British mycologists have had a major impact worldwide over the last 100 years. Commemorating the centenary of the British Mycological Society, founded in 1896, this book gives an account of the British contribution to mycology, both at professional and amateur level. A variety of distinguished British and American authors give an authoritative commentary on the present state of mycology, and on potential future developments in fields in which British mycologists made important breakthroughs.

The book is introduced by an overview of the British contribution and personal views on pioneering work on aquatic hyphomycetes, tropical mycology and the amateur contribution. Later review articles treat a number of subjects in depth such as physiology, systematics, ecology, chemistry and mapping. This unique book will be of great interest to all professional and amateur mycologists in both research and teaching.

T0215151

A CENTURY OF MYCOLOGY

A CENTURY OF MYCOLOGY

Edited by

BRIAN SUTTON
International Mycological Institute, Egham, Surrey, UK

Published for the British Mycological Society

CAMBRIDGE
UNIVERSITY PRESS

CAMBRIDGE UNIVERSITY PRESS
Cambridge, New York, Melbourne, Madrid, Cape Town, Singapore, São Paulo

Cambridge University Press
The Edinburgh Building, Cambridge CB2 8RU, UK

Published in the United States of America by Cambridge University Press, New York

www.cambridge.org
Information on this title: www.cambridge.org/9780521570565

First published 1996
This digitally printed version 2008

A catalogue record for this publication is available from the British Library

ISBN 978-0-521-57056-5 hardback
ISBN 978-0-521-05019-7 paperback

Contents

Contributors

Salomon Bartnicki-Garcia
Department of Plant Pathology, University of California, Riverside, CA 92521, USA

Arnold L. Demain
Department of Biology, Massachusetts Institute of Technology, Cambridge, MA 02139, USA

Melvin S. Fuller
Department of Botany, University of Georgia, Athens, Georgia 30602, USA and *Darling Marine Center, Walpole, Maine 04573, USA*

C. T. Ingold
12 Chiltern Close, Benson, Oxfordshire OX10 6LG, UK

D. W. Minter
International Mycological Institute, Bakeham Lane, Egham, Surrey TW20 9TY, UK

D. N. Pegler
The Herbarium, Royal Botanic Gardens, Kew, Richmond, Surrey TW9 3AE, UK

Alan D. M. Rayner
School of Biology and Biochemistry, University of Bath, Claverton Down, Bath BA2 7AY, UK

D. J. Read
Department of Animal and Plant Sciences, The University, Sheffield S10 2TN, UK

M. R. D. Seaward
Department of Environmental Science, University of Bradford, Bradford BD7 1DP, UK

Brian C. Sutton
International Mycological Institute, Bakeham Lane, Egham, Surrey TW20 9TY, UK

Roy Watling
Royal Botanic Garden, Edinburgh EH3 5LR, UK

John Webster
Department of Biological Sciences, University of Exeter, Exeter EX4 4PS, UK

Preface

When the British Mycological Society was founded in 1896, the range of techniques that mycologists could bring to bear on the investigation and testing of hypotheses and the resolution of problems were comparatively limited compared with the situation we have in 1996. Although photographic processes were well on the way to being the commercial and scientific necessity that they are now, cinematography was in its infancy, the first moving images only having been shown publicly that year. The fact that computers are available to scientific and support staff in almost every research establishment today is a very recent phenomenon. They have radically altered the ways in which mycologists analyse, organize and communicate the results of their work. Molecular techniques are little more than 20 years old, and the debate on classificatory systems now has to be viewed in the light of DNA sequencing and direct access to the fungal genome. The optical microscope, both compound and dissecting, was widely used by field- and laboratory-based mycologists, but phase contrast and interference optics had not arrived in 1896. Transmission and scanning electron microscopy were very much in the future. These are but a few of many changes which have shaped mycology between 1896 and 1996. Despite such limitations, the mycologists of 1896, and those before them, used the equipment that was at hand, and the foundations for much that we do today were established unequivocally at that time. The twentieth century has therefore been one of technical innovation, its application and exploitation.

The British Mycological Society has never been a static society; it has evolved to its present eminent but diverse status over the past 100 years. Its centenary provides not only an opportunity for looking back but also for projecting forward, and this is reflected in the scope of this centenary volume. It is of the BMS and its members and their work, and highlights just some of the major contributions that they have made to the advancement of thought and substance in mycology. The 'press' constantly bemoans the fact that the British have an enviable reputation for innovation rather than a capability for development and exploitation. This can not be levelled at the BMS or the mycologists that have been a part of it. There are many examples of initial British creativity, later consolidated and extended by the British and others. If this sounds parochial to some degree, that is the luxury afforded by the event that this volume celebrates.

British mycology and the studies it has initiated or the many breakthroughs it has made, so well documented by Webster (Chapter 1), demonstrate that the output has an historical importance in a much wider context than Britain. We are fortunate indeed to have a personal account by Ingold (Chapter 2) of how single-handedly he revived and stimulated work on aquatic hyphomycetes all over the world. To some extent, matters have come full circle with an account of the British contribution to tropical mycology by Pegler (Chapter 3); in the decades preceding the founding of the BMS, British mycologists joined others from Europe in documenting the tropical mycoflora, resulting in classifications which were predominantly eurocentric. A revival of interest in tropical aspects of mycology is provided by the recent BMS expedition to Ecuador. Another complete revolution is shown by Watling (Chapter 4), who reviews the role of the amateur in BMS activities especially in the early days, a strength which he and Webster (Chapter 1) consider vital to the future of fungal systematics in Britain and elsewhere. Chapter 5 on hyphal growth by Bartnicki-Garcia demonstrates the unique nature of hyphae in fungi. It also provides the theoretical background, only more recently realized, to the development of concepts of conidiogenesis. These are explained by Sutton (Chapter 6) and shown as being an almost uniquely British or ex-patriot British field of endeavour. Knowledge of the fungal zoospore is reviewed by Fuller (Chapter 7) who starts with the BMS Presidential Address by Waterhouse in 1962, and then considers ultrastructure and its effect on a new systematics for zoosporic fungi. The organization of the mycelium, that oft-forgotten and under-utilized component of fungi, is discussed by Rayner (Chapter 8). He treats fungi not only from the perspective of their relation to other organisms but also analyses the intellectual problems that such interconnected and individual systems pose. The stimulus to biotechnology, especially the pharmaceutical industry, that the discovery of penicillin by Fleming initiated, is reflected by Demain (Chapter 9) who deals with the regulation and function of secondary metabolites in fungi. The dependence of humankind on such activities and products of fungi is mirrored by the dependence of other organisms on fungi. One hundred years encompasses the period over which the mycorrhizal symbiosis has been recognized, and its nature and extent are discussed by Read (Chapter 10). This is followed by an account of lichens by Seaward (Chapter 11) who emphasizes the importance of advances in technology to a proper assessment of the role of lichens in the environment. Finally, Minter (Chapter 12) faces the problems of recording and mapping fungi in the longer term while at the same time coping with the rapidly changing capabilities of computers.

The Centenary Symposium *A Century of Mycology* organized by the British Mycological Society at the University of Sheffield, April, 1996, forms the basis of this volume, together with invited contributions on selected topics. I believe it inevitable that the twenty-first century and the second 100 years of the BMS will be dominated by the development of techniques that will move knowledge to levels which will be as incomprehensible and wondrous to us now as those of the twentieth century would have been to the founders of the British Mycological Society.

B. C. Sutton
2 October, 1995

1

A century of British mycology

JOHN WEBSTER

The twentieth century has been a period of unprecedented rapid change in transport, communcation, technology, scientific knowledge, agriculture, social organization and human health. I shall try to assess the scientific advances in mycology which have been initiated by British mycologists and other scientists during the past century. Others, better qualified than I, have reviewed major advances in their specialist fields in *A Century of Mycology* and as Centennial Reviews in *Mycological Research*. In a separate chapter, I shall review the history of the Society and some of the individuals and institutions which have played a part in its development.

The foundation members were mostly amateurs interested in collecting and identifying fungi, and the results of their collecting were recorded in foray lists of fungi, at first with separate lists of lichenized fungi and *Mycetozoa*. Lists of the different groups of fungi recorded for Britain have been prepared and published in the *Transactions* or by the Society. The groups now covered include *Discomycetes* (Ramsbottom, 1913; Ramsbottom & Balfour-Browne, 1951), *Phycomycetes* (Ramsbottom, 1915), *Hyphomycetes* (Wakefield & Bisby, 1941), *Pyrenomycetes* (Bisby & Mason, 1940), *Ustilaginales* (Sampson, 1940), *Uredinales* (Wilson & Bisby, 1954), Lichens (Watson, 1953), aquatic chytrids (Canter, 1953, 1955), Agarics and Boleti (Pearson & Dennis, 1948; Dennis, Orton & Hora, 1960), and *Peronosporaceae* (Francis, 1988). Such lists form an essential foundation to taxonomic work and have provided the basis of monographic treatments of British fungi, helping to make the fungus flora of Britain amongst the best understood in the world.

Taxonomy

Since the foundation members of the Society were predominantly taxonomic in their approach to mycology, it is appropriate to highlight a few of

1

the fields in fungal taxonomy which have been opened up during the century.

Hyphal analysis of polypores

The taxonomy of the *Aphyllophorales* has been revolutionized by the publication of three classical papers by Corner (1932*a*, *b*, 1953), who has been described as 'the father of the hyphal system' (Ryvarden, 1991). Corner (1932*a*) wrote 'The hyphal system of the fruit body must be considered foremost in the morphology of polypores and it will undoubtedly provide the key to a natural classification.' He described in detail the anatomy and development of the sporophore of *Polystictus xanthopus* and showed that it consisted chiefly of three distinct types of hyphae: generative, skeletal and binding. He later showed that in *Fomes levigatus* sporophores there were only two: skeletal and generative. He introduced the term trimitic for the former and dimitic for the latter. In a comparative study of four species, he also used the term monomitic for sporophores consisting of one kind of hypha only. These terms are now part of the vocabulary used in describing fruit-body anatomy of polypores (Donk, 1971). The hyphal system has been extended and modified by others (see Lentz, 1971, and Ryvarden, 1991 for references). As Corner predicted, the use of hyphal characters has greatly improved the phylogenetic classification of the genera of polypores. It has also led to the taxonomic redisposition of such gill-bearing genera as *Lentinus* whose anatomy allies them with the *Aphyllophorales* (Pegler & Young, 1983).

Conidial ontogeny

In a paper entitled 'Conidiophores, conidia and classification', Hughes (1953) significantly changed the focus of attention on conidium development and indicated its possible significance on classification of the *Deuteromycotina*. Hughes developed ideas earlier proposed by Vuillemin and by Mason. He wrote 'I believe that there are only a limited number of methods whereby conidia can develop from other cells and that morphologically related imperfect states will only be brought together when the precise methods of conidium origin take first place in the delimitation of the major groupings.' He described eight distinctive methods of development. Although his studies were based mainly on hyphomycetes, Hughes believed that 'a more rational classification in the Hyphomycetes (or Fungi Imperfecti as a whole) will be evolved only when early stages of conidium development are more closely understood.' He proposed a number of useful descriptive terms such as blastospore, porospore, meristem arthrospore and basauxic conidiophores which are now widely used. Since the publication of Hughes' paper, many studies have been made of conidial development, based not only on conventional light microscopy, but also on time-lapse cinematography, transmission and scanning electron microscopy (Cole & Samson, 1979). Hughes' ideas have since been incorporated into many taxonomic revisions of hyphomycetes, e.g. by Ellis (1971, 1976), and also of coelomycetes (Sutton, 1980, 1986). Kendrick (1981), whilst paying tribute to the importance of Hughes' concepts, was forced to the conclusion that developmental criteria cannot form the basis for a more natural classification of the conidial fungi, but that development is just one more kind of taxonomic information.

Aquatic hyphomycetes

In Chapter 2 of this book, Ingold has outlined his involvement in the study of aquatic hyphomycetes. Although a few species had been described earlier, the paper 'Aquatic hyphomycetes of decaying alder leaves' (Ingold, 1942) opened the eyes of mycologists and other aquatic biologists to an astonishing group of fungi with seemingly bizarre spore shapes which flourish abundantly on submerged leaves and twigs of riparian trees in rapidly flowing freshwater streams. The spores are large (often spanning 50 µm or more) and are of two general shapes, branched (usually tetraradiate) and sigmoid. From the outset Ingold, by painstaking studies of spore development in hanging drop culture, illustrated by clear line drawings, emphasized the importance of ontogeny. Some conidia were derived from phialides; others were holoblastic. It became necessary to erect new genera and describe many new species. Ainsworth has referred to this as a 'minor industry'. Ingold himself has certainly been industrious, having published nearly 50 papers on this group of fungi (Plunkett, 1985). Over 300 species are now known and the literature exceeds well over 1000 references. These fungi (sometimes referred to as Ingoldian fungi) are found the world over in suitable habitats. Most species can be identified with a high degree of certainty from spore morphology alone. This, coupled with the fact that the spores are readily trapped in foam, has enabled their occurrence and distribution to be studied. Ingold's ontogenetic studies showed that the two main spore shapes could result from different pathways of development, i.e. from convergent evolution (Ingold, 1966, 1975), and this has been confirmed by the demonstration that the teleomorphs of Ingoldian fungi are to be found in *Entomophthorales* and in several groups of ascomycetes and basidiomycetes (Webster, 1987, 1992). The likely reason for convergent evolution of the two types of spore is that the tetraradiate spore has a high trapping efficiency when impacted on to underwater objects, forming a tripod-like attachment with appressoria developing at the tips of the arms in contact with a surface (Webster, 1959; Read, Moss & Jones, 1991, 1992). Sigmoid spores, whilst less efficiently trapped than tetraradiates, align themselves parallel to the direction of current flow and make two points of contact (Webster & Davey, 1984).

Much is now known on their ecology (Bärlocher, 1992; Dix & Webster, 1995). In streams in temperate latitudes overhung by deciduous trees, there is an enormous pulse of activity of Ingoldian fungi related to the autumnal input of leaf litter, and estimates of conidial concentration based on millipore filtration of stream water samples may show concentrations in excess of 20000 l^{-1} (Iqbal & Webster, 1973). Interest in aquatic hyphomycetes was greatly increased by the demonstration that they play a vital role in conditioning tree leaves, making them more palatable and nutritious to aquatic invertebrates, partly by softening the tissues and also by raising the protein content (Kaushik & Hynes, 1971; Bärlocher & Kendrick, 1973 *a*, *b*; Suberkropp, 1992).

Parasites of freshwater phytoplankton

Extensive studies of chytridiaceous fungi growing as parasites of planktonic algae and cyanobacteria have been made over many years in the lakes of the English Lake District (for references see Canter, 1953, 1954; Canter & Lund, 1953; Canter-Lund & Lund, 1995). Many genera are involved. It has been shown that the parasites may play a significant role in controlling population densities of some common planktonic diatoms. *Asterionella formosa*, an abundant planktonic diatom of eutrophic lakes, is parasitized by *Rhizophydium planktonicum* agg. (Canter, 1969), and epidemics are associated with drastic reductions in the concentration of cells in suspension, and in the mean number of cells in *Asterionella* colonies (Canter & Lund, 1948, 1951, 1953).

Anaerobic rumen fungi

The discovery by Orpin (1974, 1975, 1976, 1977, 1981) that organisms previously thought to be flagellates in the guts of herbivores were obligately anaerobic chytridiomycetes came as a surprise to many mycologists because up to that time it had been generally believed that fungi were aerobic and could only increase their biomass in the presence of oxygen (Theodoru, Lowe & Trinci, 1992). Orpin was attempting to explain why the numbers of 'flagellates' found in the rumen of sheep could rise many times within 15–30 min of feeding. The ratio of the minimum (pre-feeding) to maximum concentration of 'flagellates' could vary between 15 and 296 (average 47), and if the organisms involved were dividing by binary fission, it would be necessary for them to undergo six successive divisions in 15 min. This is so improbable that another explanation was sought. It was found that fungal thalli bearing sporangia suspended in the rumen fluid were stimulated by a soluble factor in the ingested herbage to release zoospores (flagellates). Several genera are involved: the monocentric *Neocallimastix* (*Callimastix*), *Sphaeromonas* and *Piromonas* and the polycentric *Orpinomyces*. With the exception of *Neocallimastix*, the zoospores are uniflagellate, but in *Neocallimastix* there may be up to fourteen flagella. The rumen fungi are classified in the order *Spizellomycetales* (Barr, 1988; Barr *et al.*, 1989).

Ingestion of herbage is rapidly followed by release of zoospores which may reach concentrations of about 15–20000 ml^{-1} in *Neocallimastix*. Zoospore release is totally inhibited by oxygen and is stimulated by high CO_2 concentration. Oxygen is also toxic to the vegetative state. The fragments of herbage are rapidly colonized and are degraded. Rumen fungi are capable of utilizing glucose, fructose, xylose, disaccharides, polysaccharides and cellulose. Hydrolysis of pectin has been demonstrated, but it seems doubtful if lignified tissues are appreciably degraded (for references see Theodoru *et al.*, 1992; Trinci *et al.*, 1994).

In view of the toxicity of oxygen to rumen fungi, there is interest in the route by which young animals acquire these fungi. Possibly there is transmission in saliva by licking and grooming after birth, and sharing of food. It has also been shown that anaerobic fungi can be isolated from dried faeces, and there was speculation that there is an aero-tolerant survival stage (Theodoru *et al.*, 1992). It was later shown that zoosporangia can survive as resistant structures in dried faeces (Nielsen *et al.*, 1995).

Marine fungi

Interest in marine fungi was shown by British workers such as Cotton, Sutherland and Wilson (for a review see Wilson, 1960). Cotton (1907) described the symbiotic relationship between *Mycosphaerella ascophylli* and *Ascophyllum nodosum*, whilst Sutherland (1915) showed that a similar fungus *M.pelvetiae* (now regarded as a synonym of *M.ascophylli*) occurred in *Pelvetia*. Jones and his colleagues at Portsmouth have made extensive contributions to the taxonomy and ecology of marine fungi, especially by the application of scanning and transmission electron microscopical techniques to spore morphology and spore attachment. Within the *Halosphaeriaceae*, a family of marine *Pyrenomycetes*, they have revealed great variety in the ultrastructure of ascospore appendages which play a role in dispersal and subsequent attachment of the spores to their substrata (for references see Jones, Johnson & Moss, 1986; Jones, 1994). These variations in ascospore ultrastructure are of primary importance in separating taxonomically coherent genera within the family.

Fungal ecology

British mycologists have made important contributions in several fields of fungal ecology. Garrett (1950, 1956) provided a valuable conceptual framework within which to view the activities of soil fungi and fungal pathogens of roots (see below). Park (1972) and Dick (1971, 1976), have developed ideas on the role and behaviour of fungi in aquatic environments. Pugh (1980) has applied to fungal ecology the concepts of r- and k-strategies of MacArthur & Wilson, and the ideas of Grime who has classified the life strategies of higher plants into ruderals, competitors and stress-tolerant forms. Much attention has been devoted to the ecology of fungi colonizing plant remains above the level of the soil (Hudson, 1968), and to the mechanisms underlying fungal succession (Frankland, 1992). Research on the ecology of phylloplane organisms, wood decay fungi, oomycetes, coprophilous fungi, predacious fungi, root pathogens, lichens and the autecology of soil fungi have all received significant inputs from British mycologists.

It is clearly impossible to review all the fields of fungal ecology in which British mycologists have initiated major developments, and I have picked out only three.

Ecology of soil fungi

Much interest and effort have been devoted to the study of the ecology of soil fungi, and four former Presidents have addressed the Society on aspects of soil mycology (Chesters, 1949; Garrett, 1954; Burges, 1963; Pugh, 1980). Emphasis has rightly been given to the value of different techniques for studying the *activity* of fungi in soil (Warcup, 1960). A high proportion of colonies which arise on dilution plates originate from spores (Warcup, 1955b). This raises the question of the fate of spores within the soil. Dobbs & Hinson (1953) have demonstrated a widespread fungistasis (also termed mycostasis) in soils. Spores of most fungi brought into contact with soil are usually completely inhibited from germinating. The inhibitory effect can be destroyed by autoclaving, but is restored if auto-claved soil is inoculated with micro-organisms. It is also annulled by the presence of soluble nutrients. The phenomenon of soil fungistasis has received wide attention (Lockwood, 1977; Lockwood & Finonow, 1981). Intense competition for nutrients occurs in soil, and one explanation of the mechanism of soil fungistasis is that the fungus spores themselves provide local concentrations of nutrients which stimulate the development of a microbial population which effectively deprives the fungus spores of nutrients which could stimulate germination. This view is confirmed by the fact that small fungal spores are more sensitive to fungistasis than large ones with bigger nutrient reserves.

It is well known that the dilution-plate method not only over-represents fungi which can sporulate in soil, but seriously under-represents other fungi such as basidiomycetes. The use of small crumbs of soil as inoculum which ensures that all the micro-organisms in a soil sample are included was devised by Warcup (1950) in his 'soil plate method' which greatly improved the isolation of soil basidiomycetes and enabled the distribution of their mycelia to be studied (Warcup, 1959). Warcup (1955a) had the rare ability to pick out fungal hyphae from soil suspensions and to transfer them to agar medium in order for them to grow. By the use of a range of techniques, some of the mycelia could be induced to sporulate and so be identified (Warcup & Talbot, 1962). More precise methods of visualizing the course of hyphae of individual species of basidiomycete, e.g. *Mycena galopus*, in soil have been made possible by the use of anti-serum raised against the mycelium linked to a fluorescent reagent (Chard, Gray & Frankland, 1985 a, b). Autecological studies on individual species such as *M.galopus* are now possible (Frankland, 1984).

Important new concepts on the ecology of soil fungi have been promul-gated by Garrett (1950, 1956). Garrett's own research interests were on pathogenic root-infecting fungi, and he discussed their activities as part of a more general soil microflora. He developed the idea of competitive saprophytic ability, correlated with such attributes as high growth rate, dense colony growth, good enzyme equipment enabling rapid decomposi-tion of organic material, antibiotic production and ability to tolerate antibiotics produced by other organisms. He argued that the loss of competitive saprophytic ability forced certain fungi to adopt a parasitic habit, whilst others were forced into mycorrhizal associations with roots of higher plants. Griffin (1972) has written 'The original concept of competitive saprophytic ability so derived has, in my opinion, been enormously useful and has been an integrating factor in much of the literature on soil ecology.' Another valuable idea was that of inoculum potential which Garrett defined as the energy of growth of a parasite available for infection of a host at the surface of the organ to be infected. Inoculum potential can be increased by an increase in the number of infecting units or propagules per unit area of root surface, or by an increase in the nutritional status of such units. An advance in technique for studying the ability of root pathogenic fungi to compete with other soil micro-organisms was the development of the 'Cambridge method' in which pure cultures of the pathogenic fungi grown on maizemeal and sand were diluted by mixing with soil in varied proportions (Garrett, 1963). This technique has found application in other fields.

Aerobiology

Our understanding of the behaviour of fungal spores in air has been greatly influenced by the writings and experimental studies of Ingold, Gregory and Hirst. Ingold (1971) has illustrated the varied ways in which fungus spores are projected into turbulent air beyond the boundary layer of still air surrounding their sporophores. In the air they are subjected to physical forces such as wind, gravity and rainfall, and their fate depends partly on these forces, on the size, shape and surface properties of the spores themselves, and the nature of the vegetation within which they are dispersed. Experimental studies using a wind-tunnel and sticky cylindrical traps at Rothamsted Experimental Station enabled Gregory to vary wind speed, trap diameter and spore size (Gregory, 1951, 1952, 1973). Trapping (impaction) efficiency was positively correlated with wind speed and spore dimensions, larger spores being more efficiently trapped than small ones, and inversely correlated with trap diameter, narrower traps being more efficient than wider ones because they generated less turbulence. Gregory, who stressed the *functional* aspects of spore size and shape, proposed that larger spores, such as those of leaf pathogens, were adapted as *impactors*, whilst the smaller spores of fungi such as *Penicillium* were specialized as *penetrators*, not being impacted except at unusually high wind velocities.

The design and construction of a volumetric spore trap by Hirst (1952) led to a better knowledge of the concentration of spores in air, and its variation in relation to weather, time of day (Hirst, 1953) and to other variables such as season and surrounding vegetation. There is a characteristic dry air spora dominated by spores of *Cladosporium* and *Alternaria*, but at night high concentrations of basidiospores such as those of *Sporobolomyces* predominate. Rain rapidly removes many of the spores of the dry air spora, but stimulates the release of ascospores. Such studies have profound relevance to the dispersal of plant pathogens and to allergic symptoms in asthmatic patients incited by inhaling fungal spores (for references see Hirst, 1973).

The individualistic mycelium

The recognition that fungal mycelia possess individual identities and are not generally capable of indiscriminate anastomosis with mycelia of the same species to form a single co-operative unit has revolutionized our views of the function and distribution of fungal mycelia in natural habitats. Rayner (1991) has reviewed this field (see also Chapter 6). Although the phenomenon of intraspecific antagonism had been described earlier by Cayley (1923) in *Diaporthe perniciosa* (now treated as a synonym of *D.eres*) and by Campbell (1938) in *Collybia* (*Oudemansiella*) *radicata*, current interest in the phenomenon can be traced to the observations by Rayner (1976) that when isolates of *Coriolus versicolor* from decay columns in wood were paired on agar, some mycelia merged imperceptibly, whilst others failed to do so and were separated by dark zones. This phenomenon of somatic incompatibility or self – non-self recognition is very widespread, and has been clearly demonstrated amongst populations of wood-decaying basidiomycetes and ascomycetes (see for example Rayner *et al.*, 1984). *Tricholomopsis platyphylla* grows on deciduous tree stumps and forms mycelial cords extending into the surrounding litter. By pairing isolates made from basidiocarps, cords and from infected wood, it has been possible to trace the extent of *individual* mycelia, i.e. of genetically identical distinctive dikaryons, over distances of 80–100 m (Thompson & Rayner, 1982). Some mycelia showed disjunct distribution indicative of the break-up of a larger unit, possibly as a result of competition with other strains of the same fungus. Similar techniques have revealed the extent of the mycelium of *Armillaria* in coniferous forest in Washington, USA, leading to estimates that a single clone could extend over 15 ha, that its possible age was about 1500 years, with a biomass exceeding 100 t, making it possibly one of the largest and oldest living organisms (Smith, Bruhn & Anderson, 1992). In the same way, the distribution and interactions of individual mycelia of sheathing mycorrhizal fungi such as species of *Suillus* have been studied (Fries, 1987; Dahlberg, 1992).

Plant pathology

From the beginning, the Society has included members with an interest in plant pathology. The foundation members included M.C. Cooke who published *Fungoid Pests of Cultivated Plants*, and G. Massee, the first President, who also wrote two books on plant pathology and another on *Mildews, Rusts and Smuts*. His successor as President, C.B. Plowright, wrote *A Monograph of the British Uredineae and Ustilagineae*. In the first half-century of the Society's existence, a third of the Presidents were plant pathologists or had an interest in the subject, and this tradition has continued. Ainsworth (1969) has reviewed the history of plant pathology in Britain.

Compilations of lists of British plant pathogenic fungi have been sponsored by the Society, such as the *List of Common Names of British Plant Diseases* (Anon., 1929), and later editions published separately in 1934, 1944 and 1962, to be succeeded by *Names of British Plant Diseases and Causes* prepared by the British Society for Plant Pathology in 1984. These and the book by Moore (1959) have ensured that pathogens of British plants have been well-documented. During the past century several 'new' fungal pathogens have been reported or described from the British Isles, such as *Sphaerotheca mors-uvae* (American gooseberry mildew), *Phytophthora fragariae* (red-core of strawberry), *Ophiostoma ulmi* and *O.novo-ulmi* (Dutch elm disease), *Cryptostroma corticale* (sooty bark of sycamore), *Spongospora subterranea* f. sp. *nasturtii* (crook root of watercress), and many others. These occurrences reflect more intensive cultivation of certain crops and increased opportunities for entry and importation associated with improved and more rapid transport.

Although British mycologists have had a proper parochial interest in their indigenous pathogens, they have made their mark internationally in several areas such as plant disease assessment (in which the Society took an active role), forecasting, seed testing, fungicides including systemic fungicides, phytoalexins, root pathology, biological control and physiological plant pathology. Silver-leaf of fruit trees caused by *Chondrostereum purpureum* has had exhaustive investigation by Brooks and his co-workers at Cambridge. It is not possible to deal comprehensively with the major advances in plant pathology initiated by British scientists, and I have selected only five.

Breeding for resistance to plant disease

The rediscovery of Mendel's publication on inheritance in 1901 led to an important advance in plant pathology. Biffen (1905, 1907) was quick to show that rust-resistance was inherited in accordance with Mendel's laws. He made crosses between the wheat variety Rivet 'as a rule fairly immune to yellow rust' (*Puccinia striiformis*) and the susceptible variety Red King. He found that all the F_1 progeny were uniformly susceptible, but of 259 seedlings raised from them in the F_2, 64 were immune and 195 infected. 'Now the ratio 64 : 195 seems to be too close an approximation 1 : 3 to be a mere accident, and taken in conjunction with the fact that the F_1 generation was so badly attacked, it is fair proof that susceptibility and immunity are definite Mendelian characters, the former being the dominant one.' Biffen (1907) argued that the immune types would breed true to this character, whilst susceptible types should breed true in the proportion 1 : 3. On this basis he claimed (Biffen, 1931) that the solution of the most important of the problems of rust control is in sight, a prospect which Manners (1969) felt to be more distant. Control of rust diseases by genetical means, whilst still the preferred option, is much more difficult than Biffen envisaged, largely because of genetical variation within the pathogens and breakdown of resistance as novel races of the pathogens arise through recombination and selection. Nevertheless it is now obvious that Biffen's work laid a firm foundation for much that followed. 'Since that time more than a thousand papers have described the inheritance of resistance to the diseases of practically every economically important species of higher plant' (Day, 1974).

Studies in the physiology of parasitism

Brown (1915) initiated a long series of studies on the physiology of parasitism by necrotrophic plant pathogens. Preparations made from germlings of *Botrytis cinerea* caused rapid maceration of discs of tissues of turnip, beet, apple and cucumber. In addition to the macerating effect of the *Botrytis* extract, resulting from its action on the cell wall, it also brought about cell death resulting from action on the protoplast. Brown produced evidence (e.g. by heat inactivation) that the active substance was an enzyme, which he termed cytase, although it was later referred to as pectinase or protopectinase. He was unable to separate the macerating and lethal activities of the preparation and advanced two hypotheses for explaining this: a) that maceration and killing of the cells was caused by a single substance; b) that the cell walls were made permeable by the macerating enzyme to a toxic substance which kills the protoplasts. Wood (1967) has discussed several possible explanations of the relationship between maceration and killing. The simplest explanation is that once the physical barriers in the cell wall have been removed by macerating enzymes, the external membranes surrounding the protoplast are ruptured following rapid uptake of water. Support for this idea comes from the work of Tribe (1955) working with macerating enzymes of the bacterium *Erwinia aroideae* and from *B. cinerea*. Tribe found that when discs of cucumber tissue were exposed to macerating enzymes in the presence of plasmolyzing concentrations of salts and non-electrolytes, the killing effect of the enzyme preparations was greatly retarded.

Studies on the physiology of parasitism have, of course, enormously expanded since Brown's original contribution, and Brown has himself reviewed some of the later work (Brown, 1934, 1936, 1965).

Forecasting potato blight epidemics

The idea that the incidence of epidemics of late blight of potato could be related to the weather conditions had been developed by van Everdingen following a study of the outbreaks of epidemics and data on meteorological conditions in Holland. He formulated a set of rules, called the Dutch rules by Beaumont (1946), which included information on the incidence of dew, rainfall, minimum night temperature and cloudiness. Beaumont, working at Seale Hayne Agricultural College in Devon, compared the weather pattern preceding blight epidemics during an 11-year period and found that the fit between forecast and epidemic occurrence was relatively poor in that, had the Dutch rules been followed, a high proportion of false (premature) forecasts would have ensued. Nevertheless he was able to modify the Dutch rules for use under British conditions and to define a 'critical period', i.e. a combination of meteorological conditions following which a blight epidemic was likely to occur with high probability within a period of 15 days. Beaumont's critical periods were defined as a minimum temperature not less than 50°F (10°C) and a relative humidity not below 75% for at least 2 days. In practice the validity of the new rules was confined to July and August because it was found that critical periods in June were rarely immediately followed by blight epidemics. It was also found that blight epidemics were more likely to follow 'flushes' of critical periods than a single one. Large (1953) and Cox & Large (1960) found that Beaumont's new rules were valid for much of England and Wales (see also Bourke, 1970). The ability to forecast accurately the likely incidence of future blight epidemics has the obvious advantage of concentrating fungicidal spray applications to times when they are most effective, and has the added benefit that spray and tractor-wheel damage to crops, with the consequent reduction in tuber yield, can be avoided or minimized in seasons in which severe blight epidemics are unlikely or occur so late as to have relatively slight effects on crop yield.

Biological control of Heterobasidion annosum infection of pine using Peniophora gigantea

The thinning of conifer plantations leads to the exposure of the thinning stumps to infection by a range of parasitic and saprotrophic fungi. The polypore *Heterobasidion annosum* (*Fomes annosus*), the causal agent of butt-rot of pine, is capable of infecting exposed stumps by basidiospores and may remain in the stump for several years (Rishbeth, 1950, 1951). Such stumps form foci of infection for adjacent healthy trees, infection being caused by root-to-root contact underground. Rishbeth (1951, 1952) showed that the pathogen could be replaced in stumps and in root lengths by the saprotrophic *Peniophora gigantea* (*Phlebia gigantea*), and in 1963 devised practicable techniques for applying conidial suspensions of the saprotroph to freshly cut pine stumps, effectively preventing *H.annosum* from colonizing them. This technique was more effective than the application of fungicides and has resulted in considerable benefit to forestry. The mechanism of replacement of *Heterobasidion* with *Peniophora* is probably due to hyphal interference, because when colonies of the two fungi are confronted in culture, the *Heterobasidion* hyphae become vacuolated, lose opacity and turgor, and their walls become retractile. Affected cells lose their semi-permeable property, taking up the vital stain neutral red more readily than cells not in contact with *Peniophora* (Ikediugwu, Dennis & Webster, 1970; Ikediugwu, 1976).

Dutch elm disease

There have been two major epidemics of Dutch elm disease in Britain during this century. The vectors of the disease are Scolytid beetles which transmit conidia and ascospores of the fungal pathogen (Webber & Brasier, 1984). The first epidemic, caused by *Ophiostoma ulmi* (*Ceratocystis ulmi*) reached Britain in the period between 1920 and 1940, but thereafter declined in severity (Peace, 1960). A much more destructive outbreak started in the early 1970s, apparently taking its origin from importations of elm wood from Canada (Brasier & Gibbs, 1973). This second epidemic was caused by aggressive strains of an *Ophiostoma* distinguished by a more fluffy appearance and more rapid growth rate than the waxy growth of the non-aggressive form (Gibbs & Brasier, 1973). The aggressive form has since been identified as a distinct species *O. novo-ulmi* (Brasier, 1991), and it has been shown that *O.novo-ulmi* in culture secretes more of the phytotoxin cerato-ulmin than *O.ulmi* (Tegli *et al.*, 1994). Cultural characters reveal that the aggressive fungus can be separated into two races differing in their world distribution: EAN strains centred in Eurasia, and NAN strains centred in N. America (see Brasier, 1990). The original outbreak in Britain in the 1970s was associated with NAN strains only, but in 1977 outbreaks near Limerick and Cork in S.W. Ireland were caused by EAN strains. This indicates that even if NAN strains of *O.novo-ulmi* had not arrived from N. America, other aggressive EAN strains would probably have arrived later (Brasier, 1979).

Both *O.ulmi* and *O.novo-ulmi* are heterothallic, with two mating types, but each mating type is bisexual and can form ascogonia and spermatia. Attempts to cross *O.ulmi* and the EAN and NAN races of *O.novo-ulmi* amongst themselves showed that there is a reproductive barrier between them. When *O.ulmi* is the recipient, there is no barrier to accepting donor nuclei from the opposite mating type of *O.ulmi* or EAN and NAN strains of *O.novo-ulmi*. However, in reciprocal crosses with EAN or NAN as recipients, non-aggressive *O.ulmi* is rejected as a donor, i.e. there is a unidirectional barrier to hybridization. There is also a partial barrier between EAN and NAN (Brasier, 1984). In addition to the mating-type (homogenic) compatibility in *Ophiostoma*, there is also vegetative (heterogenic) incompatibility, and sampling on a world-wide scale in areas where Dutch elm disease is prevalent has shown that large numbers of vegetative-compatibility (v-c) groups exist. Even within comparatively small areas of bark in a single infected tree, a three-dimensional mosaic of mycelia of distinct v-c types occurs (Brasier, 1984; Webber, Brasier & Mitchell, 1987). This contrasts with the situation in which, in certain areas, in a number of samples, a high proportion of 'repeat' v-c groups, the so-called 'super-groups', are found. Vegetative incompatibility is believed to restrict inbreeding. It also restricts the spread of a virus-like cytoplasmic factor, the d-factor (d for disease), which occurs readily between vegetatively compatible individuals but much less frequently between vegetatively incompatible individuals. Transmission of the d-factor has a debilitating effect on the growth and reproduction of the recipient, and the possibility exists that the spread of the d-factor among populations of *O.novo-ulmi* might result in an attenuation of the disease in a similar way to the spread of hypovirulent strains of the causal organism of chestnut blight, *Endothia parasitica* (Brasier, 1983, 1984; Webber *et al.*, 1987).

Gibbs (1978) has written 'Dutch elm disease is one of the most destructive plant diseases known to man. In many host populations, the epidemic has been sustained at a very high rate until almost all the trees have been killed. . . . By killing millions of elms, Dutch elm disease has greatly disturbed man's economic and social life and struck deep at his historical and cultural traditions.' Work in Britain over the last 25 years, stimulated by the disastrous recent outbreaks, has yielded valuable insights into epidemiology, genetics of the pathogen, the replacement of an endemic non-aggressive pathogen by aggressive strains of a related fungus, and is also providing evidence of their possible origin (Brasier, 1986).

Cytogenetics

At the beginning of the century there was a considerable interest in cytology of fungi, and several review articles on fungal nuclear cytology were provided in the *Transactions* by Ramsbottom in Vols. 3, 4 and 5. Nuclear cytology in the ascomycetes was bedevilled by the supposed double nuclear fusion and double reduction in chromosome number (see, for example, Fraser, 1908) which have since been discredited. In Britain, the cytogenetics of fungi has moved forward on a broad front using *Neurospora*, *Coprinus*, *Aspergillus* and *Ustilago maydis*. Significant advances have been made in the study of complementation, heterokaryosis and cytoplasmic inheritance. Two areas of research warrant special attention.

Parasexual recombination in filamentous fungi

The phenomenon of parasexual reproduction in fungi has been extensively reviewed (e.g. by Pontecorvo, 1956; Roper, 1966; Hastie, 1981; Caten, 1981) and British mycologists, cytologists and geneticists have played a significant role in its discovery. Brierley (1931) realized that variability among conidial fungi could not be explained solely by mutation. He wrote 'The function of sexuality with its genetic consequences is, in many fungi, taken by hyphal fusion of a somatic nature,' but implied that the consequences would be 'not in heterozygotic but in heterokaryotic and heteroplasmic states' (Hastie, 1981). In fact, heterozygotic states may be involved following nuclear fusions in heterokaryons. Sansome (1946) induced the production of 'gigas' forms of *Penicillium notatum* following exposure to camphor vapour and produced evidence that they were diploid (Sansome, 1949). This led Roper (1952) to search for heterozygous diploids in *Emericella nidulans* (*Aspergillus nidulellus*), an ascomycete which reproduces asexually by means of conidia, and sexually by the formation of ascospores in cleistothecia. Roper produced forced heterokaryons between different auxotrophic mutants of *E. nidulans* which were also marked by having spore colours such as yellow or white in place of the wild type spore colour, green. Heterokaryons were able to grow on minimal medium, unlike the auxotrophic mutants. A proportion of the conidial heads formed by the heterokaryons formed diploid uninucleate conidia, and when these were plated out, diploid colonies could also be identified by their ability to grow on minimal medium. The conidia produced by these diploid colonies were green, i.e. wild type, because the mutant alleles were recessive. The diploid colonies were also able to undergo rare somatic recombination to produce further diploids homozygous for certain markers and segregating for others (Pontecorvo & Roper, 1952). Pontecorvo & Roper (1953), in explaining the occurrence of segregants and recombinants in heterozygous diploids of *E. nidulans* (and also of *Aspergillus niger* which does not reproduce sexually), proposed that somatic crossing-over occurs in diploid nuclei at the 4-strand stage of mitosis. Later Pontecorvo (1954) proposed the term parasexual reproduction, and claimed the existence of a parasexual cycle (Pontecorvo, 1956). The cycle involves the formation of a heterokaryon, fusion of unlike nuclei, multiplication of the heterozygous diploid nucleus, sorting out of a homokaryotic diploid mycelium, mitotic crossing-over and vegetative haploidization of the diploid nuclei.

It has been possible to compare genetical maps of *E. nidulans* derived from crosses involving the normal sexual process, i.e. through karyogamy, meiosis and ascospore formation, and maps based on recombination through the parasexual cycle, and there is a close degree of correspondence. Genetical analysis using parasexual recombinants is less laborious than by conventional methods. It is also possible to analyse the genetics of 'imperfect' fungi.

The ability to bring about fusion between fungal protoplasts of different genetical constitution, derived either from the same or from taxonomically distinct species of the same genus has made possible the construction of heterokaryons in fungi where this had previously not been possible. Parasexual recombination and genetic mapping have, through this technique, been achieved (for references see Peberdy, 1989).

Parasexual recombination is known to occur in a large number of filamentous fungi including some plant pathogens such as *Fusarium* and *Verticillium*. It is theoretically possible to combine the properties of different strains of economically important imperfect moulds, but progress in this field has been disappointing.

It is difficult to assess the significance of the parasexual phenomenon in nature. For it to be successful, heterokaryosis must occur between different strains of the same fungus. Whilst there is good evidence that heterokaryons do occur in the wild (Jinks, 1952; Caten, 1981), heterokaryon compatibility is under genetical control. Natural populations of *E.nidulans* and other fungi can be sub-divided into a number of heterokaryon-compatibility (h-c) groups. Compatibility is controlled by several alleles, and heterokaryons are only formed readily between strains with several h-c alleles in common (Caten & Jinks, 1966; Croft & Jinks, 1977). 'The presence of the heterokaryon incompatibility system and its mode of action would appear to rule out the possibility that parasexuality is of importance as a mechanism for recombination between h-c groups, though it could act as a means for the recombination of new genetical material which is accumulating within an h-c group' (Croft & Jinks, 1977).

Diploidy and gametangial meiosis in oomycetes

A controversial field of study, the cytology and genetics of oomycetes, has been reviewed by Shaw (1983). At the turn of the century, writers such as Wager, Trow, Stevens, Rosenburg and Ruhland had interpreted the nuclear divisions in the oogonium as meiotic, whilst others such as Berlese, Hartog, Claussen and Murphy claimed that the divisions were mitotic, meiosis taking place in the zygote, so that vegetative hyphae were haploid. The latter view prevailed until the early 1960s, influenced by the widely held view that most higher fungi were haploid. Sansome (1961, 1963), working with squash preparations of *Pythium debaryanum*, presented evidence based on nuclear counts, nuclear size and chromosomal configurations that meiosis is gametangial. This conclusion was extended to

Phytophthora and *Sclerospora* (Sansome, 1966), and it is now widely accepted that this is also true of other oomycetes (Dick & Win-Tin, 1973; Caten & Day, 1977). The conclusions based on nuclear cytology using light microscopy have been supported by ultrastructural evidence of the presence of synaptonemal complexes (axial filaments) characteristic of meiotic prophase in *Saprolegnia* (Howard & Moore, 1970) and *Achlya* (Ellzey, 1974), and by estimates of the DNA content of gametangial and vegetative nuclei in *Saprolegnia* (Bryant & Howard, 1969). The most convincing genetical confirmation for diploidy has come from studies of the inheritance of induced drug resistance in *Phytophthora drechsleri* (Shaw & Khaki, 1971) and of inheritance of induced methionine requirement in the homothallic *P.cactorum* (Elliott & MacIntyre, 1973). Since many members of the *Peronosporales* are pathogens of economically important crops, a clear understanding of their cytology and genetics is an essential prerequisite to any programme of disease control based on breeding of resistant host crops.

Medical and veterinary mycology

The history of medical and veterinary mycology has been reviewed by Ainsworth (1986). Compared with some other regions of the world, fungal diseases of man and animals have not had a serious impact in Britain. Nevertheless British mycologists and medical scientists have made a number of significant advances. Adamson, working on skin diseases at St Bartholomew's Hospital, London, showed that hairs invaded by the ringworm fungus *Microsporum audouinii* were surrounded by hyphae growing downwards along the hair shaft, stopping short at the bulb at the base of the hair. The fringe of hyphal endings at the base of the shaft has been named 'Adamson's fringe'. Adamson also developed X-ray treatment of ringworm of the scalp in children, which effectively reduced the time to cure *Tinea capitis*.

Griseofulvin

The treatment of dermatophyte infections was greatly improved by the discovery of griseofulvin, an antibiotic originally extracted from *Penicillium griseofulvum* (Oxford, Raistrick & Simonart, 1939). Brian (1960) has recounted how the same antibiotic was independently discovered by him and his colleagues in several other species of *Penicillium*, and that in low concentrations griseofulvin could cause distortion of the hyphal tips of

Botrytis allii. It was later discovered that griseofulvin could be taken up by plants and function as a systemic fungicide. This in turn led to the testing of the antibiotic against dermatophytes (see Gentles, 1958, and further references in Brian, 1960), and to the successful treatment of *Trichophyton rubrum* infections of the skin and nails, and of ringworm infection by *M. audouinii*, by oral administration without side effects (Williams, Marten & Sarkany, 1958). Griseofulvin production is now a multi-million dollar industry.

Aflatoxins

The discovery of aflatoxins had a significant input from British mycologists. 'It has been estimated that during the summer of 1960 more than 100 000 young turkeys, 14 000 ducklings, and many young pheasants and partridges in the United Kingdom died after being fed rations containing groundnut (or peanut, *Arachis hypogaea*) meal' (Ainsworth, 1986). Acute liver damage (cancer) was discovered in affected animals. As is now well known, the source of the disorder was traced to the incorporation into the feed of groundnuts infected during wet harvests in Brazil with *Aspergillus parasiticus*, a member of the *A. flavus-oryzae* group (Austwick & Ayerst, 1963; Austwick, 1978). Aflatoxin has been discovered in both animal and human food, and in parts of Africa and South-east Asia the consumption of food containing aflatoxins as a result of infection by *Aspergillus* spp. has been associated with liver cancer. Chemical analysis of aflatoxins has shown that one component, aflatoxin B_1, is carcinogenic. Ainsworth (1986) has written 'Today, twenty-five years after its discovery, study of aflatoxin still flourishes. There have been two major monographs on the aflatoxins, dozens of reviews and hundreds of papers reporting the results of research which may be consulted for further details.'

Farmer's lung disease

Farmer's lung disease, with acute respiratory symptoms associated with working with mouldy hay, is not caused by inhalation of fungus spores, but by those of the thermophilic actinomycetes *Micropolyspora faeni* and *Thermoactinomyces vulgaris*. This discovery was made by mycologists interested in aerobiology at Rothamsted Experimental Station. When mouldy hay was shaken in a wind tunnel, enormous numbers (hundreds of millions g^{-1}) of actinomycete spores were released (Gregory & Lacey, 1963; Pepys *et al.*, 1963).

Antibiotics: the discovery of penicillin

The discovery of penicillin by Fleming in 1929 has had a profound effect not only on the treatment of bacterial infections in Man, but also on the pharmaceutical industry and on research by mycologists and microbiologists searching for further sources of antibiotics. The extraordinary sequence of events which led to Fleming's discovery has been graphically re-told and reconstructed by Hare (1970). It seems likely that a spore of a powerful penicillin-producing strain of *Penicillium notatum* originated from cultures isolated by the mycologist C.J. La Touche working on the floor below Fleming's laboratory. The spore contaminated a plate inoculated with *Staphylococcus*, but the plate was not placed in an incubator. Instead, it was left on the laboratory bench whilst Fleming was away on holiday, and during this period the temperature was unusually low. The plate escaped being immersed in Lysol because it was on the top of a pile. Hare wrote 'All these events, acting in concert, brought to Fleming's notice a phenomenon which cannot, even now, be reproduced unless the conditions in which the experiment is carried out are exactly right. Had only one link in the chain been broken, Fleming would have missed his opportunity.' The effectiveness of extracts of 'mould juice' to inhibit the growth of a range of bacteria *in vitro* and *in vivo* is now well known. The purification, chemical characterization, animal testing and clinical trials which led to the successful development of penicillin have been described by Chain *et al.* (1940) and Florey *et al.* (1949). The early attempts at purification yielded a product which was far from pure, containing about 99% impurities, but by great good fortune these impurities were not toxic to experimental animals. The remarkable property of penicillin was that it was non-toxic to mammals in concentrations at which it was an effective systemic chemotherapeutant against many pathogenic bacteria. Transfer of production to USA, the use of corn-steep liquor and lactose in the fermentation, strain improvement of the original isolate of *P.notatum*, the search for other productive isolates, selection of *P.chrysogenum* as a superior producer, and bulk fermentation, have all played a part in making penicillin production cheaper (Raper, 1952). Raper wrote 'Penicillin presents a paradox. It was the first therapeutically effective antibiotic discovered: still it remains the least toxic and the most useful, despite the fact that more than 300 additional ones have been discovered. . . . Equal to its importance as a drug has been the effect of precipitating and sustaining the unprecedented search for other drugs of microbial origin'. Brian (1951) and Goldberg (1959) have listed some of the antibiotics which were later discovered, and their medical and non-medical uses, e.g. as supplements in animal food, as agents of plant disease control, and in food preservation.

Other mould metabolites

A very large number of mould metabolites have been extracted and characterized chemically. Much painstaking work in this field has been carried out by Raistrick and his co-workers in the Biochemistry Department at the London School of Hygiene and Tropical Medicine. In the period between 1929 and 1949, about 200 mould metabolic products were described (Raistrick, 1949). Amongst the substances purified and identified were griseofulvin, patulin and several anthroquinones. Raistrick raised the question of what function mould metabolic products serve in the economy of the organisms which produce them, and was unable to regard them as having no particular biological importance. He suggested that they might play a role in oxidation–reduction reactions, and also pointed out that some had strong antibacterial and antifungal properties. This is currently true of griseofulvin and patulin.

These few examples of significant work on medical and veterinary mycology do not, of course, fully represent the activities of British mycologists and reference should be made to Ainsworth (1986) for a fuller appraisal. Ainsworth pays tribute to the work of two men, E.J. Butler, first director of the International Mycological Institute, and J.T. Duncan at the London School of Hygiene and Tropical Medicine. Butler had seen to it that abstracts on medical mycology appeared in the IMI's *Review of Applied Mycology*. Later the abstracts were issued in the *Annotated bibliography of Medical Mycology*, which developed into the *Quarterly Review of Medical and Veterinary Mycology*. Great credit should also be given to Ainsworth himself, a scholarly historian of medical and veterinary mycology (Ainsworth, 1986), of plant pathology (Ainsworth, 1981) and of mycology (Ainsworth, 1976), a co-author of Ainsworth & Bisby's *Dictionary of the Fungi*, now in its eighth edition, a distinguished director of the International Mycological Institute (Aitchison & Hawksworth, 1993) and much more.

Probably because the origins and traditions of the British Mycological Society can be traced to those with a more botanical and plant pathological training or allegiance rather than to those with medical or microbiological backgrounds, medical and veterinary mycology has not featured prominently in the Society's activities. Fortunately, the subject is well catered for by the British Society for Mycopathology.

Mutualistic symbiotic associations

British mycologists have played a significant role in increasing our under-standing of mutualistic symbiotic associations. The most outstanding of these is J.L. Harley. 'He, above anyone else, put the experimental study of mycorrhizas firmly on the map.' (Smith & Lewis, 1993). His pre-war inter-ests in the ecology of beech ectomycorrhizas were followed after the war with two wide-ranging literature reviews (Harley, 1948, 1950) and books (Harley, 1959, 1969; Harley & Smith, 1983). Harley and his collaborators showed that the uptake of $^{32}PO_4$ from dilute solution by detached mycor-rhizal roots of beech was several times that of uninfected roots, and autoradiographic studies indicated that mycorrhizal root tips accumulated $^{32}PO_4$ much more rapidly than uninfected tips. Fungal sheath material dis-sected from the root core accumulated the greater proportion of the phos-phate, but if the external supply of PO_4 was diminished, the reserves accumulated by the fungus were mobilized and translocated to the host. Root dissection techniques involving the application of agar blocks con-taining radioactive sugars to the host in the form of sucrose, glucose and fructose resulted in the accumulation of radioactively labelled trehalose, mannitol and glycogen in the fungus (see Lewis, 1991).

In the endomycorrhizas of orchids, there is also an interchange of car-bohydrates between the higher plant and the mycorrhizal fungus. Experimental studies using radioactive sugars, the orchid *Dactylorchis purpurella*, and its mycorrhizal symbiont *Rhizoctonia solani*, showed that the fungus could translocate carbohydrate. When the fungus was supplied with ^{14}C-glucose, infected protocorms became labelled with the fungal sugar trehalose which later became converted to sucrose (Smith, 1966, 1967). Similar results have been obtained with *Goodyera repens* (Purves & Hadley, 1975). The higher plant can thus acquire carbohydrate at the expense of the fungus, i.e. the conventionally accepted roles of higher plant and fungus are reversed.

Harley criticized the evidence for earlier claims of seed-borne and sys-temic infection of Ericaceae by *Phoma radicis-callunae*, the supposed myc-orrhizal endosymbiont. These criticisms were amply confirmed by the isolation of an Ericaceous endophyte from macerated root segments of *Calluna vulgaris* and other members of the Ericaceae (Pearson & Read, 1973). The fungus was successfully re-inoculated into aseptically germi-nated seedlings, where it caused typical mycorrhizas. It was also shown to be widely distributed in soils even in the absence of Ericaceae. Eventually some of the isolates developed apothecia of an inoperculate discomycete, and it was named *Pezizella ericae* but has since been transferred to *Hymenoscyphus*. Although there is evidence, e.g. from transmission electron micrographs, that basidiomycete hyphae may also occur in Ericaceous roots, there is no doubt that *H.ericae* is a common and widely distributed symbiont. Experimental comparisons between mycorrhizal seedlings infected with *H.ericae* and non-mycorrhizal seedlings or seedlings associated with non-mycorrhizal soil fungi have shown that the mycorrhizal fungus has access to organic nitrogen in the soil which is not available to uninfected roots (Stribley & Read, 1980; Read, 1983).

Although vesicular–arbuscular mycorrhizal infections had been recognized and described for over a century and Peyronel had suggested that the causal fungi were related to *Endogone* (see Butler, 1939, for a review), their identity remained uncertain until Mosse (1953) reported 'fructifications of the *Endogone* type' attached to mycorrhizal strawberry plants. The fructifications were sporocarpic, consisting of clusters of chlamydospores enclosed in a loose weft of hyphae. The fungus has since been named *Glomus mosseae* in her honour. Mosse developed techniques for germinating surface-sterilized spores which she used to infect aseptically grown apple and strawberry seedlings (Mosse, 1959). Although infection of aseptic clover seedlings by this technique was possible, penetration was aided by the presence of bacteria, possibly through pectolytic action (Mosse, 1962). Several genera and many species of sporocarpic and non-sporocarpic fungi form V-A mycorrhizas. The sporocarps and chlamydospores are sufficiently large to be sieved from soil suspensions and sieved material can be used to inoculate host plants (Gerdemann & Nicolson, 1963). Comparison of mycorrhizal and non-mycorrhizal plants has shown that infected plants, especially when grown in PO_4-deficient soil grow more rapidly and take up PO_4 and water more readily than uninfected plants, effects due in part to the more extensive exploration of the soil surrounding the root than the root hairs so that the PO_4 depletion zone immediately around the root is penetrated (for references see Harley & Smith, 1983, and Read *et al.*, 1992). Because many cultivated plants are infected with V-A mycorrhizal fungi, there have been very extensive investigations of infection on crop yield.

Ecological aspects of mycorrhizal infection have also been advanced by British mycologists. Read (1984) has discussed the distribution of ericoid, sheathing and V-A mycorrhizas on a global scale, relating it to the complex of soil conditions and climate. Growth of conifer seedlings infected with sheathing mycorrhizal partners in peat in transparent chambers has permitted observation of the extent of external mycelium, and it has become obvious that plants may be directly connected by a common mycelium, sometimes in the form of strands. Through such connections, translocation of carbohydrate has been demonstrated. This may enable seedlings in nature to be nourished by older trees surrounding them.

The colonization of tree roots by sheathing mycorrhizal fungi is now known to follow a successional pattern. Evidence for this comes from two sources: the occurrence of basidiocarps around individual trees as they grow older and in stands of increasing age (Ford, Mason & Pelham, 1980; Last *et al.*, 1983; Mason *et al.*, 1982), and the identification of the fungi ensheathing root tips from morphological and anatomical features (Deacon, Donaldson & Last, 1983). The best documented examples come from birch in which early stage fungi include *Hebeloma crustuliniforme* succeeded by *Inocybe lanuginella* and they in turn by *Lactarius pubescens*. Later stage fungi include some with larger basidiocarps such as *Leccinum versipelle* and *Amanita muscaria*.

The carbohydrate interchange between fungal and autotrophic components of lichen thalli has been studied by Smith and his co-workers (see Smith, 1961, 1962, 1963, 1981; Smith, Muscatine & Lewis, 1969; Lewis & Smith, 1967). Many of their experiments made use of discs cut from intact lichen thalli which could be floated on nutrient solutions, and it proved possible with the bulkier thalli of *Peltigera polydactyla* to dissect quadrants cut from discs into an 'algal zone' (upper cortex together with the layer containing the cyanobacterium *Nostoc*) from the medulla, consisting of fungal tissue only (Harley & Smith, 1956). Exposure of *Peltigera* thalli to $H_2{}^{14}CO_3$ in the light was followed by the accumulation of organic compounds containing ^{14}C in the 'algal zone'.

Important advances followed the development of the 'inhibition technique' by Drew & Smith (1967). When lichens are incubated on solutions of $NaH^{14}CO_3$ in the light in the presence of high concentrations of the non-radioactive form of the carbohydrate normally released by the alga, the radioactive carbohydrate secreted by the alga or cyanobacterium is unable to compete for entry into the fungus with the non-radioactive form, and accumulates in the medium. The kind of carbohydrate released depends on the relationship of the photosynthetic partner. The blue-green bacterium *Nostoc* releases glucose, whilst green algal symbionts release polyols such as ribitol, erythritol and sorbitol (Richardson, Jackson Hill & Smith, 1968). Glucose released by *Nostoc* in the thallus of *Peltigera* is rapidly taken up by the fungus and converted to mannitol which cannot be metabolized by *Nostoc*. This situation closely resembles that found in beech mycorrhizas.

It has been suggested by Lewis & Harley (1965) that conversion of carbohydrate by the heterotroph to a form unavailable to the autotroph promotes the one-way flow of carbon in such symbiotic systems, i.e. it is a 'biochemical valve'.

Fungal viruses

The discovery by Gandy (1960, *a*, *b*) that the disease of cultivated mushrooms variously termed 'Watery Stipe', 'La France Disease' and 'Brown Disease' was transmissible opened up a new field, that of fungal viruses (Lemke & Nash, 1974). Gandy showed that inoculation of trays of healthy mushrooms in the mycelial stage with compost from trays containing the disease resulted in rapid transmission of the disease. In pure culture, mycelium from diseased basidiocarps is weak and appressed, whilst healthy mycelium produces more rapidly growing fluffy colonies with rhizomorphs. After anastomosis with diseased mycelium, the growth of the healthy culture is weaker. Gandy raised the question of whether a virus was involved. Hollings (1962) and Gandy & Hollings (1962) showed that this was indeed the case. They demonstrated the presence in diseased basidiocarps of more than one kind of virus-like particle. It is now known that the disease can be spread by spores from infected basidiocarps (van Zaayen, 1979).

At about the same time as viruses were demonstrated in the cultivated mushroom, virus-like particles were discovered in 'killer' strains of yeast (Bevan & Makower, 1963; see Bevan & Mitchell, 1974). Subsequent work has shown that fungal viruses occur in many fungi. Transmissible viruses in populations of fungal plant pathogens are capable of transforming virulent strains to hypovirulent ones, raising the possibility of biological control (Day & Dodds, 1974).

The mechanism of ballistospore discharge

For over a century the mechanism by which basidiospores are discharged remained a puzzle and a number of different ideas have been proposed (Webster & Chien, 1990). Although Fayod had earlier described the drop which appears at the hilar appendix of the spore shortly before discharge, the name of Buller will always be associated with it. In earlier writings and especially in his '*Researches on Fungi*', Buller (1922) drew attention to the drop which he believed contained liquid and was discharged with the spore. He wrote, 'It may be that the force of surface tension is used in some way to effect discharge.' Ingold (1939) developed the idea. He assumed that the liquid within Buller's drop had the same properties as water and calculated that the surface energy of the drop on the basidiospore of *Agaricus campestris* contains about seven times the energy required to account for the initial velocity of spore liberation. He described a way in which the potential energy of the liquid in the drop could be converted to kinetic energy. 'I think it is likely that at the moment of spore discharge, the drop excreted at the hilum (hilar appendix) flows on to the side of the spore, and while this is happening, the spore will tend to move in the opposite direction.' It is now known that a second drop of liquid, the adaxial drop, appears on the face of the basidiospore above the hilar appendix and that the two drops continue to expand until they coalesce, bringing about discharge in exactly the manner suggested by Ingold (Webster *et al.*, 1988). The expansion of the drops is not accompanied by shrinkage of the spore and is due to condensation of water vapour around hygroscopic material extruded from the spore. Recent evidence obtained by analysis of the liquid drawn off by micropipette from the unusually large Buller's drop of *Itersonilia perplexans* and by GLC analysis of the washings of basidiospores from a range of basidiomycetes indicates that the hygroscopic material consists largely of the sugar alcohol mannitol and some hexoses, and calculations based on quantitative estimates of the amount of solute in Buller's drops from *Itersonilia* show that the amounts of mannitol and hexose present match the amounts needed to permit the expansion of Buller's drop by condensation from a saturated atmosphere at the rate observed (Webster *et al.*, 1995).

Prospect for the following century

This selection of examples in which British mycologists have played a key role in the development of new subject areas shows how versatile and productive they have been during the twentieth century. The future is, however, uncertain. Concentration of funding on 'molecular' aspects of biology, and of research into a few organisms, has had a debilitating effect in Universities, research stations, and even in museums and herbaria on whole organism biology, on taxonomy and ecology. There is an alarming trend not only in Britain, but in most of the developed world, not to replace traditional mycologists with similarly qualified staff as they retire. The ability to recognize and name common fungi and their hosts is becoming increasingly rare. The motto of the Society which was printed on early issues of the *Transactions 'Recognosce notum: ignotum inspice'* now has a hollow ring, and it is probably true to say that many amateur mycologists are better qualified to recognize what is known than the professionals. This should be a matter for concern and a trend which the British Mycological Society should do its utmost to reverse if it is to uphold one of its constitutional aims – the study of mycology in all its branches.

References

Ainsworth, G.C. (1969). History of plant pathology in Great Britain. *Annual Review of Phytopathology*, **7**, 13–80.

Ainsworth, G.C. (1976). *Introduction to the History of Mycology.* Cambridge: Cambridge University Press.

Ainsworth, G.C. (1981). *Introduction to the History of Plant Pathology.* Cambridge: Cambridge University Press.

Ainsworth, G.C. (1986). *Introduction to the History of Medical and Veterinary Mycology.* Cambridge: Cambridge University Press.

Aitchison, E.M. & Hawksworth, D.L. (1993). IMI: *Retrospect and Prospect.* A Celebration of the Achievements of the International Mycological Institute 1920–1992. Wallingford, Oxon: CAB International.

Anon. (1929). List of common names of British plant diseases. *Transactions of the British Mycological Society*, **14**, 140–77.

Austwick, P.K.C. (1978). Mycotoxicoses in poultry. In *Mycotoxic Fungi, Mycotoxins, Mycotoxicoses* (ed. T.D. Wyllie and L.G. Morehouse), vol. 2, pp. 279–301. New York and Basel: Marcel Dekker.

Austwick, P.K.C. & Ayerst, G. (1963). Groundnut microflora and toxicity. *Chemistry and Industry*, **1963**, 55–61.

Bärlocher, F. (1992). (ed.) *The Ecology of Aquatic Hyphomycetes.* Berlin, Heidelberg. Springer-Verlag.

Bärlocher, F. & Kendrick, B. (1973*a*). Fungi and food preferences of *Gammarus pseudolimnaeus. Archiv für Hydrobiologie*, **72**, 501–16.

Bärlocher, F. & Kendrick, B. (1973*b*). Fungi in the diet of *Gammarus pseudolimnaeus* (Amphipoda). *Oikos*, **24**, 295–300.

Barr, D.J.S. (1988). How modern systematics relates to the rumen fungi. *BioSystems*, **21**, 351–6.

Barr, D.J.S., Kudo, H., Jakober, K.D. & Cheng, K.S. (1989). Morphology and development of rumen fungi *Neocallimastix* sp., *Piromonas communis* and *Orpinomyces bovis* gen.nov., sp.nov. *Canadian Journal of Botany*, **67**, 2815–24.

Beaumont, T.A. (1946). The dependence on the weather of the dates of outbreak of potato blight epidemics. *Transactions of the British Mycological Society*, **31**, 45–53.

Bevan, E.A. & Makower, M. (1963). The physiological basis of the killer character in yeast. In *Proceedings of the XIth International Congress of Genetics*, The Netherlands, **1**, pp. 202–203.

Bevan, E.A. & Mitchell, D.J. (1974). The killer system in yeast. In *Viruses and Plasmids in Fungi* (ed. P.A. Lemke), pp. 161–199. New York & Basel: Marcel Dekker.

Biffen, R.H. (1905). Mendel's laws of inheritance and wheat breeding. *Journal of Agricultural Science*, **1**, 4–48.

Biffen, R.H. (1907). Studies in the inheritance of disease resistance. *Journal of Agricultural Science*, **2**, 109–28.

Biffen, R.H. (1931). Presidential Address. The cereal rusts and their control. *Transactions of the British Mycological Society*, **16**, 19–37.

Bisby, G.R. & Mason, E.W. (1940). List of Pyrenomycetes recorded for Britain. *Transactions of the British Mycological Society*, **24**, 127–243.

Bourke, P.M.A. (1970). Use of weather information in the prediction of plant disease epiphytotics. *Annual Review of Phytopathology*, **8**, 345–70.

Brasier, C.M. (1979). Dual origin of recent Dutch elm disease outbreaks in Europe. *Nature, London*, **281**, 78–80.

Brasier, C.M. (1983). A cytoplasmically transmitted disease of *Ceratocystis ulmi*. *Nature, London*, **305**, 220–3.

Brasier, C.M. (1984). Inter-mycelial recognition systems in *Ceratocystis ulmi*: their physiological properties and ecological importance. In *The Ecology and Physiology of the Fungal Mycelium* (ed. D.H. Jennings and A.D.M. Rayner), pp. 451–497. Cambridge: Cambridge University Press.

Brasier, C.M. (1986). The population biology of Dutch elm disease: its principal features and some implications for other host–pathogen systems. *Advances in Plant Pathology*, **5**, 55–118.

Brasier, C.M. (1990). China and the origins of Dutch elm disease: an appraisal. *Plant Pathology*, **39**, 5–16.

Brasier, C.M. (1991). *Ophiostoma novo-ulmi* sp.nov., causative agent of current Dutch elm disease pandemics. *Mycopathologia*, **115**, 151–61.

Brasier, C.M. & Gibbs, J.N. (1973). Origin of the Dutch elm disease epidemic in Britain. *Nature, London*, **242**, 607–9.

Brian, P.W. (1951). Antibiotics produced by fungi. *Botanical Review*, **17**, 357–430.

Brian, P.W. (1960). Presidential Address. Griseofulvin. *Transactions of the British Mycological Society*, **43**, 1–13.

Brierley, W.B. (1931). Biological races in fungi and their significance in evolution. *Annals of Applied Biology*, **18**, 420–34.

Brown, W. (1915). Studies in the physiology of parasitism. I. The action of *Botrytis cinerea*. *Annals of Botany, London*, **29**, 313–48.

Brown, W. (1934). Presidential Address. Mechanism of disease resistance in plants. *Transactions of the British Mycological Society*, **19**, 11–33.

Brown, W. (1936). The physiology of host–parasite relations. *Botanical Review*, **2**, 236–81.

Brown, W. (1965). Toxins and cell-wall dissolving enzymes in relation to plant diseases. *Annual Review of Phytopathology*, **3**, 1–18.

Bryant, T.R. & Howard, K.L. (1969). Meiosis in the Oomycetes I. A microspectophotometric analysis of nuclear deoxyribonucleic acid in *Saprolegnia terrestris*. *American Journal of Botany*, **56**, 1075–83.

Buller, A.H.R. (1922). *Researches of Fungi*, **2**. London: Longmans, Green & Co.

Burges, N.A. (1963). Presidential Address. Some problems in soil microbiology. *Transactions of the British Mycological Society*, **46**, 1–14.

Butler, E.J. (1939). The occurrences and systematic position of the vesicular–arbuscular type of mycorrhizal fungi. *Transactions of the British Mycological Society*, **22**, 274–301.

Campbell, A.H. (1938). Contribution to the biology of *Collybia radicata* (Relh.) Berk. *Transactions of the British Mycological Society*, **22**, 151–9.

Canter, H.M. (1953). Annotated list of British aquatic chytrids. *Transactions of the British Mycological Society*, **36**, 278–303.

Canter, H.M. (1954). Fungal parasites of the phytoplankton. III. *Transactions of the British Mycological Society*, **37**, 111–33.

Canter, H.M. (1955). Annotated list of British aquatic chytrids. (Supplement 1). *Transactions of the British Mycological Society*, **38**, 425–30.

Canter, H.M. (1969). Studies on British chytrids. XXIX. A taxonomic revision of certain fungi found on the diatom *Asterionella*. *Journal of the Linnean Society (Botany)*, **60**, 85–97.

Canter, H.M. & Lund, J.W.G. (1948). Studies on plankton parasites I. Fluctuations in numbers of *Asterionella formosa* in relation to fungal epidemics. *New Phytologist*, **47**, 238–61.

Canter, H.M. & Lund, J.W.G. (1951). Studies on plankton parasites III. Examples of the interaction between parasites and other factors in determining the growth of diatoms. *Annals of Botany, London*, N.S., **15**, 359–71.

Canter, H.M. & Lund, J.W.G. (1953). Studies on plankton parasites II. The parasitism of diatoms with special reference to lakes in the English Lake District. *Transactions of the British Mycological Society*, **36**, 13–37.

Canter-Lund, H. & Lund, J.W.G. (1995). *Freshwater Algae: Their Microscopic World Explored*. Bristol: Biopress Ltd.

Caten, C.E. (1981). Parasexual processes in fungi. In *The Fungal Nucleus* (ed. K. Gull and S.G. Oliver), pp. 191–214. Cambridge: Cambridge University Press.

Caten, C.E. & Day, A.W. (1977). Diploidy in plant pathogenic fungi. *Annual Review of Phytopathology*, **15**, 295–318.

Caten, C.E. & Jinks, J.L. (1966). Heterokaryosis: its significance in wild homothallic Ascomycetes and Fungi Imperfecti. *Transactions of the British Mycological Society*, **49**, 81–93.

Cayley, D.M. (1923). The phenomenon of mutual aversion between monospore mycelia of the same fungus (*Diaporthe perniciosa*, Marchal) with a discussion of sex-heterothallism in fungi. *Journal of Genetics*, **13**, 353–70.

Chain, E., Florey, H.W., Gardner, A.D., Heatley, N.G., Jennings, M.A., Orr-Ewing, J. & Sanders, A.G. (1940). Penicillin as a chemotherapeutic agent. *Lancet (London)*, **ii**, 226–8.

Chard, J.M. Gray, T.R.G. & Frankland, J.C. (1985a). Purification of an antigen characteristic for *Mycena galopus*. *Transactions of the British Mycological Society*, **84**, 235–41.

Chard, J.M., Gray, T.R.G. & Frankland, J.C. (1985b). Use of an anti-*Mycena galopus* serum as an immunofluorescent reagent. *Transactions of the British Mycological Society*, **84**, 243–5.

Chesters, C.G.C. (1949). Presidential Address. Concerning fungi inhabiting soil. *Transactions of the British Mycological Society*, **32**, 197–216.

Cole, G.T. & Samson, R.A. (1979). *Patterns of Development in Conidial Fungi*. London, San Francisco, Melbourne: Pitman.

Corner, E.J.H. (1932a). The fruit body of *Polystictus xanthopus* Fr. *Annals of Botany, London*, **46**, 71–111.

Corner, E.J.H. (1932b). A *Fomes* with two systems of hyphae. *Transactions of the British Mycological Society*, **17**, 51–81.

Corner, E.J.H. (1953). The construction of polypores. I. Introduction *Polyporus sulphureus, P.squamosus, P.betulinus* and *Polystictus microcyclus*. *Phytomorphology*, **3**, 152–67.

Cotton, A.D. (1907). Notes on marine Pyrenomycetes. *Transactions of the British Mycological Society*, **3**, 92–9.

Cox, A.E. & Large, E.C. (1960). *Potato Blight Epidemics Throughout The World*. Washington, DC. Agricultural Research Service. US Department of Agriculture. Agricultural Handbook no. 174.

Croft, J.H. & Jinks, J.L. (1977). Aspects of the population genetics of *Aspergillus nidulans*. In *Genetics and Physiology of Aspergillus* (ed. J.E. Smith and J.A. Pateman), pp. 339–360. London and New York: Academic Press.

Dahlberg, A. (1992). Somatic incompatibility – a tool to reveal spatiotemporal mycelial structures of ectomycorrhizal fungi. In *Mycorrhiza in Ecosystems* (ed. D.J. Read, D.H. Lewis, A.H. Fitter and I.J. Alexander), pp. 135–140. Wallingford, Oxon: CAB International.

Day, P.R. (1974). *Genetics of Host–Parasite Interaction*. San Francisco: W.H. Freeman.

Day, P.R. & Dodds, J.A. (1974). Viruses of plant pathogenic fungi. In *Viruses and Plasmids in Fungi* (ed. P.A. Lemke), pp. 201–238. New York, Basel, Hong Kong: Marcel Dekker.

Deacon, J.W., Donaldson, S.J. & Last, F.T. (1983). Sequences and interactions of mycorrhizal fungi on birch. *Plant and Soil*, **71**, 257–62.

Dennis, R.W.G., Orton, P.D. & Hora, F.B. (1960). New check list of British Agarics and Boleti. *Transactions of the British Mycological Society*, Supplement, 1–225.

Dick, M.W. (1971). The ecology of the Saprolegniaceae in lentic and littoral muds with a general theory of fungi in the lake ecosystem. *Journal of General Microbiology*, **65**, 325–37.

Dick, M.W. (1976). The ecology of aquatic phycomycetes. In *Recent Advances in Aquatic Mycology* (ed. E.B.G. Jones), pp. 513–542. London: Elek Science.

Dick, M.W. & Win-Tin (1973). The development of cytological theory in the Oomycetes. *Biological Reviews, Cambridge*, **48**, 133–58.

Dix, N.J. & Webster, J. (1995). *Fungal Ecology*. London, Glasgow: Chapman & Hall.

Dobbs, G.C. & Hinson, W.H. (1953). A widespread fungistasis in soils. *Nature, London*, **172**, 197–9.

Donk, M.A. (1971). Multiple convergence in the Polyporaceous fungi. In *Evolution in the Higher Basidiomycetes* (ed. R.H. Petersen), pp. 393–422. Knoxville: The University of Tennessee Press.

Drew, E.A. & Smith, D.C. (1967). Studies on the physiology of lichens. VIII. Movement of glucose from alga to fungus during photosynthesis in the thallus of *Peltigera polydactyla*. *New Phytologist*, **66**, 389–400.

Elliott, C.G. & MacIntyre, D. (1973). Genetical evidence on the life-history of *Phytophthora*. *Transactions of the British Mycological Society*, **60**, 311–16.

Ellis, M.B. (1971). *Dematiaceous Hyphomycetes*. Kew, Surrey: Commonwealth Mycological Institute.

Ellis, M.B. (1976). *More Dematiaceous Hyphomycetes*. Kew, Surrey: Commonwealth Mycological Institute.

Ellzey, J.T. (1974). Ultrastructural observations of meiosis within antheridia of *Achlya ambisexualis*. *Mycologia*, **66**, 32–47.

Fleming, A. (1929). On the antibacterial action of cultures of a *Penicillium* with special reference to their use for the isolation of *B.influenzae*. *British Journal of Experimental Pathology*, **10**, 226–36.

Florey, H.W., Chain, E., Heatley, N.G., Jennings, M.A., Sanders, A.G., Abraham, E.P. & Florey, M.E. (1949). *Antibiotics. A Survey of Penicillin, Streptomycin, and Other Antimicrobial Substances from Fungi, Actinomycetes, Bacteria and Plants, volume 2, Penicillin*, pp. 631–1293. London, New York, Toronto: Oxford University Press.

Ford, E.D., Mason, P.A. & Pelham, J. (1980). Spatial patterns of sporophore distribution around a young birch tree in three successive years. *Transactions of the British Mycological Society*, **75**, 287–96.

Francis, S.M. (1988). List of Peronosporaceae reported from the British Isles. *Transactions of the British Mycological Society*, **91**, 1–62.

Frankland, J.C. (1984). Autoecology and the mycelium of a woodland litter decomposer. In *The Ecology and Physiology of the Fungal Mycelium* (ed. D.H. Jennings and A.D.M. Rayner), pp. 241–260. Cambridge: Cambridge University Press.

Frankland, J.C. (1992). Mechanisms in fungal succession. In *The Fungal Community: Its Organization and Rôle in the Ecosystem* (ed. G.C. Carroll and D.T. Wicklow), pp. 383–401. New York, Basel, Hong Kong: Marcel Dekker.

Fraser, H.C.I. (1908). Recent work on the reproduction of Ascomycetes. *Transactions of the British Mycological Society*, **3**, 100–7.

Fries, N. (1987). Somatic incompatibility and field distribution of the ectomycorrhizal fungus *Suillus luteus* (Boletaceae). *New Phytologist*, **107**, 735–9.

Gandy, D.G. (1960*a*). 'Watery stipe' of cultivated mushrooms. *Nature, London*, **185**, 482–3.

Gandy, D.G. (1960*b*). A transmissible disease of cultivated mushrooms ('Watery Stipe'). *Annals of Applied Biology*, **48**, 427–30.

Gandy, D.G. & Hollings, M. (1962). Die-back of mushrooms: a disease associated with a virus. *Report of the Glasshouse Crops Research Institute*, **1961**, 103–8.

Garrett, S.D. (1950). Ecology of the root-inhabiting fungi. *Biological Reviews*, Cambridge, **25**, 220–54.

Garrett, S.D. (1954). Presidential Address. Microbial ecology of the soil. *Transactions of the British Mycological Society*, **38**, 1–9.

Garrett, S.D. (1956). *Biology of Root-Infecting Fungi*. Cambridge: Cambridge University Press.

Garrett, S.D. (1963). *Soil Fungi and Soil Fertility*. New York: Pergamon: Macmillan.

Gentles, J.C. (1958). Experimental ringworm in guineapigs: oral treatment with griseofulvin. *Nature, London*, **183**, 256–7.

Gerdemann, J.W. & Nicolson, T.H. (1963). Spores of mycorrhizal *Endogone* species extracted from soil by wet sieving and decanting. *Transactions of the British Mycological Society*, **46**, 234–44.

Gibbs, J.N. (1978). Intercontinental epidemiology of Dutch elm disease. *Annual Review of Phytopathology*, **16**, 287–307.

Gibbs, J.N. & Brasier, C.M. (1973). Correlation between cultural characters and pathogenicity in *Ceratocystis ulmi* from Europe and North America. *Nature, London*, **243**, 381–3.

Goldberg, H.S. (1959). (ed.) *Antibiotics: Their Chemistry and Non-Medical Uses*. Princeton, Toronto, London, New York: D. van Nostrand Company.

Gregory, P.H. (1951). Deposition of air-borne *Lycopodium* spores on cylinders. *Annals of Applied Biology*, **38**, 357–76.

Gregory, P.H. (1952). Presidential Address. Fungus spores. *Transactions of the British Mycological Society*, **35**, 1–18.

Gregory, P.H. (1973). *The Microbiology of the Atmosphere*, 2nd edn. Aylesbury: Leonard Hill.

Gregory, P.H. & Lacey, M.E. (1963). Mycological examination of dust from mouldy hay associated with Farmer's lung. *Journal of General Microbiology*, **30**, 75–88.

Griffin, D.M. (1972). *Ecology of Soil Fungi*. London: Chapman & Hall.

Hare, R. (1970). *The Birth of Penicillin and the Disarming of Microbes*. London: Allen and Unwin.

Harley, J.L. (1948). Mycorrhiza and soil ecology. *Biological Reviews, Cambridge*, **23**, 127–58.

Harley, J.L. (1950). Recent progress in the study of endotrophic mycorrhiza. *New Phytologist*, **49**, 213–47.

Harley, J.L. (1959). *The Biology of Mycorrhiza*. London: Leonard Hill.

Harley, J.L. (1969). *The Biology of Mycorrhiza*, 2nd edn. London: Leonard Hill.

Harley, J.L. & Smith, D.C. (1956). Sugar absorption and surface carbohydrase activity of *Peltigera polydactyla* (Neck.) Hoffm. *Annals of Botany*, London, N.S., **20**, 513–43.

Harley, J.L. & Smith, S.E. (1983). *Mycorrhizal Symbiosis*. New York, London: Academic Press.

Hastie, A.C. (1981). The genetics of conidial fungi. In *Biology of Conidial Fungi* (ed. G.T. Cole and B. Kendrick), pp. 511–547. New York, London: Academic Press.

Hirst, J.M. (1952). An automatic volumetric spore trap. *Annals of Applied Biology*, **39**, 257–65.

Hirst, J.M. (1953). Changes in atmospheric spore content: diurnal periodicity and the effects of weather. *Transactions of the British Mycological Society*, **36**, 375–93.

Hirst, J.M. (1973). Presidential Address. A trapper's line. *Transactions of the British Mycological Society*, **61**, 205–13.

Hollings, M. (1962). Viruses associated with a die-back disease of cultivated mushroom. *Nature, London*, **196**, 962–5.

Howard, K.L. & Moore, R.T. (1970). Ultrastructure of oogenesis in *Saprolegnia terrestris*. *Botanical Gazette*, **131**, 311–36.

Hudson, H.J. (1968). The ecology of fungi on plant remains above the soil. *New Phytologist*, **67**, 837–74.

Hughes, S.J. (1953). Conidiophores, conidia and classification. *Canadian Journal of Botany*, **31**, 577–659.

Ikediugwu, F.E.O. (1976). The interface in hyphal interference by *Peniophora gigantea* against *Heterobasidion annosum*. *Transactions of the British Mycological Society*, **66**, 291–6.

Ikediugwu, F.E.O., Dennis, C. & Webster, J. (1970). Hyphal interference by *Peniophora gigantea* against *Heterobasidion annosum*. *Transactions of the British Mycological Society*, **54**, 307–9.

Ingold, C.T. (1939). *Spore Discharge in Land Plants*. Oxford: Clarendon Press.

Ingold, C.T. (1942). Aquatic hyphomycetes of decaying alder leaves. *Transactions of the British Mycological Society*, **25**, 339–417.

Ingold, C.T. (1966). The tetraradiate aquatic fungal spore. *Mycologia*, **58**, 43–56.

Ingold, C.T. (1971). *Fungal Spores, Their Liberation and Dispersal*. Oxford: Clarendon Press.

Ingold, C.T. (1975). Hooker Lecture 1974. Convergent evolution in aquatic fungi: the tetraradiate spore. *Biological Journal of the Linnean Society*, **7**, 1–25.

Iqbal, S.H. & Webster, J. (1973). Aquatic hyphomycete spora of the River Exe and its tributaries. *Transactions of the British Mycological Society*, **61**, 331–46.

Jinks, J.L. (1952). Heterokaryosis: a system of adaptation in wild fungi. *Proceedings of the Royal Society*, **B140**, 83–106.

Jones, E.B.G. (1994). Presidential Address 1992. Fungal adhesion. *Mycological Research*, **98**, 961–81.

Jones, E.B.G., Johnson, R.G. & Moss, S.T. (1986). Taxonomic studies on the Halosphaeriaceae – philosophy and rationale for the selection of characters in the delineation of genera. In *The Biology of Marine Fungi* (ed. S.T. Moss), pp. 211–229. Cambridge: Cambridge University Press.

Kaushik, N.K. & Hynes, H.B.N. (1971). The fate of dead leaves that fall into streams. *Archiv für Hydrobiologie*, **68**, 465–515.

Kendrick, B. (1981). The history of conidial fungi. In *Biology of Conidial Fungi* 1 (ed. G.T. Cole and B. Kendrick), pp. 3–18. New York, London: Academic Press.

Large, E.C. (1953). Potato blight forecasting in England and Wales. *Plant Pathology*, **2**, 1–15.

Last, F.T., Mason, P.A., Wilson, J. & Deacon, J.W. (1983). Fine roots and sheathing mycorrhizas: their formation, function and dynamics. *Plant and Soil*, **71**, 9–21.

Lemke, P.A. & Nash, C.H. (1974). Fungal viruses. *Bacteriological Reviews*, **38**, 29–56.

Lentz, P.L. (1971). Analysis of modified hyphae as a tool in taxonomic research in the higher basidiomycetes. In *Evolution in the Higher Basidiomycetes* (ed. R.H. Petersen), pp. 99–127. Knoxville: The University of Tennessee Press.

Lewis, D.H. (1991). Presidential Address 1989. Fungi and sugars – a suite of interactions. *Mycological Research*, **95**, 897–904.

Lewis, D.H. & Harley, J.L. (1965). Carbohydrate physiology of mycorrhizal roots of beech. III. Movements of sugars between host and fungus. *New Phytologist*, **64**, 256–69.

Lewis, D.H. & Smith, D.C. (1967). Sugar alcohols (polyols) in fungi and green plants. I. Distribution, physiology and metabolism. *New Phytologist*, **66**, 143–84.

Lockwood, J.L. (1977). Fungistasis in soils. *Biological Reviews*, Cambridge, **52**, 1–43.

Lockwood, J.L. & Finonow, A.B. (1981). Responses of fungi to nutrient-limiting conditions and to inhibitory substances in natural habitats. In *Advances in Microbial Ecology* (ed. M. Alexander), pp. 1–61.

Manners, J.G. (1969). Presidential Address. The rust diseases of wheat and their control. *Transactions of the British Mycological Society*, **52**, 177–86.

Mason, P.A., Last, F.T., Pelham, J. & Ingleby, K. (1982). Ecology of some fungi associated with an ageing stand of birches (*Betula pendula* and *Betula pubescens*). *Forest Ecology and Management*, **4**, 19–37.

Moore, W.C. (1959). *British Parasitic Fungi*. Cambridge: Cambridge University Press.

Mosse, B. (1953). Fructifications associated with mycorrhizal strawberry roots. *Nature, London*, **171**, 974.

Mosse, B. (1959). The regular germination of resting spores and some observations on the growth requirements of an *Endogone* sp. causing vesicular–arbuscular mycorrhiza. *Transactions of the British Mycological Society*, **42**, 273–86.

Mosse, B. (1962). The establishment of vesicular–arbuscular mycorrhiza under aseptic conditions. *Journal of General Microbiology*, **27**, 509–20.

Nielsen, B.B., Wei-Yun Zhu, Trinci, A.P.J. & Theodoru, M.K. (1995). Demonstration of zoosporangia of anaerobic fungi on plant residues recovered from faeces of cattle. *Mycological Research*, **99**, 471–4.

Orpin, C.G. (1974). The rumen flagellate *Callimastix frontalis*: does sequestration occur? *Journal of General Microbiology*, **84**, 395–8.

Orpin, C.G. (1975). Studies on the rumen flagellate, *Neocallimastix frontalis*. *Journal of General Microbiology*, **91**, 249–62.

Orpin, C.G. (1976). Studies on the rumen flagellate, *Sphaeromonas communis*. *Journal of General Microbiology*, **94**, 270–80.

Orpin, C.G. (1977). The rumen flagellate *Piromonas communis*: its life history and invasion of plant material in the rumen. *Journal of General Microbiology*, **99**, 107–17.

Orpin, C.G. (1981). Isolation of cellulolytic phycomycete fungi from the caecum of the horse. *Journal of General Microbiology*, **123**, 287–96.

Oxford, A.E., Raistrick, H. & Simonart, P. (1939). Studies in the biochemistry of micro-organisms. 60. Griseofulvin, a metabolic product of *Penicillium griseofulvum* Dierckx. *Biochemical Journal*, **33**, 240–8.

Park, D. (1972). On the ecology of heterotrophic micro-organisms in fresh water. *Transactions of the British Mycological Society*, **58**, 291–9.

Peace, T.R. (1960). *The Status and Development of Elm Disease in Britain*. Bulletin **33**, pp. 44. London: Forestry Commission.

Pearson, A.A. & Dennis, R.W.G. (1948). Revised list of British Agarics and Boleti. *Transactions of the British Mycological Society*, **31**, 145–90.

Pearson, V. & Read, D.J. (1973). Biology of mycorrhiza in the Ericaceae. I. The isolation of the endophyte and synthesis of mycorrhizas in aseptic culture. *New Phytologist*, **72**, 371–9.

Peberdy, J.F. (1989). Fungi without coats – protoplasts as tools for mycological research. *Mycological Research*, **93**, 1–20.

Pegler, D. & Young, T.W.K. (1983). Anatomy of the *Lentinus* hymenophore. *Transactions of the British Mycological Society*, **80**, 469–82.

Pepys, J., Jenkins, P.A., Festenstein, G.N., Gregory, P.H., Lacey, M.E. & Skinner, F.A. (1963). Farmer's lung. Thermophilic actinomycetes as a source of farmer's lung hay antigen. *Lancet*, **ii**, 607–11.

Plunkett, B.E. (1985). Professor Cecil Terence Ingold, C.M.G., D.Sc., F.L.S., Hon.D.Litt. (Ibadan), Hon.D.Sc. (Exeter), Hon.D.C.L. (Kent). In *Contributions to Mycology. A Tribute to Professor C.T. Ingold on his Eightieth Birthday* (ed. M.W. Dick, D.N. Pegler and B.C. Sutton). *Botanical Journal of the Linnean Society*, **91**, 1–375.

Pontecorvo, G. (1954). Mitotic recombination in the genetic systems of filamentous fungi. *Caryologia* (Supplement), **6**, 192–200.

Pontecorvo, G. (1956). The parasexual cycle in fungi. *Annual Review of Microbiology*, **10**, 393–400.

Pontecorvo, G. & Roper, J.A. (1952). Genetic analysis without sexual reproduction by means of polyploidy in *Aspergillus nidulans*. *Journal of General Microbiology*, **6**, vii (Abstract).

Pontecorvo, G. & Roper, J.A. (1953). Diploids and mitotic recombination. *Advances in Genetics*, **5**, 218–33.

Pugh, G.J.F. (1980). Presidential Address. Strategies in fungal ecology. *Transactions of the British Mycological Society*, **75**, 1–14.

Purves, S. & Hadley, G. (1975). Movements of carbon compounds between partners in orchid mycorrhiza. In *Endomycorrhizas* (ed. F.E. Sanders, B. Mosse & P.B. Tinker), pp. 173–194. London, New York: Academic Press.

Raistrick, H. (1949). A region of biosynthesis. *Proceedings of The Royal Society, A*, **199**, 141–68.

Ramsbottom, J. (1913). A list of British species of Discomycetes arranged according to Boudier's system, with a key to the genera. *Transactions of the British Mycological Society*, **4**, 343–81.

Ramsbottom, J. (1915). A list of British species of Phycomycetes etc., with a key to the genera. *Transactions of the British Mycological Society*, **5**, 304–17.

Ramsbottom, J. & Balfour-Browne, F.L. (1951). List of Discomycetes recorded from the British Isles. *Transactions of the British Mycological Society*, **34**, 38–137.

Raper, K.B. (1952). A decade of antibiotics in America. *Mycologia*, **44**, 1–59.

Rayner, A.D.M. (1976). Dematiaceous hyphomycetes and narrow dark zones in decaying wood. *Transactions of the British Mycological Society*, **67**, 546–9.

Rayner, A.D.M. (1991). The challenge of the individualistic mycelium. *Mycologia*, **83**, 48–71.

Rayner, A.D.M., Coates, D., Ainsworth, A.M., Adams, T.J.H., Williams, E.N.D. & Todd, N.K. (1984). The biological consequences of the individualistic mycelium. In *The Ecology and Physiology of the Fungal Mycelium* (ed. D.H. Jennings and A.D.M. Rayner), pp. 509–540. Cambridge: Cambridge University Press.

Read, D.J. (1983). The biology of mycorrhiza in the Ericales. *Canadian Journal of Botany*, **61**, 985–1004.

Read, D.J. (1984). The structure and function of the vegetative mycelium of mycorrhizal roots. In *The Ecology and Physiology of the Fungal Mycelium* (ed. D.H. Jennings and A.D.M. Rayner), pp. 215–240. Cambridge: Cambridge University Press.

Read, D.J., Lewis, D.H., Fitter, A.H. & Alexander, I.J. (1992). (eds) *Mycorrhizas in Ecosystems*. Wallingford, Oxon: CAB International.

Read, S.J., Moss, S.T. & Jones, E.B.G. (1991). Attachment studies of aquatic hyphomycetes. *Philosophical Transactions of the Royal Society of London B*, **334**, 449–57.

Read, S.J., Moss, S.T. & Jones, E.B.G. (1992). Attachment and germination of conidia. In *The Ecology of Aquatic Hyphomycetes* (ed. F. Bärlocher), pp. 135–151. Berlin, Heidelberg: Springer-Verlag.

Richardson, D.H.S., Jackson Hill, D. & Smith, D.C. (1968). Lichen physiology. XI. The role of the alga in determining the pattern of carbohydrate movement between lichen symbionts. *New Phytologist*, **67**, 469–86.

Rishbeth, J. (1950). Observations on the biology of *Fomes annosus*, with particular reference to East Anglian pine plantations. I. The outbreaks of disease and ecological status of the fungus. *Annals of Botany*, London, N.S. **14**, 365–83.

Rishbeth, J. (1951). Observations on the biology of *Fomes annosus*, with particular reference to East Anglian pine plantations. II. Spore production, stump infection, and saprophytic activity in stumps. *Annals of Botany*, London, N.S., **15**, 1–21.

Rishbeth, J. (1952). Control of *Fomes annosus* Fr. *Forestry*, **24**, 41–50.

Rishbeth, J. (1963). Stump protection against *Fomes annosus*. III. Inoculation with *Peniophora gigantea*. *Annals of Applied Biology*, **52**, 63–77.

Roper, J.A. (1952). Production of heterozygous diploids in filamentous fungi. *Experientia*, **8**, 14–15.

Roper, J.A. (1966). Mechanisms of inheritance. 3. The parasexual cycle. In *The Fungi: an Advanced Treatise* **2** (ed. G.C. Ainsworth and A.S. Sussman), pp. 589–617. New York, London: Academic Press.

Ryvarden, L. (1991). *Genera of Polypores. Nomenclature and Taxonomy. Synopsis Fungorum*, **5**, pp. 363.

Sampson, K. (1940). List of British Ustilaginales. *Transactions of the British Mycological Society*, **24**, 294–311.

Sansome, E. (1946). Induction of 'gigas' forms of *Penicillium notatum* by treatment with camphor vapour. *Nature, London*, **157**, 843.

Sansome, E. (1949). Spontaneous mutation in standard and 'gigas' forms of *Penicillium notatum* strain 1249 B 21. *Transactions of the British Mycological Society*, **32**, 305–14.

Sansome, E. (1961). Meiosis in the oogonium and antheridium of *Pythium debaryanum* Hesse. *Nature, London*, **191**, 827–8.

Sansome, E. (1963). Meiosis in *Pythium debaryanum* Hesse and its significance in the life-history of the Biflagellatae. *Transactions of the British Mycological Society*, **46**, 63–72.

Sansome, E. (1966). Meiosis in the sex organs of the Oomycetes. In *Chromosomes Today*, I (ed. C.D. Darlington and K.R. Lewis), pp. 77–83. Edinburgh: Oliver and Boyd.

Shaw, D.S. (1983). The Peronosporales. A fungal geneticist's nightmare. In *Zoosporic Plant Pathogens. A Modern Perspective* (ed. S.T. Buczacki), pp. 85–121. London, New York: Academic Press.

Shaw, D.S. & Khaki, I.A. (1971). Genetical evidence for diploidy in *Phytophthora*. *Genetics Research*, **17**, 75–84.

Smith, D.C. (1961). The physiology of *Peltigera polydactyla* (Neck). Hoffm. *Lichenologist*, **1**, 209–26.

Smith, D.C. (1962). The biology of lichen thalli. *Biological Reviews, Cambridge*, **37**, 537–70.

Smith, D.C. (1963). Experimental studies of lichen physiology. In *Symbiotic Associations, Society for General Microbiology Symposium 13* (ed. R. Dubos and A. Kessler), pp. 31–50. Cambridge: Cambridge University Press.

Smith, D.C. (1981). Presidential Address. The symbiotic way of life. *Transactions of the British Mycological Society*, **77**, 1–8.

Smith, D.C. & Lewis, J.L. (1993). John Laker Harley. *Obituary Notices of Fellows of the Royal Society*, 159–175.

Smith, D.C., Muscatine, L. & Lewis, D.H. (1969). Carbohydrate movement from autotrophs of heterotrophs in parasitic and mutualistic symbiosis. *Biological Reviews*, **44**, 17–90.

Smith, M.L., Bruhn, J.N. & Anderson, J.B. (1992). The fungus *Armillaria bulbosa* is amongst the largest and oldest living organisms. *Nature, London*, **356**, 428–31.

Smith, S.E. (1966). Physiology and ecology of orchid mycorrhizal fungi with reference to seedling nutrition. *New Phytologist*, **65**, 488–99.

Smith, S.E. (1967). Carbohydrate translocation in orchid mycorrhizal fungi. *New Phytologist*, **66**, 371–8.

Stribley, D.P. & Read, D.J. (1980). The biology of mycorrhiza in the Ericaceae. VII. The relationship between mycorrhizal infection and the capacity to utilise simple and complex organic nitrogen sources. *New Phytologist*, **86**, 365–71.

Suberkropp, K. (1992). Interactions with invertebrates. In *The Ecology of Aquatic Hyphomycetes* (ed. F. Bärlocher), pp. 118–134. Berlin, Heidelberg: Springer-Verlag.

Sutherland, G.K. (1915). New marine fungi on *Pelvetia*. *New Phytologist*, **14**, 33–42.

Sutton, B.C. (1980). *The Coelomycetes*. Kew, Surrey: Commonwealth Mycological Institute.

Sutton, B.C. (1986). Presidential Address. Improvisations on conidial themes. *Transactions of the British Mycological Society*, **86**, 1–38.

Tegli, S., Comparini, C., Giannetti, C. & Scala, A. (1994). Effect of temperature on growth and cerato-ulmin production of *Ophiostoma novo-ulmi* and *O.ulmi*. *Mycological Research*, **98**, 408–12.

Theodoru, M.K., Lowe, S.E. & Trinci, A.P.J. (1992). Anaerobic fungi and the rumen ecosystem. In *The Fungal Community: Its Organization and Role in the Ecosystem* (ed. G.C. Carroll and D.T. Wicklow), pp. 43–72. New York, Basel, Hong Kong: Marcel Dekker.

Thompson, W. & Rayner, A.D.M. (1982). Spatial structure of a population of *Tricholomopsis platyphylla* in a woodland site. *New Phytologist*, **92**, 103–14.

Tribe, H.T. (1955). Studies in the physiology of parasitism. XIX. On the killing of plant cells by enzymes from *Botrytis cinerea* and *Bacterium aroideae*. *Annals of Botany*, London, N.S., **19**, 351–368.

Trinci, A.P.J., Davies, D.R., Gull, K. & Lawrence, M.I. (1994). Anaerobic fungi in herbivorous animals. *Mycological Research*, **98**, 129–52.

van Zaayen, A. (1979). Mushroom viruses. In *Viruses and Plasmids in Fungi* (ed. P. Lemke), pp. 239–324. New York, Basel, Hong Kong: Marcel Dekker.

Wakefield, E.M. & Bisby, G.R. (1941). List of Hyphomycetes recorded for Britain. *Transactions of the British Mycological Society*, **25**, 49–126.

Warcup, J.H. (1950). The soil-plate method for isolation of fungi from soil. *Nature, London*, **166**, 117.

Warcup, J.H. (1955a). Isolation of fungi from hyphae present in soil. *Nature, London*, **175**, 953–4.

Warcup, J.H. (1955b). On the origin of colonies of fungi developing on soil dilution plates. *Transactions of the British Mycological Society*, **38**, 298–301.

Warcup, J.H. (1959). Studies on basidiomycetes in soil. *Transactions of the British Mycological Society*, **42**, 45–52.

Warcup, J.H. (1960). Methods for isolation and estimation of activity of fungi in soil. In *The Ecology of Soil Fungi. An International Symposium* (ed. D. Parkinson and J.S. Waid), pp. 3–21. Liverpool: Liverpool University Press.

Warcup, J.H. & Talbot, P.H.B. (1962). Ecology and identity of mycelia isolated from soil. *Transactions of the British Mycological Society*, **45**, 495–518.

Watson, W. (1953). *Census Catalogue of British Lichens*. Cambridge: Cambridge University Press.

Webber, J.F. & Brasier, C.M. (1984). The transmission of Dutch elm disease: a study of the processes involved. In *Invertebrate–Microbial Interactions* (ed. J.M. Anderson, A.D.M. Rayner and D.W.H. Walton), pp. 271–306. Cambridge: Cambridge University Press.

Webber, J.F., Brasier, C.M. & Mitchell, A.G. (1987). The role of the saprophytic phase in Dutch elm disease. In *Fungal Infection of Plants* (ed. G.F. Pegg and P.G. Ayres), pp. 298–313. Cambridge: Cambridge University Press.

Webster, J. (1959). Experiments with spores of aquatic hyphomycetes. Sedimentation and impaction on smooth surfaces. *Annals of Botany, London,* N.S., **23**, 595–611.

Webster, J. (1987). Convergent evolution and the functional significance of spore shape in aquatic and semi-aquatic fungi. In *Evolutionary Biology of the Fungi* (ed. A.D.M. Rayner, C.M. Brasier and D. Moore), pp. 191–201. Cambridge: Cambridge University Press.

Webster, J. (1992). Anamorph–teleomorph relationships. In *The Ecology of Aquatic Hyphomycetes* (ed. F. Bärlocher), pp. 99–117. Berlin, Heidelberg: Springer-Verlag.

Webster, J. & Chien, C-Y. (1990). Ballistospore discharge. *Transactions of the Mycological Society of Japan*, **31**, 301–15.

Webster, J. & Davey, R.A. (1984). Sigmoid conidial shape in aquatic fungi. *Transactions of the British Mycological Society*, **83**, 43–52.

Webster, J., Davey, R.A., Smirnoff, N., Fricke, W., Hinde, P., Tomos, D. & Turner, J.C.R. (1995). Mannitol and hexoses are components of Buller's drop. *Mycological Research*, **99**, 833–8.

Webster, J., Proctor, M.C.F., Davey, R.A. & Duller, G.A. (1988). Measurement of the electrical charge on some basidiospores and an assessment of two possible mechanisms of ballistospore propulsion. *Transactions of the British Mycological Society*, **91**, 193–203.

Williams, D.I., Marten, R.H. & Sarkany, I. (1958). Oral treatment of ringworm with griseofulvin. *Lancet*, **ii**, 259–266.

Wilson, I.M. (1960). Marine fungi: a review of the present position. *Proceedings of the Linnean Society of London*, **171**, 53–70.

Wilson, M. & Bisby, G.R. (1954). List of British Uredinales. *Transactions of the British Mycological Society*, **37**, 61–86.

Wood, R.K.S. (1967). *Physiological Plant Pathology*. Oxford and Edinburgh: Blackwell Scientific Publications.

2

My involvement with aquatic hyphomycetes

C.T. INGOLD

In the final year (1925–26) of my degree course in botany at Queen's University, Belfast, I specialized in mycology, although no member of staff was even remotely mycological. My first post-graduate experience was at Imperial College, London, following an advanced course equally divided between mycology and plant physiology. I was seduced by the latter, and, on returning to Queen's in 1927 worked for the PhD degree on the systems in the sap of plants buffering against changes in pH. The years 1927–30 seem in retrospect to have been frustrating and sterile. In 1930, I was appointed lecturer in the University of Reading with special responsibility for plant physiology, but in fact undertook more than half of all the teaching in the Department of Botany! At this stage in my life, I still had strong leanings towards mycology and even ecology.

At Reading I met Walter Buddin, a great friendly bear of a man, who for many years was Treasurer of the British Mycological Society and, on more than one occasion, firmly refused nomination as President. He introduced me and Lilian Hawker, then researching in the Botany Department on the statolith theory of geotropism, to the joys of fungal forays and was instrumental in us both joining the Society in 1932.

On progressing from Reading to University College, Leicester in 1937, I was still not sure if my heart was in plant physiology or in mycology, in spite of a growing concern with the dispersal of fungal spores. Somehow in Leicester, I developed a passing interest in the biology of Swithland Reservoir, and became excited by the chytrids attacking planktonic algae there. It was a great pleasure to discover *Endocoenobium eudorinae*, surely one of the more beautiful of these abundant microfungi. For me this opened a new chapter in mycology.

In 1938 we moved to Cropstone, north of Leicester and on the edge of

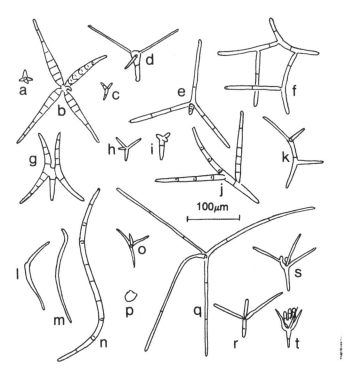

Fig. 2.1. Conidia of aquatic hyphomycetes. a, *Clavatospora stellata*; b, *Flabellospora* sp.; c, *Volucrispora graminea*; d, *Clavariopsis aquatica*; e, *Lemonniera aquatica*; f, *Varicosporium elodeae*; g, *Triposporina* sp.; h, *Clavatospora longibrachiata*; i, *Heliscus lugdunensis*; j, *Tricladium splendens*; k, *T. articulatum*; l, *Lunulospora curvula*; m, *Flagellospora curvula*; n, *Anguillospora longissima*; o, *Alatospora acuminata*; p, *Margaritispora aquatica*; q, *Tetrachaetum elegans*; r, *Articulospora tetracladia*; s, *Tetracladium marchalianum*; t, *T. setigerum* (from Ingold, 1942).

Charnwood Forest. A couple of hundred yards from our new home was the Donkey Brook, a little, alder-lined, babbling brook. There I naturally sought chytrids and, indeed, found a few, including a new species (*Chytridium lecithii*) parasitizing a shelled rhizopod in the surface scum which accumulated upstream of barriers of twigs. To my utter amazement the scum also contained in abundance many kinds of most extraordinary fungal spores. They were relatively large and colourless. Most were branched and of these the majority were tetra-radiate, but several were long and worm-like with a curvature in more than one plane. Only a very few were of a more conventional form (Fig. 2.1).

For several months I continued to find these spores in scum and especially in cakes of foam formed in the brook after heavy rain. It was only after long search that their source was tracked down to fungi on submerged decaying alder leaves in the bed of the stream.

When such a leaf, brown, softening and fast becoming a skeleton, was washed, submerged in a Petri dish and viewed next day under a dissecting microscope, it was normally found to bear rich crops of conidial fungi especially on its petiole and main veins. The short conidiophores grew into the water and the conidia, when mature, were at once liberated giving an abundant deposit on the bottom of the dish. A single leaf usually bore miniature forests of several species of these fungi. Clearly their conidia were produced, liberated and dispersed below water. Other fungi were rarely seen, although isolated sporangia of *Phytophthora* sp. on short sporangiophores were occasionally observed. Because of their abundance, it seemed likely that these fungi played a major role in the decay of the leaves.

It was fairly easy to isolate them in culture in nutrient agar. Most produced distinctive colonies, but few sporulated on the solid surface of the jelly. However, when strips of a colony were submerged in shallow water, sporulation usually occurred as on the leaves.

In studying most species, I followed the development of individual conidia in hanging-drops of water. One example from my first paper is here reproduced (Fig. 2.2). This approach to the study of sporulation has remained with me into old age and has proved rewarding in my work in the 1980s on basidium formation following teliospore formation in smut fungi. It is part of my basic philosophy of mycology that fungi are best studied as living, growing organisms.

I had no idea as to the identity of these fungi, but John Ramsbottom, with his phenomenal knowledge of mycological literature, drew my attention to Karling's paper on *Tetracladium* (Karling, 1935) and this led me to De Wildeman's three papers (De Wildeman, 1893, 1894, 1895).

It was apparent that, although a few had been described rather inadequately, most would have to be accommodated in new taxa. In my original paper (Ingold, 1942) eight genera were erected, namely: *Margaritispora*, *Articulospora*, *Tetrachaetum*, *Alatospora*, *Tricladium*, *Anguillospora*, *Flagellospora* and *Lunulospora*, all of which have withstood the test of time as accepted hyphomycete genera.

In having no training in, or aptitude for, taxonomy, I was at a great disadvantage. Further, the library at University College, Leicester was almost completely devoid of mycological literature. In partial solution of this problem, I visited Birmingham on a number of occasions and C.G.C. Chesters, then a lecturer there, was extremely helpful and especially introduced me to E.W. Mason's work on conidial fungi in which the distinction

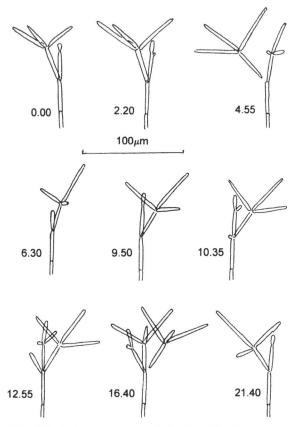

Fig. 2.2. *Articulospora tetracladia.* Conidiophore drawn at intervals over a period of 21 h 40 min (from Ingold, 1942).

between phialospores and other types of conidia was discussed. I soon found that some of my aquatic hyphomycetes produced phialoconidia in succession from a phialide, but in others this was not so. Indeed, there appeared to be pairs of genera (*Alatospora* and *Tetrachaetum*, *Clavatospora* and *Clavariopsis*, *Flagellospora* and *Anguillospora*) with the first member of each pair forming phialospores and the second, in spite of an otherwise matching ontology, producing conidia of a fundamentally different kind. Much later, *Lemonniera* and *Actinospora* were seen to represent a similar pair.

A year or two after finding these remarkable fungi, I gave a preliminary account of them to a meeting of the British Mycological Society. Two points from the ensuing discussion are vivid in my memory. John Ramsbottom remarked that he hoped I would indicate on a map exactly

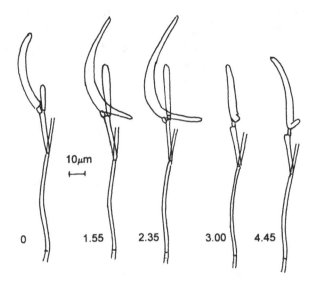

Fig. 2.3. *Lunulospora curvula.* Conidium development over a period of 4 h 45 min. Only one of the two conidium-producing apices is figured. First conidium shed at 2.45 h (from Ingold, 1942).

where they had been found as the situation might be unique and that such a distinctive fungal flora might not occur elsewhere. The other contributor to the discussion, a member from Cambridge who shall remain nameless, said it was clear that my pictures were of stellate hairs from decayed leaves and were not spores at all!

I soon began to look in other brooks and at different kinds of leaf. These fungi were there in profusion in all fast-flowing woodland streams and on every sort of leaf from dicotyledonous trees and shrubs, especially willow, oak and hawthorn. Few, however, grew on the leaves of beech and no conidial fungi were seen on pine and larch needles in the bed of a stream.

This distinctive fungal flora was to be found throughout the year, but was especially well-developed in October and November following the annual leaf-fall. Most species could, however, be collected at any time with the exception of *Lunulospora curvula* (Fig. 2.3), which seemed to be restricted to the warmer months.

Early in my studies, I found that examination of persistent foam gave an immediate picture of the species present in a stream, for most could be identified with reasonable certainty by their conidia alone. However, if a sample collected in a specimen tube was examined after the lapse of a few hours, many conidia were found to have germinated. It was, therefore, necessary to fix samples on collection by the addition of a trace of form-acetic alcohol. It is one of the outstanding features of aquatic hyphomycetes that, while their conidia are suspended in water or trapped in foam, no germination occurs, but within an hour or two of contact with a firm surface, germ-tubes develop.

In 1944 I moved to my Chair in London and for the next 33 years lived in Kent. Throughout this period my active interest in aquatic hyphomycetes continued, but most of my research energy was directed towards problems of spore discharge, and I never involved any research student in these fungi. However, in many parts of the world work on them blossomed, and they became the subject of post-graduate studies in a number of countries, including America, Sweden, Russia, Hungary, Czechoslovakia, Ireland, Japan, Malaysia, New Zealand and South Africa. This particular fungal flora was clearly of world-wide occurrence and significance.

It may be of interest to tell two stories about my involvement with these fungi in Kent. In 1952 *Actinospora megalospora* (Fig. 2.4) was described (Ingold, 1952). This had relatively enormous conidia, easily visible with a hand lens, which had very occasionally been seen in foam from the Donkey Brook, although no reference to them had been made in my original paper. Their attachment to conidiophores had not been observed, and nothing was known of their development. It was a delight to find this fungus growing sparingly on a twig in a stream at Sevenoaks. The pattern of development resembled that of *Lemonniera*, but the conidia were not formed from phialides. Although a pure culture was obtained, I could not induce it to sporulate. Many years later this was re-isolated by E. Descals and, in a spectacular film (Webster, Descals & Nawawi, 1975), the unusual development pattern of its huge conidia was recorded. This fungus was found in the week my son was born and the original idea was to call the genus *Timomyces*, but in the end my courage failed.

My second story involves the genus *Gyoerffyella*. E.A. Ellis and I had figured an extraordinary conidium (Fig. 2.5), rather like a Catherine wheel, from scum in a ditch at Wheatfen Broad, Norfolk (Ingold & Ellis, 1952). Later Petersen (Petersen, 1962), working in the USA, found the whole fungus on submerged decaying leaves and named it *Ingoldia craginiformis*. However, it transpired that the conidia had been seen long ago, had been considered to be a planktonic alga and, as such, given a valid name (Marvanová, Marvan & Ruzicka, 1967).

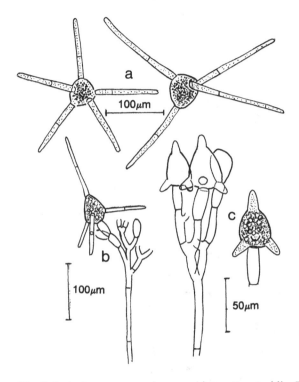

Fig. 2.4. *Actinospora megalospora.* Above, two conidia. Below left, conidiophore with only one conidium shown, and only part of its branch system which is spread out to show the manner of the branching. Below right, part of a conidiophore; eight developing conidia were present but only three are shown; the middle conidium (extreme right), drawn fifteen minutes later, illustrates the characteristic cytoplasmic contents (after Ingold, 1952).

The conidium of *Gyoerffyella* is most elegant, a feature shared by many others in the aquatic spora. To emphasize this general elegance, I could not refrain, in my Hooker Lecture to the Linnean Society (Ingold, 1977), from including a somewhat irrelevant picture that had already formed one of my annual Christmas cards. It is again reproduced here (Fig. 2.6). In the main it is an absolutely faithful copy of a Minoan plate from Crete (*ca* 1800 BC). Two figures are worshipping a god whom I have removed, substituting conidia of some aquatic hyphomycetes. Both elements unite in a ballet of dancing harmony.

But that is not the end of my *Gyoerffyella* story. In 1960 we moved, still in Kent, to a small bungalow, and the financial operations involved caused me great, but completely unnecessary, worry. Further, the move took place

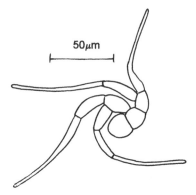

Fig. 2.5. *Gyoerffyella craginiformis.* Conidium from ditch scum at Wheatfen Broad (after Ingold & Ellis, 1952).

Fig. 2.6. Exact copy of a Minoan plate from Crete (*ca* 1800 BC) with central figure removed and replaced by camera lucida drawings of conidia of aquatic hyphomycetes.

in mid-winter during a prolonged period of intense frost. I was at a low ebb. Our new garden had a small lily pond and in it, growing on submerged decaying rose leaves, I found a beautiful new species of *Gyoerffyella* (Ingold, 1964). It was a life-saver.

Constantly and abundantly present in foam from numerous streams in widely scattered parts of Britain, were the two-celled, clove-shaped conidia of *Heliscus lugdunensis.* Also in most samples were spores of much the same shape, but somewhat larger and always unicellular. I was to see them again in Gull Lake, Michigan, USA in 1965. They were always few and far between, and, unlike the conidia of *H.lugdunensis*, did not appear capable of germination. Further, their cytoplasm was quite unlike that of other aquatic hyphomycetes, and, on staining, each spore was seen to have a single, relatively large nucleus, while in most fungi nuclei are annoyingly small. It occurred to me that they might be cysts of a rhizopod and not fungal at all. Many years later, in a foam sample collected in Scotland, unusually rich in these spores, one was seen that had produced a substantial hyphal germ-tube. About the same time John Webster and his Exeter colleagues discovered that they were entomophthoraceous conidia from fungi parasitizing blackfly larvae in a Devon stream. Their investigation of these organisms is an exciting chapter in modern mycology (Webster, Sanders & Descals, 1978).

In the 1950s and 1960s, I made many visits overseas, particularly to the West Indies and Africa (East, West and Central), in connection with the special involvement of the University of London in the development of university institutions in the colonial territories that were becoming independent. I was centrally and deeply involved in this activity which occupied most of my time abroad, although aquatic hyphomycetes were not completely neglected. In one of my visits to Jamaica, I was able to spend almost a week with Harry Hudson, then a lecturer in the University College of the West Indies, collecting decaying leaves from the delightful streams cascading down the Blue Mountains. The leaves were subsequently examined at leisure in the laboratory on the College campus. A joint paper resulted in which, among other matters, we described *Jaculispora submersa* (Hudson & Ingold, 1960), conidia of which I was later occasionally to find in foam from British streams.

In Jamaica I was impressed by the rich growth of *Lunulospora curvula* on the submerged decaying leaves, a situation later seen in both Africa and India. As already remarked, this species in Britain is to be seen mainly in the warmer months. Two other species, namely *Triscelophorus monosporus* and *Flagellospora penicillioides*, rather rare in Britain, were common in Jamaica and also in Africa.

In the course of several visits to Uganda, I examined foam from a number of localities there. My hosts were keen to show me elephant, zebra, buck and, of course, lion and were rather disappointed when I asked instead to be taken to forests with streams flowing through them.

The spora of a tropical stream, as revealed by examination of scum and foam, is in general similar to, but in detail different from, that seen in temperate regions. Were I to be transported to a part of the world without knowledge of its locality, I could make a very rough guess as to its latitude by examining the spora of a suitable stream, but would not know, however, it if were north or south of the equator. The aquatic spora of New Zealand is very similar to that of Britain.

I had assumed that, as with most other hyphomycetes, these aquatic conidial fungi were anamorphs of ascomycetes. The first proof of such a connection came when a *Nectria* developed in a culture of *Flagellospora penicillioides* (Ranzoni, 1956). Since then many more ascigerous stages have come to light, mainly as a result of the work of John Webster. In revealing these connections, I have played no part. For a short time I hopefully studied some submerged aquatic ascomycetes, but they provided no suggestive conidial stages. Being defeated, mainly by my ignorance of ascomycete taxonomy, this line of enquiry was abandoned.

In Nigeria I examined samples of foam from a number of forest streams in and around Ibadan. In two of these a relatively large tetra-radiate conidium was seen in which each long arm had septa with clamp-connections (Fig. 2.7). This was the first indication that some members of this aquatic spora might be basidiomycetes. Many years on, the fungus producing this particular kind of spore was discovered in Queensland, Australia and named *Ingoldiella hamata* (Shaw, 1972) and about the same time I was to see its conidia in a stream in Southern India. More recently the teleomorph (*Sistotrema hamatum*) has come to light (Nawawi & Webster, 1982).

In foam from the River Dee at Banchory, Scotland, close to a famous salmon leap, I had found abundant conidia of a form new to me. Each was basically tetra-radiate consisting of a bent main axis and two laterals. Midway between their points of origin was a tiny lateral knob whose true nature I failed to recognize. Shortly afterwards Miura (1974) found these conidia in Japan and recognized the knob as a clamp-connection. Soon I found the conidia in another Scottish river (Fig. 2.7) and was able to confirm this for myself.

These conidia were also picked out of foam, collected in North Wales, by the Exeter workers and transferred to nutrient agar. In the resulting cultures the teleomorph developed which was definitely identified as *Leptosporomyces galzinii*, a terrestrial basidiomycete of wide distribution especially on many kinds of coniferous wood (Nawawi, Descals & Webster, 1977).

On retiring in 1972 and still living in Kent, my interest in aquatic hyphomycetes persisted and I prepared a small illustrated guide to them as

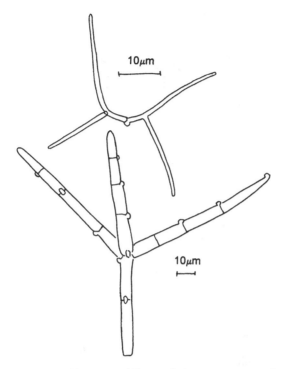

Fig. 2.7. Above, conidium of *Taeniospora gracilis* from Dundonnell River, Wester Ross, Scotland (after Ingold, 1974). Below, conidium of *Sistotrema hamatum* (perfect stage of *Ingoldiella hamata*) from Nigeria (after Ingold, 1959).

No. 30 in the series of Scientific Publications produced by the Freshwater Biological Association (Ingold, 1975). It is now out of date, but it may have helped to alert freshwater ecologists to the importance of these fungi. They seem to be the principal colonizers and degraders of submerged dicotyledonous trees and shrubs, and such leaves would appear to be the main input of organic matter into streams and small rivers flowing through areas of natural woodland in temperate regions. Felix Bärlocher has recently (Bärlocher, 1992) edited a volume on '*The Ecology of Aquatic Hyphomycetes*'.

For me, serious work on aquatic hyphomycetes ceased in 1977 on leaving Kent and coming to live in Oxfordshire. Not only were there no suitable local streams, but also, working at home, I could hardly compete with the splendid studies on these fungi under way elsewhere especially in Exeter. For the past 17 years, therefore, I have been concerned with quite

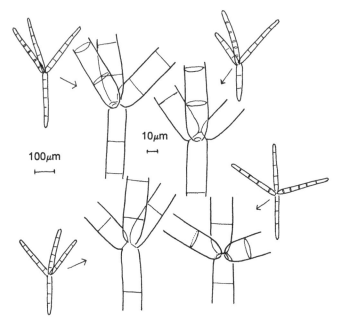

Fig. 2.8. Four conidia of an unknown species from Sezibra Falls, Uganda drawn under low power, together with high power details of their central regions (after Ingold, 1958).

different microfungi. However, on the occasional autumn foray, it has been a joy to look at samples of foam. It was especially pleasing, during the Irish foray to Roscrae in 1989, to see conidia of *Actinospora megalospora* in abundance in a little brook at Killarney.

There are a number of aquatic hyphomycetes known only as conidia in foam. In due course, some of these will, doubtless, be associated with the whole fungus either by picking them out individually and culturing them, or by discovering them attached to conidiophores on submerged decaying leaves. One in particular may be mentioned here. Its conidia are even larger than those of *Actinospora megalospora*. I have seen them in foam both from Uganda and from Swaziland. Each conidium has four divergent, septate arms about 300 μm long and 20 μm broad (Fig. 2.8).

Aquatic hyphomycetes have added spice to my life as they have to fellow mycologists scattered over the world, and it would be sad if students of mycology, at all levels, did not continue to have an opportunity to collect and look at these beautiful fungi in the living state at some stage in their training.

References

Bärlocher, F. (ed.) (1992). *The Ecology of Aquatic Hyphomycetes*. Springer-Verlag: Berlin.

De Wildeman, E. (1893). Notes mycologiques Fasicle 2. *Annales de Société Belge de Microscopie*, **17**, 35–68.

De Wildeman, E. (1894). Notes mycologiques Fasicle 3. *Annales de la Société Belge de Microscopie*, **18**, 135–61.

De Wildeman, E. (1895). Notes mycologiques Fasicle 6. *Annales de la Sociétée Belge de Microscopie*, **19**, 191–232.

Hudson, H.J. & Ingold, C.T. (1960). Aquatic hyphomycetes from Jamaica. *Transactions of the British Mycological Society*, **43**, 469–78.

Ingold, C.T. (1942). Aquatic hyphomycetes of decaying alder leaves. *Transactions of the British Mycological Society*, **25**, 339–417.

Ingold, C.T. (1952). *Actinospora megalospora* n.sp. an aquatic hyphomycete. *Transactions of the British Mycological Society*, **35**, 66–70.

Ingold, C.T. (1958). Aquatic hyphomycetes from Uganda and Rhodesia. *Transactions of the British Mycological Society*, **41**, 109–14.

Ingold, C.T. (1959). Aquatic spora of Omo Forest, Nigeria. *Transactions of the British Mycological Society*, **42**, 479–85.

Ingold, C.T. (1964). A new species of *Ingoldia* from Britain. *Transactions of the British Mycological Society*, **47**, 103–7.

Ingold, C.T. (1974). Foam spora from Britain. *Transactions of the British Mycological Society*, **63**, 487–97.

Ingold, C.T. (1975). *Guide to Aquatic Hyphomycetes*. Freshwater Biological Association Scientific Publication No. 30, Ambleside, Cumbria.

Ingold, C.T. (1977). Hooker Lecture 1974. Convergent evolution in aquatic fungi: the tetraradiate spore. *Biological Journal of the Linnean Society*, **7**, 1–25.

Ingold, C.T. & Ellis, E.A. (1952). On some hyphomycete spores including those of *Tetracladium maxilliformis*, from Wheatfen. *Transactions of the British Mycological Society*, **35**, 158–61.

Karling, J.S. (1935). *Tetracladium marchalianum* and its relations to *Asterothrix*, *Phycastrum* and *Cerasteria*. *Mycologia*, **27**, 478–95.

Marvanová, L., Marvan, P. & Ruzicka, J. (1967). *Gyoerffyella* Kol 1928, a genus of hyphomycetes. *Persoonia*, **5**, 29–44.

Miura, K. (1974). Stream spora in Japan. *Transactions of the Mycological Society of Japan*, **15**, 289–308.

Nawawi, A. (1973). Two clamp-bearing aquatic fungi from Malaysia. *Transactions of the British Mycological Society*, **61**, 521–8.

Nawawi, A., Descals, E. & Webster, J. (1977). *Leptosporomyces galzinii*, the basidial state of a branched conidium from fresh water. *Transactions of the British Mycological Society*, **68**, 31–6.

Nawawi, A. & Webster, J. (1982). *Sistostrema hamatum* sp. nov. the teleomorph of *Ingoldiella hamata*. *Transactions of the British Mycological Society*, **78**, 287–91.

Petersen, R.H. (1962). Aquatic hyphomycetes from North America. I. Aleuriosporae (Part I) and key to the genera. *Mycologia*, **54**, 117–51.

Ranzoni, F.V. (1956). The perfect stage of *Flagellospora penicillioides*. *American Journal of Botany*, **43**, 13–17.

Shaw, D. (1972). *Ingoldiella hamata* gen. et sp. nov., a fungus with clamp-connections from a stream in North Queensland. *Transactions of the British Mycological Society*, **59**, 255–9.

Webster, J., Descals, E. & Nawawi, A. (1975). Conidium development in aquatic hyphomycetes I. 16 mm film. University of Exeter.

Webster, J., Sanders, P.F. & Descals, E. (1978). Tetraradiate aquatic propagules in two species of *Entomophthora*. *Transactions of the British Mycological Society*, **70**, 472–9.

3

Advances in tropical mycology initiated by British mycologists

D.N. PEGLER

Introduction

Knowledge of the geographical distribution of fungi has grown from the long tradition of naturalists and mycologists to publish lists and, often at the same time, to describe the fungi found in localities where they live and regions which they visit. Such studies do much to retain the interest in fungi of successive generations of workers who by their current listing update classifications, correct past errors, and add to primary mycological census which is still so very incomplete. The earliest British contribution to our knowledge of tropical–subtropical fungi is probably to be found in *Synopsis Methodica Stirpium Britannicarum* (1690) by John Ray, which records the plants and fungi observed by Sir Hans Sloane when based in Jamaica, in which three fungi (species of *Auricularia* Bull., *Agaricus* L., possibly *Amanita* Pers.) were listed. Koenig, around 1770–1780 was collecting fungi in Ceylon (Sri Lanka), but these were not looked at until Berkeley examined them at the British Museum (Natural History) in 1842.

Regional descriptive mycology commenced at the beginning of the seventeenth century (Clusius), following on from the early herbal classifications of Barbaro (1454–1495) and Mattioli (1560). Taxonomic decisions, such as there were, were very much confined to Europe, particularly Germany, France and Italy, and this situation persisted up to the beginning of the nineteenth century. The first comprehensive list of English fungi was that by the Rev. Miles Joseph Berkeley (1836) (Fig. 3.1), which included much original contribution, with later volumes on British fungi by Mordecai Cubitt Cooke (1871) and George Massee (1892–95). Robert Greville first described fungi from Scotland in 1821 and went on to produce the six volumes of the *Scottish Cryptogamic Flora* (1823–28), whilst *Mycologia Scotica* was compiled by the Rev. John Stephenson in 1879.

Figs 3.1–3.5. Early British contributors to tropical mycology. Fig. 3.1. Rev. Miles Joseph Berkeley (1803–1889). Fig. 3.2. George Henry Kendrick Thwaites (1812–1882). Fig. 3.3. Mordecai Cubitt Cooke (1825–1914). Fig. 3.4. Edwin J. Butler (1874–1943). Fig. 3.5. Elsie Maud Wakefield (1886–1972). (Photographs provided by Royal Botanic Gardens, Kew.)

Throughout the nineteenth century, European mycologists did, however, examine increasing numbers of fungi from other parts of the world, collected as the results of voyages of exploration. Sir William Hooker was amongst the first to receive material. In particular, Hooker

(1816, 1822) described the Humboldt & Bonpland specimens from South America. Soon, however, he was handing over all fungal material to Berkeley. Similar collections were being sent to Fries in Sweden, Montagne and Léveillé in France, and Klotzsch in Germany. These workers began to describe exotic fungi and papers appeared with ever greater frequency.

M.J. Berkeley and the early collectors

Berkeley was particularly active, working mainly from collections regularly passed on by W. Hooker and then his son, Sir Joseph Dalton Hooker, the first Directors of the Royal Botanic Gardens at Kew. The earliest published accounts of exotic fungi by Berkeley were from Tasmania, then Van Diemen's Land (1839*a*), and also those collected by Charles Darwin on the 'Beagle' expedition (1840, 1842*a*). In 1844, Berkeley initiated his published series of the 'Decades of Fungi', which continued until 1856. Some 620 taxa were described within these 'Decades' from Australia, Ceylon, India, Jamaica, Java, North and South America (Peru, Rio Negro), and the Philippine Islands. Berkeley produced two major, tropical, floristic works, the 'Fungi Cubenses', in association with the American, M.A. Curtis (1869*a*, *b*), and the 'Fungi of Ceylon' (1846*b*, 1868, 1871*a*, 1871*b*, 1872, 1875, 1877), often in association with his close English collaborator, C.E. Broome.

In addition, during his lifetime, Berkeley was responsible for mycological accounts from the West Indies (1839*b*, 1876*a*), Borneo (1852*d*), Brazil (1839*b*, 1843*b*, 1877, 1879), Fiji (1842*c*), Hong Kong (1842*c*), India (1839*b*), Kerguelen Is. (1876*b*), Marian Is. (1876*b*), Mauritius (1839*b*), Mexico (1867), Moluccas (1878), New Guinea (1842*c*), New Hebrides (1878), New Ireland (1842*c*), Peru (1839*b*), Philippine Islands (1842*b*, 1878), St Domingo (1852*c*), South Africa (1843*a*, 1878), Tahiti (1878), Tonga (1878), Tristan da Cunha (1876*a*), and Zanzibar (1885).

Over the second half of the nineteenth century, journeys were increasingly undertaken throughout the developing countries, and botanists were frequently attached to exploration expeditions. Studies similar to those of Berkeley were initiated by Hennings in Berlin, and by Montagne in Paris. Berkeley actively exchanged both information and specimens with Montagne and Fries, and these were amongst the first to appreciate the importance of examining and comparing authentic material.

Much credit for the considerable British contribution to tropical mycology must be given to the early collectors and enthusiasts, who had either emigrated or, more often, had taken up official appointments in the 'developing countries' of the British Empire. Two individuals are especially worthy of mention, Gardner and Thwaites. George Gardner (1812–1849) qualified as a surgeon but his passion lay with botany. In 1836, he travelled to Brazil from where he sent back to England some 60000 botanical specimens, representing 3000 species. Most of these arrived at the British Museum (Natural History) but the fungi were forwarded to Berkeley, who promptly described them as new species in his 'Decades'. In 1844, Gardner was appointed to the position of Superintendent of the Botanic Garden at Peradeniya in Ceylon (Sri Lanka), having first visited Madras and the Nilgiri Hills in Southern India, and remained until his death in 1849. Although he travelled widely over the wet zone of the island, much of his collecting was restricted to or nearby Peradeniya, near Kandy, especially the Hantane Hills. He sent 135 fungal collections to Berkeley, who was able to describe (1846*b*, 1847*b*) the species in some detail, because the specimens were accompanied by a small, octavo volume of water-colour paintings. It is this kind of first hand observation, that did so much to assist the systematists back in Britain.

The second major collector was also closely linked with the Royal Botanic Garden at Peradeniya. George Henry Kendrick Thwaites (1812–1882) (Fig. 3.2) was a naturalist who published little, but corresponded extensively with the botanists of his time. He was respected to the extent that Montagne dedicated the diatom genus, *Thwaitesia*, Joseph Hooker proposed the genus *Kendrickia* (*Melastomataceae*), and Massee proposed the genus *Thwaitesiella* (*Stereaceae*) in his honour. In 1849, he succeeded Gardner as Superintendant at Peradeniya, a position which he held until 1880. He collected fungi, lichens and mosses, which were sent to specialists in England. He divided each number, retaining part of the collection at Peradeniya, sending the remainder to Berkeley who, with the aid of C.E. Broome, produced the early account of the *Fungi of Ceylon*. Later numbered sets were prepared for distribution. This work was based upon a total of 1200 collections, from which were described 403 species of agarics, with 305 providing the type-specimens to new species. Thwaites employed numerous collectors to assist him and, perhaps more importantly, most of the collections were accompanied by excellent water-colour paintings. These paintings were executed by a local artist, W. de Alwis, and careful copies were made for Kew by Cecilia Jane Berkeley, a daughter of the Rev. M.J. Berkeley. The originals were returned to Peradeniya, where they are currently housed at the Division of Plant Pathology of the Central Agricultural Research Institute at Gunnoruwa. The copies may be found in a bound volume at Kew.

To the present day, much of what is known of tropical agaricology rests on the information provided by these early collections. The collections led to investigative visits to Sri Lanka by Beccari in 1865, Cesati in 1879, Ferdinand von Höhnel in 1907, and Pegler in 1974. As Superintendant of the Gardens, Thwaites also had to perform many duties as a plant pathologist, and in 1869, it was he who sent material of the cause of the Coffee Leaf Disease, *Hemileia vastatrix*, for Berkeley to describe in 1869 (Berkeley, 1869*a*, see also Berkeley, 1872).

M.C. Cooke and G. Massee

Extensive material was gathered by other collectors and forwarded to mycologists in Britain to produce regional accounts. Both M.C. Cooke (Fig. 3.3) and G. Massee, employed as mycologists by the Royal Botanic Gardens in Kew, produced accounts for Australia (Cooke, 1880*a,b*, 1881*b,c*, 1882*a–d*, 1883*a–c*, 1885*a*, 1886*a,c*, 1887*a–d*, 1888*a–c*, 1889*a*, *d–f,h*, 1890*a,c,f,h–j*, 1891*a–d*, 1893*a–c*, 1894; Massee, 1892*c*, 1893, 1898*c*, 1899*b*, 1901*a*, 1910*b*, 1911, 1912*a*, 1913) from the collections of F. von Mueller, Bailey and Rodway; British Guiana and the West Indies (Cooke, 1878*a*, 1879*a*, 1881*a*, 1884*d*, 1889*b*, 1891*e*; Massee, 1892*a–c*, 1894, 1898*c*, 1899*a*, 1901*b*, 1907*b*, 1909*b*, 1910*b,c*, 1912*a*, 1913) from the collections of Jenman, Bartlett, Hart and Roses; India (Cooke, 1876*a,b*, 1878*b*, 1879*a*, 1880*a*, 1881*a*, 1882*e*, 1884*a,b*, 1885*a*, 1886*a*, 1887*e–g*, 1888*d–g*, 1889*g*, 1890*l*, 1893*a*; Massee, 1898*b,c*, 1899*b*, 1901*a*, 1906, 1907*b*, 1910*b*, 1912*a,b*) from the collections of King, Kurz and Burkill; Madagascar, Mauritius and Uganda (Cooke, 1881*a*, 1890*b,d*; Massee, 1891, 1894) from the collections of Scott Elliott; New Zealand (Cooke, 1884*a*, 1885*a*, 1886*b,c*, 1890*g,m*; Massee, 1898*a,c*, 1907*a*) from Colenso collections; Peninsula & Malaysia (Cooke, 1884*c*, 1885*a,b*, 1887*g*, 1888*e*; Massee, 1898*c*, 1899*b*, 1901*a*, 1906, 1907*b*, 1908*a,b*, 1909*a*, 1910*a,b*, 1911, 1912*a,c*, 1913, 1914*a,b*) from the collections of Ridley and Burkill; and for South Africa (Cooke, 1879*b*, 1880*b*, 1881*a*, 1882*e*, 1884*b*, 1885*a*, 1887*g*, 1888*e,f*, 1889*c*, 1890*k*, 1891*e*; Massee, 1908*c*, 1911) from the collections of Kalchbrenner, MacOwan and Medley Wood. Other fungi were described from the Andaman Is, Brazil, Burma, Japan, Java (Cooke, 1890*e*), Malaccas Is (Cooke, 1882*g*), New Guinea (Cooke, 1886*d*), Philippine Is (Massee, 1907*c*), Samoa, Sarawak, Socotra Is (Cooke, 1882*f,h*).

Plant pathology and the 'Overseas Empire'

Much of the British contribution to tropical mycology, however, was still undertaken within the United Kingdom. From the middle of the nineteenth century onwards, the importance of studying plant pathogens was increasingly realized, particularly of those that arose through the planting and production of new crops in the developing countries of the 'Overseas Empire'. A series of short tours of duty was undertaken by British mycologists and plant pathologists. In 1880, Harvey Marshall Ward was appointed as the Government Cryptogamist in Ceylon, and made a special study of *Hemileia* rust, which had so devastated the coffee plantations of that country. In 1887, A.E. Shipley was sent to Bermuda to investigate onion disease and, in 1897, J.B. Carruthers became Government Mycologist in Ceylon to work on cacao canker before moving on in 1905 to the Federation of Malay States, where he contributed so much of our knowledge on the diseases of tea, rubber, cacao, and coconut. In 1890, Daniel McAlpine arrived in Melbourne, Australia, to become Vegetable Pathologist, this being the first full-time appointment in applied mycology. McAlpine's contribution was considerable. The production of a compilation account on the fungi of Australia by Cooke (1892) stimulated McAlpine to produce the first local census of fungi (1895), and he then went on to publish outstanding monographs on the diseases of citrus fruits (1899), stone fruits (1902), rusts (1906), smuts (1910), and potato diseases (1912). He, together with N. Cobb in New South Wales, who investigated the fungus diseases of sugar-cane, laid the foundation for plant pathology studies in the overseas British Empire.

A major British contributor, both to tropical mycology and plant pathology, was Thomas Petch and, once again, Ceylon played a major role. Petch succeeded Carruthers in 1905 as Director and Mycologist at Peradeniya, where he spent the next 20 years, contributing so much to our knowledge of the diseases of rubber (Petch, 1908, 1921) and tea (Petch, 1923). At the same time he promptly set about revising previous fungal records and published innumerable accounts both of the macrofungi and, especially, on entomogenous fungi. In 1950, with the help of G.R. Bisby, he was able to publish a detailed list of the 'Fungi of Ceylon', which remains unrivalled to the present day. In addition, he contributed considerably to our knowledge on the biology of tropical fungi, including that of thread-blights (1915, 1925), termite fungi (1913a,b), and root-disease fungi (1910, 1928). The early part of the twentieth century saw major additions to the knowledge of the geographical distribution of both pathogenic and saprobic fungi by plant pathologists. An understanding of the distribution of pathogens was essential for the provision of appropriate import and export regulations and quarantine procedures. Inspectors of plant diseases were established in Jamaica, British Guiana (Guyana), Ceylon and the

Gold Coast (Ghana). Nowell (1923) published the 'Diseases of Crop Plants in the Lesser Antilles'. Plant pathologists could often contribute to a knowledge of the mycota of an area. This was particularly true for India and 'British Africa'. In South Africa, I.B. Pole Evans became Chief of the Division of Botany and Plant Pathology for the Union, and was able to establish an excellent local herbarium. Major contributions were provided by W.J. Dowson in Kenya, W. Small in Uganda, R.H. Bunting in Ghana, and F.C. Deighton in Sierra Leone.

Edwin J. Butler (Fig. 3.4) became the first Imperial Mycologist when he was sent to India in 1906, and soon after the Agricultural Research Institute at Pusa was established, where research was carried out on the diseases of tea, cereals, oil-seeds, spice crops, fibres, palms and sugar-cane. He made major additions to the knowledge of Indian fungi, carrying on from the early pioneering work of D.D. Cunningham and A. Barclay, two Indian army medical officers, and the *Fungi of India* was published in 1931, aided by G.R. Bisby. The contributions to the development of economic mycology in the British 'Overseas Empire' were summarised by Butler (1929).

It soon became apparent that this rapid expansion in the fields of economic mycology and plant pathology, together with the need for phytosanitary regulations, required a centralized facility which could provide international advice. Following the success of the Imperial Bureau of Entomology, a similar Bureau of Mycology was proposed in 1918, and commenced work in 1920. The first Director was E.J. Butler, who formulated the role of the Bureau and published a policy statement in 1921. The aims were to be of use to economic mycologists in the Empire Overseas, to accumulate and distribute information on all plant disease matters, and to assist in the identification of pathogens. In addition the need was seen to index useful literature and to produce periodical abstracting journals, to build up a host-plant index, and to carry out critical studies of parasitic fungi. It was sited close to the Royal Botanic Gardens, Kew for reasons of close collaboration. From relatively humble beginnings the original Bureau has today become the International Mycological Institute, one of four scientific institutes of CAB International (CABI), and is now based at Egham, Surrey. It publishes *Distribution Maps of Plant Diseases, Index of Fungi, Bibliography of Systematic Mycology* and, since 1943, the research publications of *Mycological Papers, Phytopathological Papers*, and *Systema Ascomycetum*. In addition it maintains departments for Bacteriology, Culture Collections and Industrial Mycology, and organizes a number of training courses, under its present Director, Professor D.L. Hawksworth.

The production of the *Mycological Papers* and *Phytopathological Papers* provided the means for taxonomists and plant pathologists to publish regional accounts for tropical countries, including the following: Africa (Cejp & Deighton, 1969); Ghana (Gold Coast) (Hughes, 1952, 1953); Kenya (Nattrass, 1961); Malawi (Nyasaland) (Bisby & Wiehe, 1953; Corbett, 1964; Peregrine & Siddiqi, 1972; Sutton, 1993); Malaysia (Johnston, 1960); Sierra Leone (Deighton, 1944, 1960, 1969; Deighton & Pirozynski, 1972; Hansford & Deighton, 1948); Sudan (Boughey, 1946, 1947; Tarr, 1963); Tanzania (Tanganyika) (Ebbels & Allen, 1979; Pirozynski, 1972; Riley, 1960; Wallace & Wallace, 1949, 1953); Uganda (Hansford, 1946); Zambia (Northern Rhodesia) (Riley, 1956); India (Cannon & Minter, 1986); Fiji (Firman, 1972); Malaysia (Thompson & Johnston, 1953); Sabah (Williams & Liu, 1976); Sarawak (Turner, 1971); West Indies (Baker & Dale, 1948, 1951); Barbados (Norse, 1974); Jamaica (Dale, 1955*b*; Larter & Martyn, 1943); Trinidad (Baker, 1955; Dale, 1955*a*).

Production of mycofloras and monographs

British taxonomic contributions to tropical mycology remained strong throughout the twentieth century. Elsie Wakefield (Fig. 3.5), who succeeded Massee at the Royal Botanic Gardens, Kew, following her appointment in 1910, spent some months visiting, and exploring the mycota of, the West Indies. Her interest in the tropics led to the continuation of Massee's 'Fungi exotici' series, in which she described taxa from India (1922), Malay States (1918), Nigeria (1912, 1914, 1917*a*, *b*, 1920*a*), South Africa (1922, 1927) and Uganda (1918, 1920*b*, 1922). Indeed, her Presidential address to the British Mycological Society in 1930, was entitled 'Fungi exotici, past work and present problems'. Together with Petch, it was Wakefield who increasingly involved the British Mycological Society in tropical mycology to the extent that she was able to describe the Society (1930) as the 'mycological society of the British Empire'.

Probably the most important contribution by one individual to tropical mycology has been that of Professor E.J.H. Corner, who spent many years at the Botanic Garden, Singapore, and carried on the tradition, initiated by Chipp (1921), in producing a long series of publications relating to the fungi of Malaysia (Corner, 1935, 1950*a*, 1954, 1970*a*, 1971, 1972, 1974, 1981, 1986*b,c*; Corner & Bas, 1962; Corner & Hawker, 1953; Cash & Corner, 1958). In addition, there have been a series of monographic treatments, which have included the tropical taxa, such as: cantharelloid fungi and *Trogia* (1966, 1991*a*), clavarioid fungi (Corner, 1950*b*, 1967, 1970*b*), poroid fungi (Corner, 1983, 1984, 1986*a*, 1989*a,b*, 1991*b*), and *Thelephora* (Corner, 1968). These and other publications covered North Borneo (Corner, 1964), Australasia, equatorial Africa (Corner & Heinemann, 1967), and South America.

Kew has continued to specialize in the fungi of the pantropical region and southern hemisphere over the past half-century, largely based on first-hand collecting experience in the field. Collecting expeditions to Trinidad in 1949 and Venezuela in 1958 were undertaken by R.W.G. Dennis, which resulted in a series of publications on the region, including the West Indies (Dennis, 1951*a,b*, 1952*b–d*, 1953*a–e*, 1954*b*, 1968*a*; Dennis & Brodie, 1954), and tropical America (Dennis, 1951*d,e*, 1952*a,b*, 1954*a*, 1956, 1959, 1961*c,d*, 1965, 1970; Dennis & Müller, 1965). This culminated in the *Fungus Flora of Venezuela* (Dennis, 1970), which provides an introductory guide to the whole of tropical South America, and remains the only comprehensive account for any tropical region. Additional publications dealt with Africa (1955*b*, 1961*a,b*, 1963, 1964, 1972); Australasia (1955*a*, 1971), and India (1951*c*, 1954*c*). Emphasis was given by D.A. Reid to the macrofungi found in Australasia and southern Africa, with several visits to each region, and his publications included: Australasia (Reid, 1955, 1956, 1957, 1963*a,b*, 1975*b*, 1978, 1979, 1980, 1986, 1992); southern Africa (Reid, 1965, 1973, 1974, 1975*a*; Reid & Eicker, 1991*a,b*; Reid & Guillarmod, 1988; Eicker & Reid, 1990), together with contributions on Malaysia (Reid, 1958; Oldridge, Pegler, Reid & Spooner, 1986; Dennis & Reid, 1957). His monograph on the stipitate stereoid fungi (Reid, 1965) largely deals with tropical species. Expeditions to East Africa, India and Sri Lanka and the West Indies resulted in several series of publications from D.N. Pegler, including Tropical Africa (Pegler, 1966, 1967, 1969*a,b*, 1971, 1972*b*, 1973, 1976, 1977*c*, 1982; Pegler & Piearce, 1980; Pegler & Rayner, 1969); India and Sri Lanka (Pegler, 1972*a*, 1977*a,d*); tropical America (1978, 1987*a,b*, 1988, 1990; Pegler & Fiard, 1978, 1979, 1980; Pegler & Lodge, 1990). Several agaric floras were published for these areas, including tropical east Africa (Pegler, 1977*b*), the Lesser Antilles (Pegler, 1983), and Sri Lanka (Pegler, 1986*b*). In addition, there were contributions on Australasia (Pegler, 1965, 1984*a*; Pegler, Beaton & Young, 1984*a,b*; 1985*a–d*; Pegler & Holland, 1983*a*) and Malaysia (Pegler & Vanhaecke, 1993; Oldridge, Pegler, Reid & Spooner, 1986). A monograph on the genus *Lentinus* (Pegler, 1984*b*) dealt with one of the most common of all tropical taxa. The gasteroid fungi received special attention from D.M. Dring, especially those of tropical Africa (Demoulin & Dring, 1971, 1975; Dring & Pegler, 1978), these included West Africa (Dring, 1964*a,b*; Dring &

Rose, 1977) and East Africa (Dring & Rayner, 1967). His monograph on the Clathraceae (edited posthumously by Dennis, 1980) covered many tropical taxa.

The interest initiated by Kew on tropical African macromycetes stimulated other workers, especially in eastern Africa, to consider further the mycota of the area, particularly with regard to the ethnomycological aspects, e.g. Morris (1981, 1984, 1990), Piearce (1981, 1987), Pegler & Piearce (1980), and Williamson (1973). Tropical Africa has also interested R. Watling (1973, 1974, 1992; Watling & Turnbull, 1993; Hjortstam, Ryvarden & Watling, 1993; Petersen & Watling, 1989). It is with the Australian mycota, however, that Watling has made his most significant contributions to tropical mycology (Watling, 1978; Watling & Gregory, 1986, 1988*a*,*b*, 1989*a*,*b*, 1991; Whalley & Watling, 1989; Kile & Watling, 1983, 1988; Hoiland & Watling, 1990; Margot & Watling, 1981).

Present-day activity in the tropics

In recent years, British mycologists have increasingly turned their attentions towards the tropics, in a variety of activities. John Hedger (1985, 1990; Hedger, Lewis & Gitay, 1993) has initiated pioneering work on the ecological approach to macromycete formation in tropical rain forests, whilst E.B. Gareth Jones continues to undertake extensive investigations into the tropical marine fungi, e.g. Jones (1993) and the fungi of mangroves, e.g. Hyde & Jones (1988) and A.J.S. Whalley (Whalley, 1993; Whalley, Hammelev & Taligoola, 1988; Whalley & Taligoola, 1978) is giving consideration to the Xylariaceae.

Finally, in recent years the British Mycological Society has increasingly turned its attention towards the tropics, culminating in 1992 with both a major symposium (Isaac, Frankland, Watling & Whalley, 1993) and a 'tropical rain forest expedition' to the Amazonian region of Ecuador in 1993 (Figs. 3.6, 3.7). The expedition, organized by G. Dickson and J.N. Hedger, involved some 30 individuals and undertook projects on litter decay, wood-decomposing and mycorrhizal fungi, aero-aquatic fungi, insect pathogens, and ethnomycology (Hedger & Dickson, 1993).

References

Baker, R.E.D. (1955). Species of the genus *Parodiopsis* found in Trinidad. *Mycological Papers*, **58**, 1–16.
Baker, R.E.D. & Dale, W.T. (1948). Fungi of Barbados and the Windward Islands. *Mycological Papers*, **25**, 1–26.

Fig. 3.6. British Mycological Expedition, Cuyabeno Rain Forest Reserve, July–August 1993. Loading provisions and equipment at Cuyabeno Bridge to take to the reserve. (Photograph: G. Dixon).

Fig. 3.7. British Mycological Expedition, Cuyabeno Rain Forest Reserve, July–August 1993. Expedition member at work in the Carretera principal (main road of Reserve). (Photograph: G. Dixon).

Baker, R.E.D. & Dale, W.T. (1951). Fungi of Trinidad and Tobago. *Mycological Papers*, **33**, 1–123.

Barbaro, E. (1454–1495). *Corallinum in Dioscoridem libri V*.

Berkeley, M.J. (1836). *The English Flora of Sir James Edward Smith*, 5, part 2, 1–386. London.

Berkeley, M.J. (1839*a*). Contribution towards a flora of Van Diemen's Land; from collections sent by R.W. Lawrence and Ronald Gunn. *Annals and Magazine of Natural History*, **3**, 322–7, pl. 7.

Berkeley, M.J. (1839*b*). Descriptions of exotic fungi in the collection of Sir W.J. Hooker, from memoirs and notes of J.F. Klotzsch, with additions and corrections. *Annals and Magazine of Natural History*, **3**, 375–401.

Berkeley, M.J. (1840). Notice of some fungi collected by Ch. Darwin during the expedition of H.M. Ship Beagle. *Annals and Magazine of Natural History*, **4**, 291–3

Berkeley, M.J. (1842*a*). Notice of some fungi collected by Ch. Darwin in South America and the Islands of the Pacific. *Annals and Magazine of Natural History*, **9**, 443–8.

Berkeley, M.J. (1842*b*). Enumeration of fungi collected by H. Cuming in the Philippine Islands. *Hooker's London Journal of Botany*, **1**, 447–57.

Berkeley, M.J. (1842*c*). Descriptions of fungi, collected by R.B. Hinds, Esq. principally in the islands of the Pacific. *Hooker's London Journal of Botany*, **6**, 447–57.

Berkeley, M.J. (1843*a*). Enumeration of fungi collected by Zeyher at Uitenhage. *Hooker's London Journal of Botany*, **2**, 629.

Berkeley, M.J. (1843*b*). Notices of some Brazilian Fungi. *Hooker's London Journal of Botany*, **2**, 629–41, pl. 24.

Berkeley, M.J. (1844*a*). Decades of fungi I, Philippines, South Africa. *Hooker's London Journal of Botany*, **3**, 185–94.

Berkeley, M.J. (1844*b*). Decades of fungi II, Java. *Hooker's London Journal of Botany*, **3**, 329–57.

Berkeley, M.J. (1845*a*). Decades of fungi III–VI, Australia. *Hooker's London Journal of Botany*, **4**, 42–73.

Berkeley, M.J. (1845*b*). Decades of fungi VII–X, Australia, North America. *Hooker's London Journal of Botany*, **4**, 298–315.

Berkeley, M.J. (1846*a*). Decades of fungi XI, South Africa, Jamaica, Peru. *Hooker's London Journal of Botany*, **5**, 1–6.

Berkeley, M.J. (1846*b*). Notices of three new fungi collected by Mr Gardner in Ceylon. *Hooker's London Journal of Botany*, **5**, 534–5.

Berkeley, M.J. (1847*a*). Decades of fungi XII–XIV, Ohio. *Hooker's London Journal of Botany*, **6**, 312–26.

Berkeley, M.J. (1847*b*). Decades of fungi XV–XIX, Ceylon. *Hooker's London Journal of Botany*, **6**, 479–514.

Berkeley, M.J. (1848). Decades of fungi XX, Tasmania. *Hooker's London Journal of Botany*, **7**, 572–80.

Berkeley, M.J. (1849*a*). Decades of fungi XXI–XXII, North and South Carolina. *Hooker's Journal of Botany and Kew Miscellany*, **1**, 97–104.

Berkeley, M.J. (1849*b*). Decades of fungi XXIII–XXIV, North and South Carolina. *Hooker's Journal of Botany & Kew Miscellany*, **1**, 234–9.

Berkeley, M.J. (1850*a*). Decades of fungi XXV–XXX, Sikkim. *Hooker's Journal of Botany & Kew Miscellany*, **2**, 42–51.

Berkeley, M.J. (1850*b*). Decades of fungi XXV–XXX, Sikkim. *Hooker's Journal of Botany & Kew Miscellany*, **2**, 76–88.

Berkeley, M.J. (1850*c*). Decades of fungi XXV–XXX, Sikkim. *Hooker's Journal of Botany & Kew Miscellany*, **2**, 106–12.

Berkeley, M.J. (1851*a*). Decades of fungi XXXI, Brazil. *Hooker's Journal of Botany & Kew Miscellany*, **3**, 14–21.

Berkeley, M.J. (1851*b*). Decades of fungi XXXII–XXXIII, Sikkim. *Hooker's Journal of Botany & Kew Miscellany*, **3**, 39–49.

Berkeley, M.J. (1851*c*). Decades of fungi XXXIV, Sikkim. *Hooker's Journal of Botany & Kew Miscellany*, **3**, 77–84.

Berkeley, M.J. (1851*d*). Decades of fungi XXXV, Sikkim. *Hooker's Journal of Botany & Kew Miscellany*, **3**, 167–72.

Berkeley, M.J. (1851*e*). Decades of fungi XXXVI, Sikkim. *Hooker's Journal of Botany & Kew Miscellany*, **3**, 200–6.

Berkeley, M.J. (1852*a*). Decades of fungi XXXVII–XXXVIII, Sikkim. *Hooker's Journal of Botany & Kew Miscellany*, **4**, 97–107.

Berkeley, M.J. (1852*b*). Decades of fungi XXXIX–XL, Sikkim. *Hooker's Journal of Botany & Kew Miscellany*, **4**, 130–42.

Berkeley, M.J. (1852*c*). Enumeration of some fungi from St Domingo. *Annals & Magazine of Natural History*, ser. 2, **9**, 192–203.

Berkeley, M.J. (1852*d*). Enumeration of a small collection of fungi from Borneo. *Hooker's Journal of Botany & Kew Miscellany*, **4**, 161–4.

Berkeley, M.J. (1854*a*). Decades of fungi XLI–XLIII, India. *Hooker's Journal of Botany & Kew Miscellany*, **6**, 129–43.

Berkeley, M.J. (1854*b*). Decades of fungi XLIV–XLVI, India. *Hooker's Journal of Botany & Kew Miscellany*, **6**, 161–75.

Berkeley, M.J. (1854*c*). Decades of fungi XLVII–XLVIII, India. *Hooker's Journal of Botany & Kew Miscellany*, **6**, 204–12.

Berkeley, M.J. (1854*d*). Decades of fungi XLIX–L, India. *Hooker's Journal of Botany & Kew Miscellany*, **6**, 225–35.

Berkeley, M.J. (1856*a*). Decades of fungi LI–LIV, Brazil. *Hooker's Journal of Botany & Kew Miscellany*, **8**, 129–44.

Berkeley, M.J. (1856*b*). Decades of fungi LV–LVI, Brazil. *Hooker's Journal of Botany & Kew Miscellany*, **8**, 169–77.

Berkeley, M.J. (1856*c*). Decades of fungi LVII–LVIII, Brazil. *Hooker's Journal of Botany & Kew Miscellany*, **8**, 193–200.

Berkeley, M.J. (1856*d*). Decades of fungi LIX–LX, Brazil. *Hooker's Journal of Botany & Kew Miscellany*, **8**, 233–41.

Berkeley, M.J. (1856*e*). Decades of fungi LXI–LXII, Brazil. *Hooker's Journal of Botany & Kew Miscellany*, **8**, 272–80.

Berkeley, M.J. (1867). On some new fungi from Mexico. *Journal of the Linnean Society, London*, **9**, 423–5.

Berkeley, M.J. (1869*a*). *Hemileia vastatrix* Berk. and Broome. *Gardeners Chronicle, 1869*, 1157, fig. 1.

Berkeley, M.J. (1869*b*). On a collection of fungi from Cuba. Part 2. *Journal of the Linnean Society, London*, **10**, 341–92.

Berkeley, M.J. (1872). Fungi in the coffee plantation of Ceylon. *Gardeners Chronicle, 1872*, 425.

Berkeley, M.J. (1876*a*). Enumeration of the fungi collected during the expedition of H.M.S. Challenger II. *Journal of the Linnean Society, London*, **15**, 48–53.

Berkeley, M.J. (1876*b*). Report of the fungi collected on Kerguelen's Land by the Rev. A.E. Eaton during the stay of the Transit-of-Venus Expedition. *Journal of the Linnean Society, Botany*, **15**, 221–222.

Berkeley, M.J. (1878). Enumeration of the fungi collected during the expedition of H.M.S. Challenger, III. *Journal of the Linnean Society, London*, **16**, 38–54, pl. 2.

Berkeley, M.J. (1879). Fungi Brasiliensis in provincia Rio de Janeiro a clar. Dr A. Glaziou lecti. *Videnskabelige Meddelelser fra dansk Naturhistorisk Forening i Kjøbenhavn, 1879–1880*, 31–34.

Berkeley, M.J. (1885). Notices of fungi collected in Zanzibar in 1884 by Miss R.E. Berkeley. *Annals and Magazine of Natural History*, ser. 5, **15**, 384–7.

Berkeley, M.J. & Broome, C.E. (1868). On some species of the genus *Agaricus* from Ceylon. *Transactions of the Linnean Society, London*, **27**, 149–52.

Berkeley, M.J. & Broome, C.E. (1871*a*). The fungi of Ceylon. *Journal of the Linnean Society, London*, **11**, 494–567.

Berkeley, M.J. & Broome, C.E. (1871*b*). On some species of the genus *Agaricus* from Ceylon. *Transactions of the Linnean Society, London*, **27**, 149–52, pl. 33–34.

Berkeley, M.J. & Broome, C.E. (1875). Enumeration of the fungi of Ceylon Part II, containing remainder of Hymenomycetes. *Journal of the Linnean Society, London*, **14**, 29–140, pl. 2–10.

Berkeley, M.J. & Broome, C.E. (1877). Supplement to the enumeration of fungi of Ceylon. *Journal of the Linnean Society, London*, **15**, 82–86.

Berkeley, M.J. & Cooke, M.C. (1877). The fungi of Brazil, including those collected by J.W.H. Trail esq. MA in 1874. *Journal of the Linnean Society, London*, **15**, 363–98.

Berkeley, M.J. & Curtis, M.A. (1869*a*). Fungi Cubenses, Part I, Hymenomycetes. *Journal of the Linnean Society, Botany*, **10**, 280.

Berkeley, M.J. & Curtis, M.A. (1869*b*). Fungi Cubenses, Part II, Gasteromycetes, Phycomycetes, Coniomycetes, Hyphomycetes and Ascomycetes. *Journal of the Linnean Society, Botany*, **10**, 341.

Berkeley, M.J. & Montagne, J.F.C. (1844). Decades of fungi. *Hooker's London Journal of Botany*, **3**, 329.

Berkeley, M.J. & Montagne, J.F.C. (1849). Sixième centurie de plantes cellulaires nouvelles, tant indigenes qu'exotiques. Decade 7. *Annales des Sciences Naturelles, Botanique*, sér. 3, **11**, 235–46.

Bisby, G.R. & Wiehe, P.O. (1953). The rusts of Nyasaland. *Mycological Papers*, **54**, 1–12.

Boughey, A.S. (1946). A preliminary list of plant diseases in the Anglo-Egyptian Sudan. *Mycological Papers*, **14**, 1–16.

Boughey, A.S. (1947). The effect of rainfall on plant disease distribution in the Anglo-Egyptian Sudan. *Mycological Papers*, **19**, 1–13.

Butler, E.J. (1921). The Imperial Bureau of Mycology. *Transactions of the British Mycological Society*, **7**, 168–72.

Butler, E.J. (1929). The development of economic mycology in the Empire Overseas. *Transactions of the British Mycological Society*, **14**, 1–18.

Butler, E.J. & Bisby, G.R. (1931). *The Fungi of India*. Imperial Council of Agricultural research, India.

Cannon, P.F. & Minter, D.W. (1986). The Rhytismataceae of the Indian subcontinent. *Mycological Papers*, **155**, 1–123.

Cash, E.K. & Corner, E.J.H. (1958). Malayan and Sumatran Discomycetes. *Transactions of the British Mycological Society*, **41**, 273–82.
Cejp, K. & Deighton, F.C. (1969). Microfungi III. Some African species of *Phyllosticta* and *Septoria*; new genera and species and redispositions of some Hyphomycetes, mostly African. *Mycological Papers*, **117**, 1–31.
Chipp, T.F. (1921). A list of fungi of the Malay Peninsula. *Gardens' Bulletin of the Straits Settlements*, **2**, 311–418.
Cooke, M.C. (1871). *Handbook of British Fungi, with full descriptions of all the species, and illustrations of the genera.* Vols I & II. London & New York: Macmillan & Co.
Cooke, M.C. (1876a). Some Indian Fungi. *Grevillea*, **4**, 114–18.
Cooke, M.C. (1876b). Some Indian Fungi. *Grevillea*, **5**, 14–17.
Cooke, M.C. (1878a). Cocos Palm fungi. *Grevillea*, **5**, 101–3.
Cooke, M.C. (1878b). Some Indian Fungi. *Grevillea*, **6**, 117–18.
Cooke, M.C. (1879a). Some exotic fungi. *Grevillea*, **7**, 94–6.
Cooke, M.C. (1879b). Natal fungi. *Grevillea*, **8**, 69–72.
Cooke, M.C. (1880a). Fungi of India. *Grevillea*, **8**, 93–6.
Cooke, M.C. (1880b). *Fungi australiani, imprimis e collectionibus a reverendo. J.M. Berkeley pervisis numerati, additis circa centum speciebus e collectione Baileyana a C.E. Frome examinatis et insertis circiter triginta aliis a Friesio e collectione Preissii divulgatis.* London & Melbourne (Supplementum ad vol. XI, fragmentorum phytographiae Australiae baronis F. de Müller).
Cooke, M.C. (1881a). Some exotic fungi. *Grevillea*, **9**, 97–101.
Cooke, M.C. (1881b). Australian fungi. *Grevillea*, **9**, 142–9.
Cooke, M.C. (1881c). Australian fungi. *Grevillea*, **10**, 60–4, pl. 10.
Cooke, M.C. (1882a). Australian fungi. *Grevillea*, **10**, 93–4.
Cooke, M.C. (1882b). Australian fungi. *Grevillea*, **10**, 131–6.
Cooke, M.C. (1882c). Australian fungi. *Grevillea*, **11**, 28–34.
Cooke, M.C. (1882d). Australian fungi. *Grevillea*, **11**, 57–65.
Cooke, M.C. (1882e). Exotic fungi. *Grevillea*, **10**, 121–30.
Cooke, M.C. (1882f). Fungi of Socotra. *Grevillea*, **11**, 39.
Cooke, M.C. (1882g). Three asiatic fungi. *Grevillea*, **11**, 76.
Cooke, M.C. (1882h). Diagnoses fungorum novorum (*Stereum retirugum* Cke, *Trametes socotrana* Cke) in Insula Socotra a Bayley Balfour, Coraolo Cockburn et Alexandra Scott, lectorum. *Transactions & Proceedings of the Royal Society, Edinburgh*, **11**, 456.
Cooke, M.C. (1883a). Australian fungi. *Grevillea*, **11**, 97–104.
Cooke, M.C. (1883b). Australian fungi. *Grevillea*, **11**, 145–52.
Cooke, M.C. (1883c). Australian fungi, *Grevillea*, **12**, 8–21.
Cooke, M.C. (1884a). Some exotic fungi. *Grevillea*, **12**, 85.
Cooke, M.C. (1884b). Some exotic fungi. *Grevillea*, **13**, 6–7.
Cooke, M.C. (1884c). Fungi of Perak. *Grevillea*, **13**, 1–4.
Cooke, M.C. (1884d). Demarara fungi, *Grevillea*, **13**, 32–3.
Cooke, M.C. (1885a). Some exotic fungi. *Grevillea*, **14**, 11–14.
Cooke, M.C. (1885b). Fungi of Malayan Peninsula. *Grevillea*, **14**, 43–4.
Cooke, M.C. (1886a). Some exotic fungi. *Grevillea*, **14**, 89–90.
Cooke, M.C. (1886b). Some exotic fungi. *Grevillea*, **14**, 129–30.
Cooke, M.C. (1886c). Some exotic fungi. *Grevillea*, **15**, 16–18.
Cooke, M.C. (1886d). Fungi of New Guinea. *Grevillea*, **14**, 115–18.
Cooke, M.C. (1887a). Australian fungi. *Grevillea*, **15**, 93–5.
Cooke, M.C. (1887b). Australian fungi. *Grevillea*, **15**, 97–101.

Cooke, M.C. (1887*c*). Australian fungi. *Grevillea*, **16**, 1–6.
Cooke, M.C. (1887*d*). Australian fungi. *Grevillea*, **16**, 30–3.
Cooke, M.C. (1887*e*). Some exotic fungi. *Grevillea*, **16**, 15–16.
Cooke, M.C. (1887*f*). Some exotic fungi. *Grevillea*, **16**, 25–6.
Cooke, M.C. (1887*g*). Exotic agarics. *Grevillea*, **16**, 105–6.
Cooke, M.C. (1888*a*). Australian fungi. *Grevillea*, **16**, 72–6.
Cooke, M.C. (1888*b*). Australian fungi. *Grevillea*, **16**, 113–14.
Cooke, M.C. (1888*c*). Australian fungi. *Grevillea*, **17**, 7–8.
Cooke, M.C. (1888*d*). Some exotic fungi. *Grevillea*, **16**, 69–72.
Cooke, M.C. (1888*e*). Some exotic fungi. *Grevillea*, **16**, 105–6.
Cooke, M.C. (1888*f*). Some exotic fungi. *Grevillea*, **16**, 121.
Cooke, M.C. (1888*g*). Exotic fungi. *Grevillea*, **17**, 42–3.
Cooke, M.C. (1889*a*). Australian fungi. *Grevillea*, **17**, 55–6.
Cooke, M.C. (1889*b*). Exotic fungi. *Grevillea*, **17**, 59–60.
Cooke, M.C. (1889*c*). Three Natal fungi. *Grevillea*, **17**, 70.
Cooke, M.C. (1889*d*). Two Australian fungi. *Grevillea*, **17**, 81.
Cooke, M.C. (1889*e*). New Australian fungi. *Grevillea*, **18**, 1–8.
Cooke, M.C. (1889*f*). New Australian fungi. *Grevillea*, **18**, 25–6.
Cooke, M.C. (1889*g*). Some exotic fungi. *Grevillea*, **18**, 34–5.
Cooke, M.C. (1889*h*). Some Brisbane fungi. *Grevillea*, **17**, 69–70.
Cooke, M.C. (1890*a*). New Australian fungi. *Grevillea*, **18**, 49.
Cooke, M.C. (1890*b*). Fungi of Madagaskar, collected by Mr Scott Elliot. *Grevillea*, **18**, 49–51.
Cooke, M.C. (1890*c*). New Australian fungi. *Grevillea*, **18**, 80–1.
Cooke, M.C. (1890*d*). Some exotic fungi. *Grevillea*, **18**, 86–7.
Cooke, M.C. (1890*e*). Fungi of Java. *Grevillea*, **18**, 54–6.
Cooke, M.C. (1890*f*). On *Campbellia*. *Grevillea*, **18**, 87–8.
Cooke, M.C. (1890*g*). New Zealand fungi. *Grevillea*, **19**, 1–4.
Cooke, M.C. (1890*h*). Australian fungi. *Grevillea*, **19**, 5.
Cooke, M.C. (1890*j*). Australian fungi. *Grevillea*, **19**, 44–7.
Cooke, M.C. (1890*k*). Some African fungi. *Grevillea*, **19**, 6–7.
Cooke, M.C. (1890*l*). Some Asiatic fungi. *Grevillea*, **19**, 7.
Cooke, M.C. (1890*m*). Fungi of New Zealand. *Grevillea*, **19**, 47–9.
Cooke, M.C. (1891*a*). Australian fungi. *Grevillea*, **19**, 60–2.
Cooke, M.C. (1891*b*). Two Australian fungi. *Grevillea*, **19**, 81–3.
Cooke, M.C. (1891*c*). Australian fungi. *Grevillea*, **19**, 89–92.
Cooke, M.C. (1891*d*). Australian fungi. *Grevillea*, **20**, 35.
Cooke, M.C. (1891*e*). Exotic fungi. *Grevillea*, **20**, 15–16.
Cooke, M.C. (1892). *Handbook of Australian fungi*. London: Williams & Nordgate.
Cooke, M.C. (1893*a*). Australian fungi. *Grevillea*, **22**, 36–8.
Cooke, M.C. (1893*b*). Exotic fungi. *Grevillea*, **21**, 73–5.
Cooke, M.C. (1893*c*). Fungi. *Transactions of the Royal Society of South Australia*, **16**, 15.
Cooke, M.C. (1894). Australian fungi. *Grevillea*, **22**, 67–8.
Cooke, M.C. & Kalchbrenner, C. (1880*a*). Australian fungi. *Grevillea*, **9**, 1–4.
Cooke, M.C. & Kalchbrenner, C. (1880*b*). South African fungi. *Grevillea*, **9**, 17–31.
Cooke, M.C. & Kalchbrenner, C. (1881). Natal fungi. *Grevillea*, **10**, 26–7.
Corbett, D.C.M. (1964). Supplementary list of plant diseases in Nyasaland. *Mycological Papers*, **95**, 1–15.

Corner, E.J.H. (1935). The seasonal fruiting of agarics in Malaya. *Gardens' Bulletin of the Straits Settlements*, **9**, 79–88.

Corner, E.J.H. (1950*a*). Descriptions of two luminous tropical agarics (*Dictyopanus* and *Mycena*). *Mycologia*, **42**, 423–31.

Corner, E.J.H. (1950*b*). Monograph of *Clavaria* and allied genera. *Annals of Botany Memoir*, **1**, 1–298.

Corner, E.J.H. (1954). Further descriptions of luminous agarics. *Transactions of the British Mycological Society*, **37**, 256–71.

Corner, E.J.H. (1964). Fungi (Royal Society Expedition to North Borneo). *Proceedings of the Linnean Society, London*, **175**, 40–2.

Corner, E.J.H. (1966). A monograph of the cantharelloid fungi. *Annals of Botany Memoirs*, **2**, 1–255.

Corner, E.J.H. (1967). Clavarioid fungi of the Solomon Islands. *Proceedings of the Linnean Society, London*, **178**, 91–106.

Corner, E.J.H. (1968). A monograph of *Thelephora*. *Beihefte zur Nova Hedwigia*, **27**, 1–100.

Corner, E.J.H. (1970*a*). *Phylloporus* Quél. and *Paxillus* Fr. in Malaya and Borneo. *Nova Hedwigia*, **20**, 793–822.

Corner, E.J.H. (1970*b*). Supplement to 'A monograph of *Clavaria* and allied genera'. *Beihefte zur Nova Hedwigia*, **33**, 1–299.

Corner, E.J.H. (1971). Merulioid fungi in Malaya. *Gardens' Bulletin, Singapore*, **25**, 355–81.

Corner, E.J.H. (1972). *Boletus in Malaysia*. Singapore: Government Printing Office.

Corner, E.J.H. (1974). *Boletus* and *Phylloporus* in Malaysia: further notes and descriptions. *Gardens' Bulletin, Singapore*, **27**, 1–16.

Corner, E.J.H. (1981). The agaric genera *Lentinus, Panus* and *Pleurotus*, with particular reference to the Malaysian species. *Beihefte zur Nova Hedwigia*, **69**, 1–169.

Corner, E.J.H. (1983). Ad Polyporaceas I. *Amauroderma* and *Ganoderma*. *Beihefte zur Nova Hedwigia*, **75**, 1–182.

Corner, E.J.H. (1984). Ad Polyporaceas II, *Polyporus, Mycobonia* and *Echinochaete*; III, *Piptoporus, Buglossoporus, Laetiporus, Meripilus* and *Bondarzewia*. *Beihefte zur Nova Hedwigia*, **78**, 1–219.

Corner, E.J.H. (1986*a*). Ad Polyporaceas IV, the genera *Daedalea, Flabellophora, Flavodon, Gloeophyllum, Heteroporus, Irpex, Lenzites, Microporellus, Nigrofomes, Nigroporus, Oxyporus, Paratrichaptum, Rigidoporus, Scenidium, Trichaptum, Vanderbylia* and *Steccherinum*. *Beihefte zur Nova Hedwigia*, **86**, 1–265.

Corner, E.J.H. (1986*b*). The tropical complex of *Mycena pura*. *Transactions of the Botanical Society of Edinburgh 1836–1986, 150th Anniversary Supplement*, 61–7.

Corner, E.J.H. (1986*c*). The agaric genus *Panellus* (including *Dictyopanus*) in Malaysia. *Gardens' Bulletin, Singapore*, **39**, 103–47.

Corner, E.J.H. (1989*a*). Ad Polyporaceas V. The genera *Albatrellus, Boletopsis, Gloeoporus, Grifola, Hapalopilus, Heterobasidion, Hydnopolyporus, Ischnoderma, Loweporus, Parmastomyces, Perenniporia, Pyrofomes, Stecchericium, Trechispora, Truncospora* and *Tyromyces*. *Beihefte zur Nova Hedwigia*, **96**, 1–218.

Corner, E.J.H. (1989*b*). Ad Polyporaceas VI. The genus *Trametes*. *Beihefte zur Nova Hedwigia*, **97**, 1–197.

Corner, E.J.H. (1991*a*). *Trogia* (Basidiomycetes). *Gardens' Bulletin, Singapore*, supplement 2, 1–100.

Corner, E.J.H. (1991*b*). Ad Polyporaceas VII. The xanthochroic polypores. *Beihefte zur Nova Hedwigia*, **101**, 1–177.

Corner, E.J.H. & Bas, C. (1962). The genus *Amanita* in Singapore and Malaya. *Persoonia*, **2**, 241–304.

Corner, E.J.H. & Hawker, L.E. (1953). Hypogeous fungi from Malaya. *Transactions of the British Mycological Society*, **32**, 125–37.

Corner, E.J.H. & Heinemann, L.E. (1967). Clavaires et *Thelephora. Flore Iconographique des Champignons du Congo*, fascicle 16, 309–21.

Dale, W.T. (1955*a*). New species of Uredinales from Trinidad. *Mycological Papers*, **59**, 1–11.

Dale, W.T. (1955*b*). A preliminary list of Jamaican Uredinales. *Mycological Papers*, **60**, 1–21.

Deighton, F.C. (1944). West African Meliolineae I. Meliolineae on Malvaceae and Tiliaceae. *Mycological Papers*, **9**, 1–24.

Deighton, F.C. (1960). African fungi I. *Mycological Papers*, **78**, 1–43.

Deighton, F.C. (1969). Microfungi IV, some hyperparasitic Hyphomycetes and a note on *Cercosporella uredinophila* Sacc. *Mycological Papers*, **118**, 1–41.

Deighton, F.C. & Pirozynski, K.A. (1972). Microfungi V. More hyperparasitic Hyphomycetes. *Mycological Papers*, **128**, 1–110.

Demoulin, V. & Dring, D.M. (1971). Two new species of *Scleroderma* from tropical Africa. *Transactions of the British Mycological Society*, **56**, 163–5.

Demoulin, V. & Dring, D.M. (1975). Gasteromycetes of Kivu (Zaire), Rwanda and Burundi. *Bulletin du Jardin national de Belgique*, **45**, 339–72.

Dennis, R.W.G. (1951*a*). *Lentinus* in Trinidad. *Kew Bulletin*, **5**, 321–33.

Dennis, R.W.G. (1951*b*). An earlier name for *Omphalia flavida. Kew Bulletin*, **5**, 434.

Dennis, R.W.G. (1951*c*). The genus *Trogia* in India. *Indian Phytopathology*, **4**, 119–22.

Dennis, R.W.G. (1951*d*). Species of *Marasmius* described by Berkeley from tropical America. *Kew Bulletin*, **6**, 153–63.

Dennis, R.W.G. (1951*e*). Murrill's West Indian species of *Marasmius. Kew Bulletin*, **6**, 196–210.

Dennis, R.W.G. (1952*a*). Some tropical American Agaricaceae referred by Berkeley and Montagne to *Collybia* or *Heliomyces. Kew Bulletin*, **6**, 387–410.

Dennis, R.W.G. (1952*b*). Some Agaricaceae of Trinidad and Venezuela. Part I. Leucosporae. *Transactions of the British Mycological Society*, **34**, 411–82.

Dennis, R.W.G. (1952*c*). The *Laschia* complex in Trinidad and Venezuela. *Kew Bulletin*, **7**, 437–8.

Dennis, R.W.G. (1952*d*). *Lepiota* and allied genera in Trinidad. *Kew Bulletin*, **7**, 459–99.

Dennis, R.W.G. (1953*a*). Some pleurotoid fungi from the West Indies. *Kew Bulletin*, **8**, 31–45.

Dennis, R.W.G. (1953*b*). New species of *Dictyonia* and *Coccomyces. Kew Bulletin*, **8**, 49–50.

Dennis, R.W.G. (1953*c*). Some West Indian collections referred to *Hygrophorus.* Fr. *Kew Bulletin*, **8**, 253–68.

Dennis, R.W.G. (1953*d*). Some West Indian Gasteromycetes. *Kew Bulletin*, **8**, 307–28.

Dennis, R.W.G. (1953*e*). Les Agaricales de l'Isle de la Trinidad: rhodosporae – ochrosporae. *Bulletin de la Société Mycologique de France*, **69**, 145–98.

Dennis, R.W.G. (1954*a*). Some inoperculate Discomycetes from tropical America. *Kew Bulletin*, **9**, 289–348.

Dennis, R.W.G. (1954*b*). Operculate Discomycetes from Trinidad and Jamaica. *Kew Bulletin*, **9**, 417–21.

Dennis, R.W.G. (1954*c*). *Heimiomyces* in the old world tropics. *Kew Bulletin*, **9**, 422.

Dennis, R.W.G. (1955*a*). New or interesting Queensland Agaricales. *Kew Bulletin*, **10**, 107–10.

Dennis, R.W.G. (1955*b*). Fungi from Sierra Leone. Pezizales and Helotiales. *Kew Bulletin*, **10**, 363–8.

Dennis, R.W.G. (1956). Some Xylariaceae from tropical America. *Kew Bulletin*, **11**, 401–44.

Dennis, R.W.G. (1959). Fungi Venezuelani I. *Exobasidium woroninii*. *Kew Bulletin*, **13**, 402–3.

Dennis, R.W.G. (1961*a*). Xylarioideae and Thamnomycetoideae of the Congo. *Bulletin du Jardin de l'état de Bruxelles*, **31**, 109–54.

Dennis, R.W.G. (1961*b*). A collection of *Polydiscidium* from Africa. *Bulletin du Jardin de l'état de Bruxelles*, **31**, 155–7.

Dennis, R.W.G. (1961*c*). Fungi Venezuelani III. *Kew Bulletin*, **14**, 418–58.

Dennis, R.W.G. (1961*d*). Fungi Venezuelani IV. Agaricales. *Kew Bulletin*, **15**, 67–156.

Dennis, R.W.G. (1963). Hypoxyloideae of Congo. *Bulletin du Jardin de l'état de Bruxelles*, **33**, 317–40.

Dennis, R.W.G. (1964). Further records of Congo Xylariaceae. *Bulletin du Jardin de l'état de Bruxelles*, **34**, 231–41.

Dennis, R.W.G. (1965). Fungi Venezuelani VII. *Kew Bulletin*, **19**, 231–73.

Dennis, R.W.G. (1968*a*). Some Agaricales from the Blue Mountains of Jamaica. *Kew Bulletin*, **22**, 73–85.

Dennis, R.W.G. (1970). Fungus Flora of Venezuela and adjacent countries. *Kew Bulletin Additional Series*, **3**, 1–531.

Dennis, R.W.G. (1971). A new discomycete from New Guinea. *Kew Bulletin*, **25**, 375–6.

Dennis, R.W.G. (1972). A leaf-spot of *Phyllanthus* in Kenya. *Kew Bulletin*, **26**, 45–6.

Dennis, R.W.G. & Brodie, H.J. (1954). The Nidulariaceae of the West Indies. *Transactions of the British Mycological Society*, **37**, 151–60.

Dennis, R.W.G. & Müller, E. (1965). Fungi Venezuelani VIII. Plectascales, Sphaeriales, Loculoascomycetes. *Kew Bulletin*, **19**, 357–86.

Dennis, R.W.G. & Reid, D.A. (1957). Some marasmioid fungi allegedly parasitic on leaves and twigs in the tropics. *Kew Bulletin*, **12**, 287–92.

Dring, D.M. (1964*a*). Gasteromycetes of West Tropical Africa. *Mycological Papers*, **98**, 1–64.

Dring, D.M. (1964*b*). Some aspects of the Gasteromycete flora of West Africa. *Transactions of the British Mycological Society*, **47**, 646–7.

Dring, D.M. (Ed. Dennis, R.W.G.) (1980). Contributions towards a rational arrangement of the Clathraceae. *Kew Bulletin*, **35**, 1–96.

Dring, D.M. & Pegler, D.N. (1978). New and noteworthy gasteroid relatives of the Agaricales from tropical Africa. *Kew Bulletin*, **32**, 563–9.

Dring, D.M. & Rayner, R.W. (1967). Some Gasteromycetes from eastern Africa. *Journal of the East African Natural History Society*, **26** (2), 5–46.

Dring, D.M. & Rose, A.C. (1977). Additions to the West African phalloid fungi. *Kew Bulletin*, **31**, 741–51.

Ebbels, D.L. & Allen, D.J. (1979). A supplementary list of plant diseases, pathogens and associated fungi in Tanzania. *Phytopathological Papers*, **22**, 1–89.

Eicker, A. & Reid, D.A. (1990). *Clathrus transvaalensis*, a new species from the Transvaal. South Africa. *Mycological Research*, **94**, 422–3.

Firman, I.D. (1972). A list of fungi and plant parasitic bacteria, viruses and nematodes in Fiji. *Phytopathological Papers*, **15**, 1–36.

Greville, R.K. (1821). Descriptions of seven new Scottish fungi. *Memoirs of the Wernerian Society*, **4** (1), 67.

Greville, R.K. (1823–28). *Scottish cryptogamic flora or coloured figures and descriptions of cryptogamic plants found in Scotland and belonging chiefly to the order Fungi and intended to serve as a contribution to English Botany.* 6 vols. Edinburgh: Maclachnan.

Hansford, C.G. (1946). The foliicolous ascomycetes, their parasites and associated fungi, especially illustrated by Ugandan species. *Mycological Papers*, **15**, 1–240.

Hansford, C.G. & Deighton, F.C. (1948). West African Meliolineae II. Meliolineae collected by F.C. Deighton. *Mycological Papers*, **23**, 1–79.

Hedger, J.N. (1985). Tropical agarics, resource relations and fruiting periodicity. In *Developmental Biology of Higher Fungi* (ed. Moore, D., Casselton, L.A., Wood, D.A. and Frankland, J.C.). British Myological Society, Symposium Series, **10**, pp. 41–46.

Hedger, J.N. (1990). Fungi in the tropical forest canopy. *The Mycologist*, **4**, 200–202.

Hedger, J.N. & Dickson, G. (1993). British Mycological Society tropical rainforest expedition, 1993. *The Mycologist*, **7**, 205–6.

Hedger, J.N., Lewis, P. & Gitay, H. (1993). Litter-trapping fungi in moist tropical forest. *British Mycological Society Symposium Series*, **19**, 15–35.

Hjortstam, K., Ryvarden, L. & Watling, R. (1993). Preliminary checklist of non-agaricoid macromycetes of the Korup National Park, Cameroon and surrounding area. *Edinburgh Journal of Botany*, **50**, 105–19.

Hoiland, K. & Watling, R. (1990). Some *Cortinarius* species (Agaricales) of the Cooloola Sandmass, Queensland, Australia. *Plant Systematics and Evolution*, **171**, 135–46.

Hooker, W. (1816). *Plantae cryptogamicae, quas in plaga orbis novi aequinoctiali colleagerunt Alexander de Humboldt et Aimé.* Fasc. 1, 8 fol., 4 col. pl. London.

Hooker, W. (1822). Fungi et Lichenes in C.S. Kunth, *Synopsis Plantarum quas in itinere ad plagem aequinoctialem orbis novi collererunt A. de Humboldt et Am. Bonpland* **1**, 7(fungi), 14 (lichens).

Hughes, S.J. (1952). Fungi from the Gold Coast. *Mycological Papers*, **48**, 1–91.

Hughes, S.J. (1953). Fungi from the Gold Coast II. *Mycological Papers*, **50**, 1–104.

Hyde, K.D. & Jones, E.B.G. (1988). Marine mangrove fungi. *P.S.Z.N.I.: Marine Ecology*, **9**, 15–33.

Isaac, S., Frankland, J.C., Watling, R. & Whalley, A.J.S. (1993). Aspects of tropical mycology. *British Mycological Society Symposium Series*, **19**, 325.

Johnston, A. (1960). A supplement to the host list of plant diseases in Malaya. *Mycological Papers*, **77**, 1–30.

Jones, E.B.G. (1993). Tropical marine fungi. *British Mycological Society Symposium Series*, **19**, 73–89.

Kile, G.A. & Watling, R. (1983). *Armillaria* species from south-eastern Australia. *Transactions of the British Mycological Society*, **81**, 129–40.

Kile, G.A. & Watling, R. (1988). Identification and occurrence of Australian *Armillaria* species, including *A. pallidula* sp. nov. and non-Australian tropical and Indian *Armillaria*. *Transactions of the British Mycological Society*, **91**, 305–15.

Larter, L.N.H. & Martyn (1943). A preliminary list of plant diseases of Jamaica. *Mycological Papers*, **8**, 1–15.

Margot, P. & Watling, R. (1981). Studies in Australian agarics and boletes II. Further studies in *Psilocybe*. *Transactions of the British Mycological Society*, **76**, 485–9.

Massee, G. (1891). New fungi from Madagascar, *Journal of Botany, London*, **29**, 102.

Massee, G. (1892*a*). Some West Indian fungi. *Journal of Botany, London*, **30**, 161–4.

Massee, G. (1892*b*). Some West Indian Fungi. *Journal of Botany*, **30**, 196–8.

Massee, G. (1892*c*). Notes on exotic fungi in the Royal herbarium, Kew. *Grevillea*, **21**, 1–6.

Massee, G. (1892–1895). *British Fungus Flora. A classified text-book of mycology*, 4 vols, London: Bell & Sons.

Massee, G. (1893). Australian fungi. *Grevillea*, **22**, 17–19.

Massee, G. (1894). Exotic fungi. *Grevillea*, **22**, 67–8.

Massee, G. (1898*a*). The fungus flora of New Zealand. *Transactions & Proceedings of the New Zealand Institute, Wellington*, **31**, 282.

Massee, G. (1898*b*). Tea blights. *Bulletin of Miscellaneous Information, Kew, 1898*, 105–12.

Massee, G. (1898*c*). Fungi exotici. *Bulletin of Miscellaneous Information, Kew, 1898*, 113–6.

Massee, G. (1899*a*). Cacao diseases in Trinidad. *Bulletin of Miscellaneous Information, Kew, 1899*, 1–6.

Massee, G. (1899*b*). Fungi exotici II. *Bulletin of Miscellaneous Information, Kew, 1899*, 164–85.

Massee, G. (1901*a*). Fungi exotici III. *Bulletin of Miscellaneous Information, Kew, 1901*, 150–69.

Massee, G. (1901*b*). Thallophytes of two botanical collections from Mount Roraima, British Guyana. *Transactions of the Linnean Society, London*, ser. 2, **6**, 101–2.

Massee, G. (1906). Fungi exotici IV. *Bulletin of Miscellaneous Information, Kew, 1906*, 91–4.

Massee, G. (1907*a*). The fungus flora of New Zealand Part II. *Transactions of the New Zealand Institute*, **39**, 1–49.

Massee, G. (1907*b*). Fungi exotici VI. *Bulletin of Miscellaneous Information, Kew, 1907*, 121–4.

Massee, G. (1907*c*). Philippine Myxogastres. *Philippine Journal of Science*, **2**, 113–115.

Massee, G. (1908*a*). Fungi exotici VII collected in Singapore Botanic Gardens. *Bulletin of Miscellaneous Information, Kew, 1908*, 1–6.

Massee, G. (1908*b*). Fungi exotici VIII collected in Singapore Botanic Gardens. *Bulletin of Miscellaneous Information, Kew, 1908*, 216–19.

Massee, G. (1908*c*). The South African locust fungus (*Entomophthora grylli* Fres.) *Bulletin of Miscellaneous Information, Kew, 1908*, 197–8.

Massee, G. (1909a). Fungi exotici IX. *Bulletin of Miscellaneous Information, Kew, 1909*, 204–9.

Massee, G. (1909b). Coffee diseases of the New World. *Bulletin of Miscellaneous Information, Kew, 1909*, 337–41.

Massee, G. (1910a). Fungi from Penang. *Agricultural Bulletin of the Straits and Federated Malay States*, **9**, 135.

Massee, G. (1910b). Fungi exotici XI. *Bulletin of Miscellaneous Information, Kew, 1910*, 249–53.

Massee, G. (1910c). Trinidad fungi. *Proceedings of the Agricultural Society of Trinidad & Tobago*, **10**, 87–90.

Massee, G. (1911). Fungi exotici XII. *Bulletin of Miscellaneous Information, Kew, 1911*, 223–6.

Massee, G. (1912a). Fungi exotici XIII. *Bulletin of Miscellaneous Information, Kew, 1912*, 189–91.

Massee, G. (1912b). Fungi exotici XIV. *Bulletin of Miscellaneous Information, Kew, 1912*, 253–5.

Massee, G. (1912c). Fungi exotici XV. *Bulletin of Miscellaneous Information, Kew, 1912*, 357–9.

Massee, G. (1913). Fungi exotici XVI. *Bulletin of Miscellaneous Information, Kew, 1913*, 104–5.

Massee, G. (1914a). Fungi exotici XVII. *Bulletin of Miscellaneous Information, Kew, 1914*, 72–6.

Massee, G. (1914b). Fungi exotici XIX. *Bulletin of Miscellaneous Information, Kew, 1914*, 357–9.

Mattioli, P.A. (1560). *Commentarii in sex libros Pedacii Dioscoridis Anazarbei de Medica Materia.*

McAlpine, D. (1895). *Systematic arrangement of Australian fungi, together with host index and lists of works with subject.* Victoria: Agriculture Department.

McAlpine, D. (1899). *Fungus Diseases of Citrus Trees in Australia, and their Treatment.* Melbourne, Department of Agriculture of Victoria.

McAlpine, D. (1902). *Fungus Diseases of Stone-fruit Trees in Australia, and their Treatment.* Melbourne, Department of Agriculture of Victoria.

McAlpine, D. (1906). *The Rusts of Australia, their Structure, Nature and Classification.* Victoria: Agriculture Department.

McAlpine, D. (1910). *The Smuts of Australia, Their Structure, Life-history, Treatment and Classification.* Melbourne.

McAlpine, D. (1912). *Handbook of Fungus Diseases of the Potato in Australia, and their Treatment.* Melbourne.

Morris, B. (1981). *Common Mushrooms of Malawi.* Oslo: Fungiflora.

Morris, B. (1984). Macrofungi of Malawi: some ethnobotanical notes. *Bulletin of the British Mycological Society*, **18**, 48–57.

Morris, B. (1990). An annotated checklist of the macrofungi of Malawi. *Kirkia*, **13**, 323–64.

Nattrass, R.M. (1961). Host list of Kenya fungi and bacteria. *Mycological Papers*, **81**, 1–46.

Norse, D. (1974). Plant diseases of Barbados. *Phytopathological Papers*, **18**, 1–38.

Nowell, W. (1923). *Diseases of Crop Plants in the Lesser Antilles.* London: West India Committee.

Oldridge, S.G., Pegler, D.N., Reid, D.A. & Spooner, B.M. (1986). A collection of fungi from Pahang and Negeri Sembilan (Malaysia). *Kew Bulletin*, **41**, 855–72.

Pegler, D.N. (1965). Studies on Australasian Agaricales. *Australian Journal of Botany*, **13**, 323–56.
Pegler, D.N. (1966). Tropical African Agaricales. *Persoonia*, **4**, 73–124.
Pegler, D.N. (1967). Studies on African Agaricales. *Kew Bulletin*, **21**, 499–533.
Pegler, D.N. (1969*a*). Studies on African Agaricales II. *Kew Bulletin*, **23**, 219–49.
Pegler, D.N. (1969*b*). A New Pathogenic Species of *Marasmiellus* Murr. (Tricholomataceae). *Kew Bulletin*, **23**, 523–5.
Pegler, D.N. (1971). *Lentinus* Fr. and related genera from Congo-Kinshasa (Fungi). *Bulletin du Jardin National de Belgique*, **41**, 273–81.
Pegler, D.N. (1972*a*). A revision of the genus *Lepiota* from Ceylon. *Kew Bulletin*, **27**, 155–202.
Pegler, D.N. (1972*b*). Lentineae (Polyporaceae), Schizophyllaceae et espèces lentinoides et pleurotoides des Tricholomataceae. *Flore Illustrée des Champignons d'Afrique Centrale*, fascicle 1, 5–26.
Pegler, D.N. (1973). *Skepperiella cochlearis* sp.nov., and the genus *Skepperiella* Pilàt. *Kew Bulletin*, **28**, 257–65.
Pegler, D.N. (1976). A revision of the Zanzibar Agaricales described by Berkeley. *Kew Bulletin*, **30**, 429–42.
Pegler, D.N. (1977*a*). *Pleurotus* (Agaricales) in India, Nepal and Pakistan. *Kew Bulletin*, **31**, 501–510.
Pegler, D.N. (1977*b*). A preliminary agaric flora of East Africa. *Kew Bulletin, Additional Series*, **6**, 1–615.
Pegler, D.N. (1977*c*). A new species of *Richoniella* (Hymenogastrales) from Ghana. *Kew Bulletin*, **32**, 12.
Pegler, D.N. (1977*d*). A revision of Entolomataceae (Agaricales) of India and Sri Lanka. *Kew Bulletin*, **32**, 189–220.
Pegler, D.N. (1978). *Crinipellis perniciosa* (Agaricales). *Kew Bulletin*, **32**, 731–736.
Pegler, D.N. (1982). Agaricoid and boletoid fungi (Basidiomycota) from Malawi and Zambia. *Kew Bulletin*, **37**, 255–71.
Pegler, D.N. (1983). Agaric Flora of the Lesser Antilles. *Kew Bulletin, Additional Series*, **9**, 1–688.
Pegler, D.N. (1984*a*). *Lentinus araucariae*, an Australasian member of the *L. badius* – complex. *Cryptogamie, Mycologie*, **4**, 123–8.
Pegler, D.N. (1984*b*). The genus *Lentinus*, a World Monograph. *Kew Bulletin, Additional Series*, **10**, 1–281.
Pegler, D.N. (1986*a*). *Cystangium crichtonii* sp. nov. (Russulaceae), a secotioid species from New South Wales. *Transactions of the British Mycological Society*, **86**, 181–4.
Pegler, D.N. (1986*b*). Agaric flora of Sri Lanka. *Kew Bulletin, Additional Series*, **12**, 1–519.
Pegler, D.N. (1987*a*). A revision of the Agaricales of Cuba 1. Species described by Berkeley & Curtis. *Kew Bulletin*, **42**, 501–85.
Pegler, D.N. (1987*b*). A revision of the Agaricales of Cuba 2. Species described by Earle and Murrill. *Kew Bulletin*, **42**, 853–86.
Pegler, D.N. (1988). A revision of the Agaricales of Cuba 3. Keys to families, genera and species. *Kew Bulletin*, **43**, 53–75.
Pegler, D.N. (1990). Agaricales of Brazil described by J.P.F.C. Montagne. *Kew Bulletin*, **45**, 161–77.
Pegler, D.N., Beaton, G. & Young, T.W.K. (1984*a*). Gasteroid Basidiomycota of Victoria State, Australia 1, Hydnangiaceae. *Kew Bulletin*, **39**, 499–508.

Pegler, D.N., Beaton, G. & Young, T.W.K. (1984*b*). Gasteroid Basidiomycota of Victoria State, Australia, 2, Russulales. *Kew Bulletin*, **39**, 698–9.

Pegler, D.N., Beaton, G. & Young, T.W.K. (1985*a*). Gasteroid Basidiomycota of Victoria State, Australia 3, Cortinariales. *Kew Bulletin*, **40**, 167–204.

Pegler, D.N., Beaton, G. & Young, T.W.K. (1985*b*). Gasteroid Basidiomycota of Victoria State, Australia 4, *Hysterangium. Kew Bulletin*, **40**, 435–44.

Pegler, D.N., Beaton, G. & Young, T.W.K. (1985*c*). Gasteroid Basidiomycota of Victoria State, Australia 5–7, *Kew Bulletin*, **40**, 573–98.

Pegler, D.N., Beaton, G. & Young, T.W.K. (1985*d*). Gasteroid Basidiomycota of Victoria State, Australia 8–9. *Kew Bulletin*, **40**, 827–42.

Pegler, D.N. & Fiard, J.P. (1978). *Hygrocybe* Sect. *Firmae* in Tropical America. *Kew Bulletin*, **32**, 297–312.

Pegler, D.N. & Fiard, J.P. (1979). Taxonomy and ecology of *Lactarius* (Agaricales) in the Lesser Antilles. *Kew Bulletin*, **33**, 601–28.

Pegler, D.N. & Fiard, J.P. (1980). New taxa of *Russula* from the Lesser Antilles. *Mycotaxon*, **12**, 92–6.

Pegler, D.N. & Holland, A. (1983*a*). *Hebeloma victoriense* and the genus *Metraria. Transactions of the British Mycological Society*, **80**, 157–60.

Pegler, D.N. & Lodge, D.J. (1990). Hygrophoraceae of Luquillo Mountains of Puerto Rico. *Mycological Research*, **94**, 443–56.

Pegler, D.N. & Piearce, G. (1980). The edible fungi of Zambia. *Kew Bulletin*, **35**, 475–91.

Pegler, D.N. & Rayner, R.W. (1969). A contribution to the Agaric Flora of Kenya. *Kew Bulletin*, **23**, 347–412.

Pegler, D.N. & Vanhaecke, M. (1993). *Termitomyces* in South-east Asia. *Kew Bulletin*, **49**, 717–36

Peregrine, W.T.H. & Siddiqi, M.A. (1972). A revised and annotated list of plant diseases in Malawi. *Phytopathological Papers*, **16**, 1–51.

Petch, T. (1908). Die Pilze von *Hevea brasiliensis. Zeitschrift für Pflanzenkrankheiten*, **18**, 81–2.

Petch, T. (1910). Root diseases of *Acacia decurrens. Circular of the Agricultural Journal of the Royal Botanic Gardens, Ceylon*, **5** (10), 89–94.

Petch, T. (1913*a*). Termite fungi: a résumé. *Annals of the Royal Botanic Gardens, Peradeniya*, **5**, 303–41.

Petch, T. (1913*b*). White ants and fungi. *Annals of the Royal Botanic Gardens, Peradeniya*, **5**, 389–93.

Petch, T. (1915). Horse-hair blights. *Annals of the Royal Botanic Gardens, Peradeniya*, **6**, 43–68.

Petch, T. (1921). *The Diseases and Pests of the Rubber Tree.* London: Macmillan & Co.

Petch, T. (1923). *The Diseases of the Tea Bush.* London: Macmillan & Co.

Petch, T. (1925). Thread blights. *Annals of the Royal Botanic Gardens, Peradeniya*, **9**, 1–46.

Petch, T. (1928). Tropical root disease fungi. *Annals of the Royal Botanic Gardens, Peradeniya*, **13**, 238–253.

Petch, T. & Bisby, G.R. (1950). The Fungi of Ceylon. *Peradeniya Manual 6*, 1–111.

Petersen, R.H. & Watling, R. (1989). New or interesting *Ramaria* taxa from Australia. *Notes of the Royal Botanic Garden, Edinburgh*, **46**, 144–59.

Piearce, G.D. (1981). Zambian mushrooms – customs and folklore. *Bulletin of the British Mycological Society*, **15**, 139–42.

Piearce, G.D. (1987). The genus *Termitomyces* in Zambia. *The Mycologist*, **1**, 111–16.

Pirozynski, K.A. (1972). Microfungi of Tanzania I. Miscellaneous fungi on oil palm. II. New Hyphomycetes. *Mycological Papers*, **129**, 1–64.

Reid, D.A. (1955). New or interesting records of Australian Basidiomycetes. *Kew Bulletin*, **10**, 631–648.

Reid, D.A. (1956). New or interesting records of Australian Basidiomycetes II. *Kew Bulletin*, **11**, 535–40.

Reid, D.A. (1957). New or interesting records of Australian Basidiomycetes III. *Kew Bulletin*, **12**, 127–43.

Reid, D.A. (1958). A new species of *Thelephora* from Malaya. *Kew Bulletin*, **13**, 227–8.

Reid, D.A. (1963*a*). Fungi Venezuelani VI (New or interesting records of Australian Basidiomycetes IV). *Kew Bulletin*, **16**, 437–45.

Reid, D.A. (1963*b*). New or interesting records of Australian Basidiomycetes V. *Kew Bulletin*, **17**, 267–308.

Reid, D.A. (1965). A Monograph of the Stipitate Stereoid Fungi. *Beihefte zur Nova Hedwigia*, **18**, 1–382.

Reid, D.A. (1973). A reappraisal of type and authentic specimens of Basidiomycetes in the Van der Byl herbarium, Stellenbosch. *Journal of South African Botany*, **39**, 141–78.

Reid, D.A. (1974). A reappraisal of type and authentic material of the larger Basidiomycetes in the Pretoria herbarium. *Bothalia*, **11**, 221–30.

Reid, D.A. (1975*a*). Type studies of the larger Basidiomycetes described from southern Africa. *Contributions from the Bolus Herbarium*, **7**, 1–255.

Reid, D.A. (1975*b*). A new species of *Coryneliospora* on the fruits of a *Rapanea* from New Guinea. *Kew Bulletin*, **30**, 17–18.

Reid, D.A. (1978). New species of *Amanita* from Australia. *Victorian Naturalist*, **95**, 47–9.

Reid, D.A. (1979). *Tremelloscypha* and *Papyrodiscus*, two new genera of basidiomycetes for Australasia. *Sydowia, Beihefte*, **8**, 332–4.

Reid, D.A. (1980). A monograph of the Australian species of *Amanita*. *Australian Journal of Botany, supplementary series*, **8**, 1–96.

Reid, D.A. (1986). New or interesting records of Australian Basidiomycetes VI. *Transactions of the British Mycological Society*, **86**, 429–40.

Reid, D.A. (1992). The genus *Elmerina* (Tremellales) with accounts of two species from Queensland, Australia. *Persoonia*, **14**, 465–74.

Reid, D.A. & Eicker, A. (1991*a*). South African fungi: the genus *Amanita*. *Mycological Research*, **95**, 80–95.

Reid, D.A. & Eicker, A. (1991*b*). A taxonomic survey of the genus *Montagnea*, with special reference to South Africa. *South African Journal of Botany*, **57**, 161–70.

Reid, D.A. & Guillarmod, A.J. (1988). *Marasmius titanosporus*, a new species from the Eastern Cape, South Africa. *Transactions of the British Mycological Society*, **91**, 707–9.

Riley, E.A. (1956). A preliminary list of plant diseases in Northern Rhodesia. *Mycological Papers*, **63**, 1–28.

Riley, E.A. (1960). A revised list of plant diseases in Tanganyika Territory. *Mycological Papers*, **75**, 1–42.

Stephenson, J. (1879–1885). Mycologia Scotica. *Scottish Naturalist* vols. 6–8.

Sutton, B.C. (1993). Mitosporic fungi from Malawi. *Mycological Papers*, **167**, 1–93.

Tarr, S.A.J. (1963). A supplementary list of Sudan fungi and plant diseases. *Mycological Papers*, **85**, 1–31.

Thompson, A. & Johnston, A. (1953). A host list of plant diseases in Malaya. *Mycological Papers*, **52**, 1–38.

Turner, G.J. (1971). Fungi and plant disease in Sarawak. *Phytopathological Papers*, **13**, 1–55.

Wakefield, E.M. (1912). Nigerian fungi. *Bulletin of Miscellaneous Information, Kew, 1912*, 141–4.

Wakefield, E.M. (1914). Nigerian Fungi II. *Bulletin of Miscellaneous Information, Kew, 1914*, 253–61.

Wakefield, E.M. (1917*a*). Fungi exotici. XXIII. *Bulletin of Miscellaneous Information, Kew, 1917*, 308–14.

Wakefield, E.M. (1917*b*). Nigerian Fungi III. *Bulletin of Miscellaneous Information, Kew, 1917*, 105–11.

Wakefield, E.M. (1918). Fungi exotici XXIV. *Bulletin of Miscellaneous Information, Kew, 1918*, 207.

Wakefield, E.M. (1920*a*). Diseases of oil palm in West Africa. *Bulletin of Miscellaneous Information, Kew, 1920*, 306–8.

Wakefield, E.M. (1920*b*). Fungi exotici XXV. Notes on Ugandan fungi II Microfungi. *Bulletin of Miscellaneous Information, Kew*, 1920, 289–300.

Wakefield, E.M. (1922). Fungi exotici XXVI. Notes on Ugandan fungi. *Bulletin of Miscellaneous Information, Kew, 1922*, 161–5.

Wakefield, E.M. (1927). The genus *Cystopus* in South Africa. *Bothalia*, **2**, 242–6.

Wakefield, E.M. (1930). Presidential address. Fungi exotici: past work and present problems. *Transactions of the British Mycological Society*, **15**, 12–31.

Wallace, G.B. & Wallace, M.M. (1949). A list of plant diseases of economic importance in Tanganyika Territory. *Mycological Papers*, **26**, 1–25.

Wallace, G.B. & Wallace, M.M. (1953). A supplement to a list of plant diseases of economic importance in Tanganyika Territory, *Mycological Papers*, **51**, 1–7.

Watling, R. (1973). New species of Bolbitiaceae (Agaricales) from Zaire. *Bulletin du Jardin Botanique de Belgique*, **43**, 187–92.

Watling, R. (1974). Bolbitiaceae. *Flore Illustrée des Champignons d'Afrique Centrale*, Fascicle 3, 55–71.

Watling, R. (1992). *Armillaria* Staude in the Cameroon Republic. Persoonia, **14**, 483–91.

Watling, R. (1978). Studies on Australian agarics and boletes 1. Some species of *Psilocybe. Notes of the Royal Botanic Gardens, Edinburgh*, **36**, 199–210.

Watling, R. & Gregory, N.M. (1986). Observations on the boletes of the Cooloola Sandmass, Queensland and notes on their distribution in Australia. Part I, introduction and keys. *Proceedings of the Royal Society of Queensland*, **97**, 97–128.

Watling, R. & Gregory, N.M. (1988*a*). Observations on the boletes of the Cooloola Sandmass, Queensland and notes on their distribution in Australia. Part 2A. Smooth-spored taxa – introduction, keys and references. *Proceedings of the Royal Society of Queensland*, **99**, 45–63.

Watling, R. & Gregory, N.M. (1988*b*). Observations on the boletes of the Cooloola Sandmass, Queensland and notes on their distribution in Australia. Part 2B. Smooth-spored taxa of the family Gyrodontaceae and the genus *Pulveroboletus. Proceedings of the Royal Society of Queensland*, **99**, 65–75.

Watling, R. & Gregory, N.M. (1989*a*). Observations on the boletes of the Cooloola Sandmass, Queensland and notes on their distribution in Australia. Part 2C. Smooth-spored taxa – Strobilomycetaceae. *Proceedings of the Royal Society of Queensland*, **100**, 13–30.

Watling, R. & Gregory, N.M. (1989*b*). Observations on the boletes of the Cooloola Sandmass, Queensland and notes on their distribution in Australia. Part 2D. Smooth-spored taxa – Boletaceae, Xerocomaceae. *Proceedings of the Royal Society of Queensland*, **100**, 31–47.

Watling, R. & Gregory, N.M. (1991). Observations on the boletes of the Cooloola Sandmass, Queensland and notes on their distribution in Australia. Part I, lamellate taxa. *Edinburgh Journal of Botany*, **48**, 353–91.

Watling, R. & Turnbull, E. (1993). Boletes from south and east-central Africa, 1. *Edinburgh Journal of Botany*, **49**, 343–61.

Whalley, A.J.S. (1993). Tropical Xylariaceae: their distribution and ecological characters. In *Aspects of Tropical Mycology* (ed. Isaac, S., Frankland, J.C., Watling, R. and Whalley, A.J.S.). British Mycological Society Symposium Series, **19**, 103–9.

Whalley, A.J.S., Hammerlev, D. & Taligoola, H.K. (1988). Two new species of *Hypoxylon* from Nigeria. *Transactions of the British Mycological Society*, **90**, 139–41.

Whalley, A.J.S. & Taligoola, H.K. (1978). Species of *Hypoxylon* from Uganda. *Transactions of the Royal Botanical Society of Edinburgh* (Cryptogamic Centenary Symposium), **42**, 93–98.

Whalley, A.J.S. & Watling, R. (1989). *Versiomyces cahuchucosus* gen. & sp. nov. from Queensland, Australia. *Notes of the Royal Botanic Garden, Edinburgh*, **45**, 401–4.

Williams, T.H. & Liu, P.S.W. (1976). A host list of plant diseases in Sabah, Malaysia. *Phytopathological Papers*, **19**, 1–67.

Williamson, J. (1973). Preliminary list of some edible fungi of Malawi. *Journal of the Society of Malawi*, **26**, 15–27.

4

The amateur contribution within the Society

ROY WATLING

At the time of the formation of the Society, three groups dedicated to the study of fungi were in existence, a very active element within the umbrella of the Yorkshire Naturalists' Union (Blackwell, 1961*b*) (Fig. 4.1), a smaller but very influential group originally under the leadership of Dr Bull based in Herefordshire (Anon., 1873; Dennis, 1975; Ainsworth, 1976) – the Woolhope Club (Fig. 4.2), and a strong group north of the 'border', the Scottish Cryptogamic Society which although merged with the Botanical Society of Edinburgh, celebrated its centenary 20 years ago (Noble, 1978; Watling, 1986) (Fig. 4.3). Although professionals such as

Fig. 4.1. Yorkshire Naturalists' Union fungus foray at Maltby 1903. Charles Crossland and Alfred Clarke in foreground left and right respectively. James Needham to the middle left.

81

Fig. 4.2. The Hereford Fungus Festival 1874. Cartoon of members and activities of the Woolhope Club by Worthington G. Smith. *The Graphic* 17 × 1874.

Fig. 4.3 Botanical Society of Edinburgh (Cryptogamic Section) foray at Edzell 1958. Back row: Professor J.A. Macdonald, P. Austwick, P.D. Orton and R. Johnstone. Front row: Florence Greig, Dr Mary Noble, Dr D. Downie and Professor D.M. Henderson.

M.C. Cooke and G. Massee and later foreign visitors such as Dr. G. Atkinson from Cornell University (Fig. 4.4) and Profs M.J. de Seynes and M. Cornu of Paris (Fig. 4.5) attended, these groups were amateur based.

There was therefore already a large body of workers and the time was ripe for a national society to be formed. This came about at the Yorkshire Naturalists' Union foray based in Selby in 1896 when the concept of a national society was formally floated. In fact, the ideas began to gel at the now famous tea party (Blackwell, 1961*a*, 1966) held at the Huddersfield home of Alfred Clarke in 1895, a night school teacher, photographer and chemist (Blackwell, 1966; Watling & Sykes, 1992) who recorded the event. Present at that historic meeting were M.C. Cooke, the daddy of them all (Ramsbottom, 1915*a,b*; English, 1978) (Fig. 4.6); C. Rea, a Woolhopean and Worcestershire barrister (Ramsbottom, 1948*b*) and especially invited to the meeting; Charles Crossland, a Halifax butcher (Ramsbottom, 1917*a*) and his companion James Needham, a Hebden Bridge dyer (Crossland, 1913). G. Massee, a Yorkshireman who drifted into mycology from farming, art college and agricultural work abroad (Ramsbottom, 1917*b*; Watling, 1982; Ainsworth, 1992*a*), also attended and the YNU was represented by the Rev. William Weeks Fowler of Normanton, who, whilst not primarily a mycologist, was very sympathetic to the cause and

Fig. 4.4. Mycologists at the Yorkshire Naturalists' Union foray, Hemsley 1903. From left to right: Charles Crossland, Alfred Clarke, George Massee, W.N. Cheesman and Professor George T. Atkinson, visiting from Cornell University.

Fig. 4.5. Cartoon of Mycological activities and personalities. *The Gardeners' Magazine* 17 × 1877 by Worthington G. Smith. 1. Mycologists spilling precious collections. 2. Gathering *Fistulina* (Berkeley depicted). 3. Collecting bog fungi. 4, 5. Going to the collecting sites. 6. Truffle-hunting. 7. M.J. de Seynes, Paris. 8. Professor M. Maxime Cornu, Paris.

Fig. 4.6. Yorkshire Naturalists' Union foray at Huddersfield 1895 at which the national society was spawned. Back row: George Massee, Reverend William Weeks Fowler and James Needham. Front row: Charles Crossland, Mordecai Cubitt Cooke and Carleton Rea. Photograph by Alfred Clarke.

often lectured on the necessity to study the little known groups. He was, however, a founder member of the British Mycological Society. This meeting has been documented by Blackwell (1961*a*, 1966).

The Society was originally formed around fieldwork, the so-called forays, and the amateurs have been a cornerstone of these activities and their expansion ever since. They are still an important facet of the Society, although the membership has expanded still further over the years and embraced the newer disciplines within the subject as they developed. But the larger fungi were, and remain, in the domain of the amateur.

The Huddersfield group was, in fact, building on a long tradition of amateurs which had been under way. One immediately thinks of the Rev. M.J. Berkeley, the father of British Mycology whose publications (Fig. 4.7) and authority in this country were second to none (Ainsworth, 1987*a*); he later collaborated with C.E. Broome (Ainsworth, 1989) who during his lifetime amassed an enormous herbarium including many specimens of his particular interest, viz. truffles. In Scotland there was Berkeley's mentor Robert Kay Greville (Watling, 1990*a*); he was William Hooker's friend who later became Director of the Royal Botanic Gardens, Kew. Both Berkeley

GREVILLEA.   *Frontispiece*

THE REV. M. J. BERKELEY, M A.

Fig. 4.7. Reverend Miles Joseph Berkeley from *Grevillea* frontispiece, Father of British Mycology.

and Greville, judging from their publications and correspondence, had a network of amateur collectors, and they were competent illustrators in their own right. The former's influence spread far from these shores as time went on, and his works on exotic fungi are still called upon today.

Even before these multi-dimensional naturalists were active James Bolton, an industrialist and publican, had catalogued and illustrated the fungi of the Halifax district in Yorkshire (Watling & Seaward, 1981; Watling, 1990*c*). His books were republished in German (Fig. 4.8) and recently a symposium based at the Liverpool Museum was held on his life and works. J. Sowerby also produced fine descriptions and illustrations of fungi still used as references today (Ainsworth, 1988). The early Society's founders therefore really had to live up to high standards!

And they did! There were 50 founder members, the greatest proportion being amateurs or representing amateurs (Dublin Naturalists Club; Louth Antiquarian Society and the Woolhope Club). One of them was Charles Plowright, who succeeded George Massee, the first President, for the period 1898–9. Plowright was able to successfully marry mycology and a very active and distinguished medical career, which saw him move from a house surgeon to medical officer for 32 years for Freebridge, Lynn Rural District Council. He first devoted himself to pyrenomycetes for which he is honoured in two genera of *Dothideales* (*Plowrightia* Sacc. and *Plowrightiella* (Sacc.) Trott. (= *Sydowia* Bres.)) (Anon., 1910). Indeed, he described 296 species with his friend W. Phillips of Shrewsbury; his book on discomycetes is still extremely useful. He had a prolific output, publishing regularly in the *Gardeners' Chronicle*, but it was in the study of rusts and smuts he is most known, carrying out experimentation to determine the heteroeceous nature of some rust species. In fact, this was a lead followed by many amateurs since Plowright's time, e.g. H.T. Soppitt of Halifax (Anon., 1898). Plowright made a large collection, now deposited in the University Herbarium at Birmingham, Edgbaston. He was one of the first to document poisoning by *Amanita phalloides* Vaill.: Fr. in Britain (Plowright, 1907). In parallel to many mycologists of his time he was a very keen archaeologist; he was made Hunterian Professor of Comparative Anatomy and Physiology of the Royal College of Surgeons 1890–94. Another physician practising at about the same time was Somerset Hastings, the author of *Toadstools at Home*, 1906 (Ainsworth, 1991).

Forayers consider the autumn foray of the BMS a 'highlight of the year', a quote I've heard many times, and those amateurs attending on many occasions make most interesting records. In fact for many, the autumn and spring forays substitute as their annual holiday breaks. The Rev. William L.W. Eyre, Rector of Swarraton and Vicar of Northington, Hants., enjoyed his mycology also and this was reflected in the content and title of his Presidential Address (Eyre, 1903) 'Mycology as an instrument of recreation'. The Rev. Eyre of course is well known to many in that his

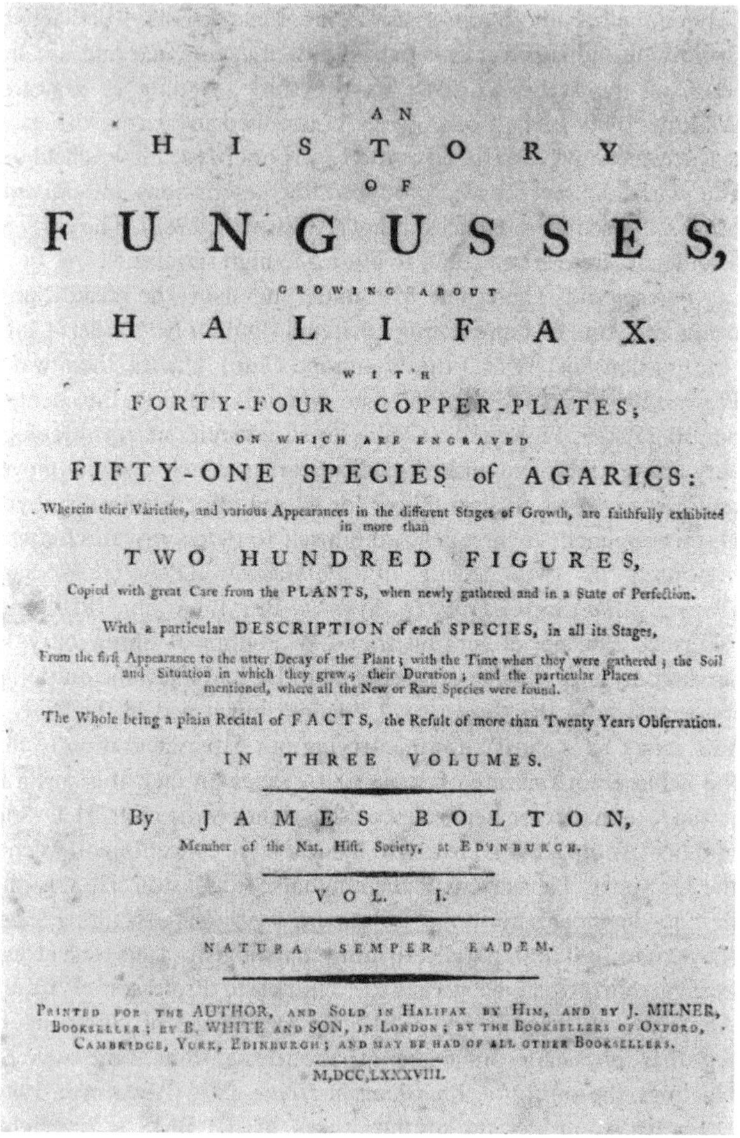

AN

HISTORY

OF

FUNGUSSES,

GROWING ABOUT

HALIFAX.

WITH

FORTY-FOUR COPPER-PLATES;

ON WHICH ARE ENGRAVED

FIFTY-ONE SPECIES of AGARICS:

Wherein their Varieties, and various Appearances in the different Stages of Growth, are faithfully exhibited in more than

TWO HUNDRED FIGURES,

Copied with great Care from the PLANTS, when newly gathered and in a State of Perfection.

With a particular DESCRIPTION of each SPECIES, in all its Stages,

From the first Appearance to the utter Decay of the Plant; with the Time when they were gathered; the Soil and Situation in which they grew; their Duration; and the particular Places mentioned, where all the New or Rare Species were found.

The Whole being a plain Recital of FACTS, the Result of more than Twenty Years Observation.

IN THREE VOLUMES.

By JAMES BOLTON,

Member of the Nat. Hist. Society, at EDINBURGH.

VOL. I.

NATURA SEMPER EADEM.

PRINTED FOR THE AUTHOR, AND SOLD IN HALIFAX BY HIM, AND BY J. MILNER, BOOKSELLER; BY B. WHITE AND SON, IN LONDON; BY THE BOOKSELLERS OF OXFORD, CAMBRIDGE, YORK, EDINBURGH; AND MAY BE HAD OF ALL OTHER BOOKSELLERS.

M,DCC,LXXXVIII.

Fig. 4.8. Title page to James Bolton's *An History of Fungusses growing about Halifax* Vol. 1 1788; the only classic British mycological text to be translated into a second language.

name forms the epithet for *Melanophyllum eyrei* (Mass.) Singer (as *Schulzeria eyrei* Mass.), a beautiful although rare, small, green-blue gilled member of the *Agaricaceae*. Rea (1922), in fact, introduced the new generic name *Glaucospora* for it. He is also celebrated in a jelly fungus (*Bourdotia* = *Basidiodendron*; Wakefield, 1913) and a poroid fungus (*Poria* = *P. hypolateritius* Berk. fide Lowe = *Schizopora paradoxa* (Fr.) Donk *sensu lato*; Bresadola, 1911). He was very enthusiastic but was happy to send collections of interesting fungi to his many correspondents as opposed to describing new entities himself. He (Eyre apud Smith & Rea, 1903) did, however, describe *Lepiota grangei* (Eyre) Lange (as *Schulzeria*) another small lepiotoid agaric. Eyre illustrates how keen amateurs can find very interesting fungi, often missed by the professional, by concentrating on small areas. In so doing they help the progress of mycology by publishing lists of the fungi of their locality, and emphasize their ecology as Eyre did. Eyre was a member of the Woolhopeians in the later days of their active mycological period along with Worthington G. Smith (Dennis, 1975). These two helped to bridge the gap between the era of Berkeley and the early days of the Society.

W.G. Smith became a book illustrator after studying as an architect, and his talents were soon recognized by the *Gardeners' Chronicle* (Fig. 4.9). He followed Plowright in offering many mycological contributions including an account of original work on potato blight and by illustrating fungi in his own special unique way. He illustrated many botanical articles with great skill (Anon., 1873) especially from 1875–1910 in addition to the manual issued by the British Museum to describe the Sowerby models therein (Smith, 1898) and became known as 'Worth Gee Up Smith'. In fact, Smith published several books on topics ranging from diseases in field crops to early man and the history of the counties of England; he was an enthusiastic antiquarian. These publications were in addition to a supplement to Berkeley's *Outlines of British Fungology* (Smith, 1891) and *Synopsis of British Basidiomycetes* (Smith, 1908*b*), the forerunner of C. Rea's book, but still of great use recognizing therein genera which today, under different names, are used in classification, e.g. *Togaria, Deconica* (although subsumed by Rea years later) but *Deconica* reappeared briefly in the *New Check List* (Dennis, Orton & Hora, 1960). These books are illustrated but it is for his informal montages of forays depicting mycologists and their activities that he is remembered more by some; his satirical illustrations of political activities are also thought-provoking. One illustration depicting him illustrating plants in a greenhouse at the top of a ladder accompanies his obituary in the *Gardeners Chronicle* (Nov. 3rd

Fig. 4.9. The Perth and Hereford Fungus Meetings. The foundation of the Scottish Cryptogamic Society. 1. Sir Thomas Moncreiffe, President of the Scottish Cryptogamic Society. 2. Collecting in inclement weather. 3. Gathering mushrooms. 4, 5. Hats and mushrooms. 6. Buchanan White. 7. Reverend John Stevenson. 8, 9. Circles (mushrooms and stones) in Perth and Hereford, respectively. 10. Headquarters of the Woolhopeians with Dr Bull represented by a head of a bull and Reverend M.J. Berkeley blessing his followers. 11. Examination of potato blight fungus. 12. Finding earthstars. 13. Muddy path to Sir John Cotterell's seat. From *The Graphic* 12 xi 1875 by Worthington G. Smith.

1917; 181); it encapsulates his humour, draftsmanship and biological knowlege. He was awarded the Knightian Medal and Veitch Medals of the Royal Horticultural Society and was President of the BMS in 1904 (Smith, 1920; Ainsworth, 1990). It is a shame that the link between mycology and the RHS is so much poorer these days, but the celebratory exhibition of the Society will draw us together again. Smith demonstrates how an amateur can add so much to a subject. Many taxonomists will tell you that one can only get to grips with a fungus when you start to draw its features.

In 1906 the first of a father/daughter team became President. Arthur Lister was a wine merchant in London and took an interest in myxomycetes, a subject close to the heart of many amateurs even now. In fact, the subject in this country has been dominated by amateurs. He published the first edition of the British Museum's account of *Mycetozoa*, which in subsequent editions incorporated more and more of his daughter Gulielma's work (Lister, A., 1894; Lister, G., 1925). Her input to the subject was recognized as fundamental and she became President in 1912 and again 20 years later. The Listers were the pioneers in British myxomycete studies (Ainsworth, 1993).

Carleton Rea, as indicated earlier, was a barrister and had enough private means to be a continual amateur mycologist with a wide circle of friends who collected for him and with him, e.g. W.B. Allen (Rea, 1923), leading as one might expect to an incredible knowledge of our native mycota. He was President in 1907 and 1921 but also held many official positions in the Society. His contributions have been fully explored by Pearson (1947) and Ramsbottom (1948b). In addition to his articles on larger fungi in the *Transactions*, often illustrated with colour pictures, he is best known for his compilation of the *British Basidiomycetae* (Rea, 1922), a most useful publication even today! He is commemorated in the nasty tasting *Hygrocybe reae* (Maire) J. Lange and *Leptonia reae* Maire for which a modern description is required desperately. Rea's 1908 Presidential Address covered the use of basidia and spores in classification, something still at the front of enquiry (see Wells, 1994).

C. Rea's wife, Emma Amy, became President in 1915. She was a very accomplished artist and assisted her husband in the documentation of his fungi. Examples are found in the early volumes of the *Transactions*, e.g. *Lactarius theiogalus* (Bull.: Fr.) S.F. Gray in Vol. 3 Plate 3. In fact it was the Rev. Eyre who made money available to the Society in order to publish plates in its publications. A whole portfolio of illustrations was built up of beautifully executed paintings and is now at Kew.

It was Emma Amy who in fact instructed her daughter, later to become Violet Astley-Cooper, in painting, and she too was dragooned into painting fungi for her father. Violet regularly attended the YNU forays where she exhibited her 'father's' paintings and on numerous occasions kept members entertained with anecdotes especially on very wet mornings before leaving for the foray or after being rained off. Her mother's Presidential Address is a mine of information on ancient and classical fungal illustrations and many would do well to read it (Rea, 1915).

Whilst Rea was playing an important role in organizing the young society and documenting the British basidiomycota in a formal way, E.W. Swanton, Curator at the Educational Museum, Haslemere, was spreading the gospel (Ainsworth, 1987b). His Presidential Address in 1916 was on 'Education in Mycology'; rather different in content from the documents seen today on how mycology is being lost from our tertiary education and research programmes, but none the less sound reading (Swanton, 1916). My early development was considerably helped by reading Swanton's book entitled *Fungi and How to Know Them* (1909). I think it and Rolfe & Rolfe's *Romance of the Fungi* (1925) had lasting effects on me. Certainly, Swanton helped and encouraged many would-be mycologists and was a colleague of many well-known names. A.A. Pearson for example was a great friend living in Haslemere, although with Yorkshire business connections.

Mycology has had its full support from the clergy, Rev. M.J. Berkeley, Rev. W.L.W. Eyre and in 1918 the Reverend, later the Very Rev. David Paul became President. He was based in Roxburgh and was one of the few formal Scottish links which the Society had. Although the BMS visited Scotland on several occasions for forays, few Scots ventured south, probably because they were generally amateurs with only limited amounts of time and money to travel long distances, e.g. Rev. Stevenson; J.H. Trail, Professor of Botany in Aberdeen, was a link at the professional level. The Rev. Paul was a very enthusiastic amateur and held the post of Secretary of the Cryptogamic Society of Scotland for many years and not only organized the Society but badgered the forayers to supply lists of fungi, etc. His own interests were in the macromycetes but his religious background no doubt stood him in excellent stead to give as his Presidential Address the early history of mycology. It offers a full spectrum of thoughts from his wide experience and makes very interesting and essential reading (Paul, 1918). Amateurs following Paul's example have been a mainstay in the Society's organization: C. Crossland, for instance, treasurer at the foundation of the Society published several new taxa especially with his friend

George Massee. He was stimulated to take up mycology very late in life by his daughter's interest in botany (Ramsbottom, 1917*b*). The year prior to Paul, Annie Lorraine Smith was President for a second time, the earlier year being 1906 (Smith, 1907, 1917*a*). Smith, born in Dumfriesshire, was a governess and an official assistant at the British Museum (Nat. Hist.) for 46 years. She worked as an 'unpaid' assistant. The result of her labours was several publications ranging from microfungi and blocked drains to W.G. Smith, in addition to the BM *Handbook on Lichens* (Smith, 1921), although she did not produce the first volume as widely stated of the monograph as that had been carried out by Crombie (1894). Lichens are more universally considered part of the fungi as a whole only representing a trophic state, and much to the horror of some traditionalists have been incorporated even within the framework of fungal classification. Smith was keen to integrate lichens into mycology and published a short note to this effect 'Lichenology a new departure' (Smith, 1917*b*); foray lists during her time even included lichens, something lost for a while! She in effect produced the only workable keys for British lichens until Elif Dahl (1952) produced a key to macrolichens; the crustose forms waited another generation.

After the First World War, mycology became more professional, something commented upon by Ramsbottom (1926) in the obituary to W.N. Cheesman, although certain exceptions may be mentioned, e.g. H.C.W. Hawley who was preparing an account of British pyrenomycetes cut short by his untimely death (Ramsbottom, 1924) and reported again by F.T. Brooks in his Presidential Address (Brooks, 1924). Cheesman was really an all-round amateur mycologist. He was brought up in Lincolnshire, went to London and then returned to Selby where he was to play a pivotal role in the Society's founding. He was a draper by trade and a very active general mycologist examining both micro- and macro-forms. Cheesman hosted the second official foray held in Yorkshire in 1884 (Blackwell, 1961*a*; Watling, 1982) which ultimately led to the Yorkshire Nat. Union Mycological Section in 1892, still very active today. It was the same man that catalysed the elements brought together at Selby in 1896 at which the fate of British mycology was sealed. Cheesman was also an archaeologist and active member of the British Association. In fact it was in connection with the latter he travelled to Canada, South Africa, Australia and New Zealand. The first visit was based on Winnipeg and he gathered fungi in the Rocky Mountains (Cheesman, 1910), later published as a list with many new records for the New World. I remember as a young man reading an intriguing paper by Thomas Gibbs of Sheffield describing fungi on dung collected in S. Africa; it was part of an article by Cheesman on fungi he had collected at the BA meeting in 1908 in which both authors added many new records (Cheesman, 1909). It described a technique which has been successfully used in many schools and colleges to demonstrate the fungal succession on dung. It is very pleasing to know that the study of coprophilous fungi continued in Sheffield but this time under the eye of our present President, John Webster, in the University Botany Department, and it was here and on the YNU forays that the BMS *Keys to Fungi on Dung* (Richardson & Watling, 1968 and now in its 3rd edition) were born. Cheesman was a very interesting and observant man and it is a shame that due to illness he was unable to give a Presidential Address. His name will continue, however, because he gave 100 guineas to the Society to start a fund which would allow students to attend forays, something of which he was a very keen supporter (Ramsbottom, 1926).

An account of the early informative years has been given by Ramsbottom (1948*a*) but it took a little time after the Second World War before the amateur element became a strong working force again. The Society, however, soon realized the potential of such members. This spurred the launching of a Newsletter first edited by J.T. Palmer, an active amateur interested in gasteromycetes and latterly the *Sclerotiniaceae*. This publication expanded in coverage and subsequent editors were professionals but the amateur element was strongly maintained and was the source of the results of the forays – the foray lists. The amateurs were still the main participants of the forays. The *Newsletter* became the *Bulletin* and the present-day *Mycologist* is its direct descendant. One only needs to look at an issue to see the articles and photographs by amateurs and for amateurs. The foray lists are no longer published in the journal but are available as printouts. This resulted directly from the introduction of computer recording, again initially by amateurs interested in the subject, especially Bill Moodie, and later it was put on a more professional basis. The Society's database, a most valuable resource and increasing in value annually, is largely the result of the Society's amateur army.

The database is under the keen eye of the foray committee which evolved to be a mixture of amateur and professional elements; it was their duty to select and organize the spring and autumn forays and coordinate a series of day forays in different parts of the country. The committee also tried to encourage other organizations to hold forays and publicized them together in an autumn programme.

The programme included the arrangements for the Yorkshire Naturalists' Union forays which had gone from strength to strength, the amateur organizers attracting employees of the Royal Botanic Gardens, Kew, Commonwealth Mycological Institute (IMI) and Edinburgh. The records so accumulated over such a long period of time (100 years; Blackwell, 1961*b*) and painstakingly catalogued by W.G. Bramley, have been edited by Watling, Richardson and Preece, 1970 (Bramley, 1985). The enormous contribution made by Bramley was recognized by the Society when it awarded him the Society's Associates medal in 1990 (Preece & Watling, 1992) (Fig. 4.10). Bramley concentrated on rusts and smuts, undoubtedly because his interest was aroused by his seeing their effects on a daily basis as a farmer by occupation. However, he made many valuable records of pyrenomycetes also and had a good eye for *Aphyllophorales*.

The YNU success contrasted with the decline of the Woolhope Club and the merging of the Cryptogamic Society of Scotland with the Botanical Society of Edinburgh in 1921, now the Botanical Society of Scotland (Watling, 1986). However, meetings are still regularly held and records maintained in the library of the Royal Botanic Garden, Edinburgh. Whilst the Scottish amateur element slipped from prominence, in other parts of the country interest was blossoming. Thus under the guidance of Dr Nancy Montgomery recording of fungi for the Midlands commenced. A new era had begun and has continued, finally stimulating the production of the *Fungus Flora* of *Warwickshire* (Clark, 1980*a*). This was largely based on the records of local amateurs through their great energies. M.C. Clark and Reg. Evans, who was also a leading member of that Warwickshire team, were awarded the Society's Associates medals in 1988, Clark for his dedicated work with the micromycetes (Clark, 1980*b*), and Evans with parallel determination but particularly in the field of resupinate fungi.

Two other amateur mycologists have played an important role in the study of larger fungi in Britain, A.A. Pearson (Orton, 1954; Ainsworth, 1992*b*) (Fig. 4.11) and P.D. Orton. The former became President of the Society in 1931 and 1952 and the latter was awarded the Society's Associates medal in 1989. A.A. Pearson was a Yorkshire businessman who travelled widely in Europe and commenced a series of papers of selected genera, *British Boleti* (1946), *Russula* (1948) and *Lactarius* (1950); all were subsequently reprinted twice. Two further papers in the same format were completed by Pearson's long-time friend Dr R.W.G. Dennis, viz. *Inocybe* (1968) and *Mycena* (1969). They had collaborated in Pearson's life-time to produce a checklist of *British Agarics and Boleti* (Pearson & Dennis, 1948). Pearson became an Honorary member; he was very popular with those attending the forays. It was with the same mycologist from Kew that

Fig. 4.10. Glass goblet presented to Willis G. Bramley, BMS Honorary Associate, by his friends on the occasion of his 90th birthday: 15 × 1987. The glass engraving depicts the spatterdashed agaric (= *Collybia peronata*) taken from Bolton's Plate 58 (in *History of Fungusses* . . . Vol. 2). Photograph by C. Stephenson.

Fig. 4.11. Arthur Anselm Pearson 1874–1954. President 1931 and 1952. Author of several papers on macromycetes including with Dr R.W.G. Dennis, RBG Kew the *Check List of British Agarics and Boleti*. Belfast foray 1945. With him left to right: W.C. Moore, E.C. Large, Geoffrey C. Ainsworth, P. O'Connor, A.E. Muskett, E.N. Carrothers, Mme M. Le Gal (from Paris), K. St G. Cartwright, Miss T. Gogol, Miss E.M. Blackwell and J. Colhoun. President on far right demonstrating a find.

P.D. Orton collaborated, along with Dr F.B. Hora, to produce the *New Check List* (Dennis, Orton & Hora, 1960). Hora incidentally was a professional physiologist and became President in 1958. To Orton, a school teacher by occupation, we are also indebted for such important papers as *Cortinarius* I and II (Orton, 1955, 1958) which followed Pearson's format and appeared in the same journal, *The Naturalist*, undoubtedly as a result of the strong mycological following in this Yorkshire organization. Orton's Magnum Opus however must be a large paper in the *Transactions* which accompanied the *New Check List*; 300 new species were described or new records to Britain offered (Orton, 1960). Since then, Orton has continued to produce papers at regular intervals in a parallel way to the 1960 article; they appear in the *Kew Bulletin, Notes of the Royal Botanic Garden Edinburgh* and the Society's *Transactions*. A.A. Pearson had strong links with the French Mycological Society which today still has a very high proportion of amateur members with the accompanying strong interest in macromycetes (see Pearson, 1953).

Parker-Rhodes conducted forays for amateurs at Flatford Mill and computed the results of species collections, one of the first studies in biodiversity (Parker-Rhodes, 1955)! He was a statistician by training but a very active mycologist, and his enthusiasm was transferred to the many who attended his forays. There are many amateurs today who are greatly indebted to him for introducing them to fungi during his long-running course at Flatford, an encouragement which is still with them today. Parker-Rhodes was also noted for his researches on the fungi of Skokholm Island published in the Society's *Transactions* and in the *New Phytologist*.

The amateurs have not neglected the micro-forms; Clark has already been mentioned in connection with the Warwickshire flora, but his interest was firstly in the *Myxomycetes*, a group commonly adopted by amateurs, and then latterly the discomycetes. In the latter he followed W.D. Graddon, a businessman in ceramics who contributed enormously to our understanding of many obscure cryptic disc-fungi (Graddon, 1957); he became President in 1956. Like Pearson he had strong links with our sister Society in France, and like Pearson was a guiding hand to the present author.

One only has to look through the herbaria at the two national Botanic Gardens and at the International Mycological Institute to see the huge numbers of collections of fungi housed there made by these amateurs and the many not mentioned.

The Society has been indebted to all those foray committee secretaries, many who were amateurs, e.g. Audrey Thomas, not only for overseeing the committee but participating in Council and making sure the amateur membership was fairly represented. The Foray Committee gave rise to the Foray, Systematics and Structure Committee, which has now, more logically, become the Foray and Conservation Committee because conservation has developed into such an important area. Meantime, the November taxonomy meeting has developed as a feature of the Society's calendar and workshops at the forays reflect an earlier period of the Society when public lectures and displays were featured. The educational aspect of the forays has now become well established, and a recent innovation is the organization of forays abroad to link up with mycologists of like interests, e.g. Norway, Canary Is., Spain.

Since 1965 the author has been involved with the North American Mycological Association, including the North West Key Council and during his presidency of the British Society floated the idea of a parallel activity in Britain. The system was that amateurs would select a group to work on, produce a key and circulate it amongst their friends knowing that the hard-pressed professionals were there to give assistance but not simply to act as an identificatory postal service. Jack Marriott took this on and is at the helm of a very successful group; it has now expanded with facets more than our wildest ideas. Its mimeographed newsletter *Keys* is now in its eighth part with interesting notes, keys to *Clavaria*, checklists etc., in addition to expanding out into manuals for the amateur. A joint meeting was held last year with the Key Council and UK amateurs based at Kindrogan, Perthshire.

Probably one of the most influential books published at the start of the 1980s was *Mushrooms and Other Fungi of Great Britain and Europe* by Roger Phillips (1981). Although a photographer and originally not a mycologist, he set out to depict our native fungi. He soon became a mycologist, like many others being bitten by the bug, and joined the Society. Many amateurs of the day and subsequently will be indebted to him for years to come. Many books by others followed his publication but generally had less impact; subsequently Roger Phillips expanded his interests here and abroad.

The 'foray box' as it is affectionately known contains books relevant to the identification of fungi and selected tomes are taken to the forays. Several people have donated books but it was expanded with some exceedingly important volumes from the library of the late Wing Commander Schofield. Right up until his death, Schofield regularly attended the forays and was a great encouragement to young people. One strong link he made was with Richard Jennings who was particularly active in the application of conservation to fungi and guided the Society in its first hesitant steps in this direction. It is very fitting that some of Richard's legacy has been put towards the production of information boards at the first British Cryptogamic Sanctuary at Dawyck in the Scottish Borders. Jennings was awarded the Society's Associate medal posthumously in 1990 (Watling, 1990*b*).

More recently the Society has seen the setting-up of area study groups throughout the country recording fungi, preparing databases and organizing workshops under the framework of Fungus Groups, e.g. South East Scotland Region Group, Mid-Yorkshire Fungus Group. These groups ensure an active future for amateur mycologists.

In 1987 the Society held a meeting on Tropical mycology which spawned the *Tropical Mycology Newsletter* and from there on the Society and the world suddenly became serious about tropical fungi. An expedition to Ecuador sponsored by the Society was organized by an amateur, dental practitioner by training, G. Dickson. He brought together professionals, amateurs and students from the United Kingdom and abroad, so maintaining the long tradition of the Society of blending its membership resources.

The amateurs' work rarely appears in the citation indices but it will continue to be an important facet in the understanding of our natural heritage not only of the organisms but how and where they occur. They have made and continue to make important contributions following a long tradition and I have only been able to mention a few. We now await with excitement the next President from the amateur element of the Society to follow on from W.D. Graddon.

With the continued deliberate move by academic institutes away from systematics into molecular techniques, bioengineering and allied areas, future input by the amateur will become paramount. Biodiversity studies and conservation both in Britain and abroad will become more heavily dependent on amateur participation as time goes on. At best in many groups good keys and descriptive works are available so that the dedicated amateur can play an important or even more important part in the future development of mycology within and outwith the Society. The contributions of the Society's amateurs abroad similarly will be a very necessary facet of the future, but here an expansion of the 'Keys' concept will undoubtedly be a prerequisite.

Already the era has commenced with the Forestry Commission contacting the Society to carry out biodiversity studies; the bulk of the field work is being conducted by our amateurs. My professional colleagues and are heartened by this new role for the amateur continuing the tradition which founded the Society.

References

Ainsworth, G.C. (1976). *Introduction to the History of Mycology*. Cambridge: Cambridge University Press.

Ainsworth, G.C. (1987*a*). British Mycologists 1. M.J. Berkeley (1803–89). *The Mycologist*, **1**, 126.

Ainsworth, G.C. (1987*b*). British Mycologists 2. E.W. Swanton (1870–1958). *The Mycologist*, **1**, 172–3.

Ainsworth, G.C. (1988). British Mycologists 5. J. Sowerby (1757–1822). *The Mycologist*, **2**, 125.

Ainsworth, G.C. (1989). British Mycologists 9. C.E. Broome. *The Mycologist*, **3**, 189.

Ainsworth, G.C. (1990). British Mycologists 10. Worthington G. Smith (1835–1917). *The Mycologist*, **4**, 32.

Ainsworth, G.C. (1991). British Mycologists 16. Somerville Hastings. *The Mycologist*, **5**, 149.

Ainsworth, G.C. (1992*a*). British Mycologists 17. G.E. Massee (1850–1917). *The Mycologist*, **6**, 13.

Ainsworth, G.C. (1992*b*). British Mycologists 19. A.A. Pearson (1874–1954). *The Mycologist*, **6**, 130.

Ainsworth, G.C. (1993). British Mycologists 23. Arthur (1830–1908) and Gulielma (1860–1949) Lister. *The Mycologist*, **7**, 134.

Anon. (1873). The Fungus Festival at Hereford. *The Graphic*, 15 Nov., 1873.

Anon. (1898). Mr H.T. Soppitt. *Transactions of the British Mycological Society*, **1**, 83–5.

Anon. (1910). Charles Bagge Plowright MD FRCS Hon JP. *Transactions of the British Mycological Society*, **1**, 231–2.

Blackwell, E.M. (1961*a*). Link with Past Yorkshire Mycologists. *The Naturalist*, **877**, 53–66.

Blackwell, E.M. (1961*b*). Fungus forays in Yorkshire and the history of the Mycological Committee. *The Naturalist*, **871**, 163–8.

Blackwell, E.M. (1966). Alfred Clarke, His Herbarium and the birth of the British Mycological Society. *News Bulletin of the British Mycological Society*, **25**, 13–17.

Bramley, W.G. (1985). *A Fungus Flora of Yorkshire* (ed. R. Watling, M.J. Richardson and T. Preece). Leeds.

Bresadola, G. (1911). *Poria eyrei. Transactions of the British Mycological Society*, **3**, 264.

Brooks, F.T. (1924). Presidential Address. Some present day aspects of mycology. *Transactions of the British Mycological Society*, **9**, 14–32.

Cheesman, W.N. (1909). A contribution to the flora of South Africa – with a note on coprophilous fungi by Thomas Gibbs. *Journal of the Linnean Society of London*, **38**, 408–17.

Cheesman, W.N. (1910). A contribution to the mycologic flora and the Mycetozoa of the Rocky Mountains. *Transactions of the British Mycological Society*, **3**, 267–76.

Clark, M.C. (ed.) (1980*a*). *A Fungus Flora of Warwickshire*. Cambridge.

Clark, M.C. (1980*b*). Non-lichenized discomycetes recorded in Britain in recent years. *Bulletin of the British Mycological Society*, **14**, 24–56.

Crombie, J.M. (1894). *A monograph: Lichens found in Britain*. London: British Museum.

Crossland, C. (1913). In memoriam: James Needham. *The Naturalist*, 294–8.

Dahl, E. (1952). *Analytical Keys to British Macrolichens*. Mimeographed.

Dennis, R.W.G. (1954). *Inocybe. The Naturalist*, 117–40.

Dennis, R.W.G. (1955). *Mycena. The Naturalist*, 41–63.

Dennis, R.W.G. (1975). Dr Bull's drawings and the Herefordshire Flora. *Kew Bulletin*, **30**, 541–2.

Dennis, R.W.G., Orton, P.D. & Hora, F.B. (1960). The new Check List of British Agarics and Boleti. *Transactions of the British Mycological Society*, **43**, (Suppl.), 1–225.

English, M.P. (1978). Mordecai Cubitt Cooke, the father of the Queckett Club. *Microscopy*, **33**(6), 329–38.

Eyre, W.L.W. (1903). Presidential Address: Mycology as an instrument of recreation. *Transactions of the British Mycological Society*, **2**, 49–53.

Gibbs, T. (1909). In Cheesman, W.N. A contribution to the mycology of S. Africa. With a note on coprophilous fungi. *Journal of the Linnean Society of London*, **38**, 408–17.

Graddon, W.D. (1957). Presidential Address. Approach to Discomycetes. *Transactions of the British Mycological Society*, **40**, 1–8.

Hastings, S. (1906). *Toadstools at Home*. London.

Lister, A. (1894). *A Monograph of the Mycetozoa*. London: British Museum.

Lister, G. (1925). *A Monograph of the Mycetozoa*. 3rd edn. London: British Museum.

Noble, M. (1978). The Cryptogamic Society of Scotland – 1875–1935–1975. *Transactions of the Botanical Society of Edinburgh*, **42**, (Suppl.), 11–17.

Orton, P.D. (1954). Arthur Anselm Pearson (1874–1954). *Transactions of the British Mycological Society*, **37**, 321–323.

Orton, P.D. (1955). The genus *Cortinarius* I. *Myxacium* and *Phlegmacium. The Naturalist*. London, 1–80.

Orton, P.D. (1958). The genus *Cortinarius* II. *Inoloma* and *Dermocybe. The Naturalist* Suppl., 81–149.

Orton, P.D. (1960). New Check List of Agarics & Boleti. Part III. Notes on genera and species in the list. *Transactions of the British Mycological Society*, **43**, 154–439.

Paul, D. (1918). Presidential Address. On the earlier study of fungi in Britain. *Transactions of the British Mycological Society*, **6**, 91–103.

Parker-Rhodes, A.F. (1955). Statistical aspects of fungus forays. *Transactions of the British Mycological Society*, **38**, 283–90.

Pearson, A.A. (1946). Notes on the Boleti. *The Naturalist*, 85–99.

Pearson, A.A. (1947). Carleton Rea: obituary. *Transactions of the Worcestershire Naturalists Club* 1945–47, 78–81.

Pearson, A.A. (1948). The Genus *Russula. The Naturalist*, 1–24.

Pearson, A.A. (1950). The Genus *Lactarius. The Naturalist*, 81–99.

Pearson, A.A. (1953). Presidential Address. Hommage aux Mycologues Français. *Transactions of the British Mycological Society*, **36**, 1–12.

Pearson, A.A. & Dennis, R.W.G. (1948). Revised List of British Agarics and Boleti. *Transactions of the British Mycological Society*, **31**, 145–90.

Phillips, R. (1981). *Mushrooms and Other Fungi of Great Britain and Europe*. Pan Books, London.

Plowright, C.B. (1907). Six fatal cases of poisoning by *Amanita phalloides* (Vaill.) Fr. *Transactions of the British Mycological Society*, **3**, 25–26.

Preece, T. & Watling, R. (1992). Willis G. Bramley. *The Mycologist*, **6**, 202.

Ramsbottom, J. (1915a). Mordecai Cubitt Cooke (1825–1914). *Transactions of the British Mycological Society*, **5**, 69–185.

Ramsbottom, J. (1915b). William Leigh Williamson Eyre (1841–1914). *Transactions of the British Mycological Society*, **5**, 185–6.

Ramsbottom, J. (1916). John William Ellis (1857–1916). *Transactions of the British Mycological Society*, **5**, 462–4.

Ramsbottom, J. (1917a). Charles Crossland (1844–1916). *Transactions of the British Mycological Society*, **5**, 466–7.

Ramsbottom. J. (1917b). George Edward Massee (1850–1917). *Transactions of the British Mycological Society*, **5**, 469–73.

Ramsbottom, J. (1924). Henry Cusak Wingfield Hawley (1876–1923). *Transactions of the British Mycological Society*, **9**, 241–3.

Ramsbottom, J. (1926). William Norwood Cheesman (1847–1925). *Transactions of the British Mycological Society*, **11**, 1–4.

Ramsbottom, J. (1948a). The British Mycological Society. *Transactions of the British Mycological Society*, **30**, 1–12.

Ramsbottom, J. (1948*b*). Carleton Rea. *Transactions of the British Mycological Society*, **30**, 180–5.

Rea, C. (1922). *British Basidiomycetae*. Cambridge.

Rea, C. (1923). William Beriah Allen (1875–1922). *Transactions of the British Mycological Society*, **8**, 191–2.

Rea, E.A. (1915). Presidential Address. Notes on fungus illustrations. *Transactions of the British Mycological Society*, **5**, 211–28.

Richardson, M.J. & Watling, R. (1968). Keys to Dung Fungi. *Bulletin of The British Mycological Society*.

Rolfe, R.T. & Rolfe, F.W. (1925). *The Romance of the Fungus World*. London.

Smith, A.L. (1907). Presidential Address. Microfungi: a historical sketch. *Transactions of the British Mycological Society*, **3**, 18–25.

Smith, A.L. (1917*a*). Presidential Address: The relation of fungi to other organisms. *Transactions of the British Mycological Society*, **6**, 17–31.

Smith, A.L. (1917*b*). Lichenology, a new departure. *Transactions of the British Mycological Society*, **6**, 32.

Smith, A.L. (1920). Worthington G. Smith as mycologist. *Transactions of the British Mycological Society*, **6**, 65–7.

Smith, A.L. (1921). *British Lichens*. London: British Museum.

Smith, A.L. & Rea, C. (1903). Fungi new to Britain. *Transactions of the British Mycological Society*, **2**, 31–40.

Smith, Worthington G. (1891). *Supplement to the Outlines of British Fungology by Berkeley*. London.

Smith, Worthington G. (1898). *Guide to Sowerby's Models of British Fungi*. London: British Museum.

Smith, Worthington G. (1908*a*). *Guide to Sowerby's Models of British Fungi in the Department of Botany, British Museum* (Nat. Hist.). London: British Museum.

Smith, Worthington G. (1908*b*). *Synopsis of British Basidiomycetes*. London: British Museum.

Swanton, E.W. (1909). *Fungi and How to Know Them*. London.

Swanton, E.W. (1916). Presidential Address. Education in mycology. *Transactions of the British Mycological Society*, **5**, 381–407.

Wakefield, E.M. (1913). Some New British Hymenomycetes. *Transactions of the British Mycological Society*, **5**, 126–37.

Watling, R. (1982). The British Mycological Society: The Yorkshire connection. *The Naturalist*, **107**, No 963, 121–9.

Watling, R. (1986). 150 years of Paddockstools: a history of agaric ecology and floristics in Scotland. *Transactions of the Botanical Society of Edinburgh*, **45**, 1–42.

Watling, R. (1990*a*). British Mycologists 11. Robert Kay Greville (1794–1866). *The Mycologist*, **4**, 73.

Watling, R. (1990*b*). Obituary: Richard Douglas Jennings (1935–1989). *The Mycologist*, **4**, 99.

Watling, R. (1990*c*). British Mycologists 12. James Bolton (1750–1799). *The Mycologist*, **4**, 143.

Watling, R. & Seaward, M.R.D. (1981). James Bolton, mycological pioneer. *Archives Natural History*, **10**, 89–100.

Watling, R. & Sykes, M. (1992). Alfred Clarke (1848–1925). *The Mycologist*, **6**, 197–8.

Wells, K. (1994). Jelly fungi: then and now! *Mycologia*, **86**, 18–48.

5

The hypha: unifying thread of the fungal kingdom

SALOMON BARTNICKI-GARCIA

Introduction

Nothing else defines fungi more clearly and universally than their ability to produce long tubular cells – hyphae – by a process of tip growth. Branching of the hyphae generates a mycelium, the characteristic soma of the vast majority of fungi, and the structure that makes these organisms superbly suited for a lifestyle devoted to decomposition and absorption of dead or live organic matter. The anatomical and physiological features of the mycelium have given fungi ecological advantages that explain their evolutionary success and ubiquity. These were the features invoked by Whittaker (1969) to justify elevating the Fungi to kingdom status.

Hyphae and apical growth: the basis for morphological and physiological unity in the fungi

Throughout the fungal kingdom, hyphae are similar in form and function. This unity provides the common denominator to tie together in one kingdom all groups of organisms that we call fungi. Although differences exist in the hyphae from major classes of fungi, namely in details such as septation (Moore & McAlear, 1962; Bracker, 1967), cell wall chemical composition (Bartnicki-Garcia, 1968, 1987a; Ruiz-Herrera, 1992), and apical vesicle organization (Grove & Bracker, 1970; Lopez-Franco & Bracker, 1995), hyphae lack the morphological diversity needed to distinguish the tens of thousands of species of fungi currently recognized. Reproductive structures, instead, contain a rich variety of morphological features that have allowed taxonomists to differentiate up to 69000 species of fungi (Hawksworth, 1991). Against a background of enormous reproductive diversity, we find in fungi an overriding unity in somatic development. Much of this similarity stems from the fact that hyphae elongate by a common, vesicle-based, mechanism of apical growth. Despite differences in the structure and organization of vesicles in the tip region and in the chemical composition of cell wall polymers, the principles of hyphal morphogenesis seem to be the same throughout the fungi.

Apical growth and hyphal morphogenesis

How does a fungus manage to grow a tubular cell, so efficiently, so precisely? This deceptively simple question has intrigued mycologists and physiologists from around the world for more than a century. What we know about hyphae today is the result of an international effort to which members of our celebrated, centenarian society have been major contributors. Although the ultimate secret of tip growth is still concealed in the beautiful intricacy of every fungal hypha, a veritable treasure of knowledge about fungal cytology, physiology, and biochemistry has been collected in the course of searching for this elusive secret.

The following is a brief account of some of the findings and concepts collected during the last hundred years that brought us to our current understanding of hyphal biology. A discussion of current models of fungal tip growth is included.*

Hyphae grow mainly at their apices

Reinhardt's (1892) discovery is the seminal point of our present understanding of hyphal growth. His pioneering studies revealed that hyphal elongation occurred at the apex (Figs 5.1 A, B, C). This conclusion was confirmed (Smith, 1923; Castle, 1958; Grove, Bracker & Morré, 1970) and extended by increasingly sophisticated methodology including labelling with fluorescent antibodies (Marchant & Smith, 1968), and autoradiography to discern the exact pattern of deposition of cell wall polymers (Bartnicki-Garcia & Lippman, 1969, 1977; Katz & Rosenberger, 1970; Gooday, 1971; Van der Valk & Wessels, 1977; Fèvre & Rougier, (1982). There seems to be universal agreement that a sharp gradient of wall growth exists at the tip, with its maximum at the apical pole and decreasing rapidly toward the subapical region.

---

* For additional and/or different views on the subject, the reader is directed to recent reviews by Gow (1994), Heath (1994), and Sietsma & Wessels (1994). A more general review by Harold (1990) contains a highly recommended account, provocative and insightful, of current thinking on the basis of microbial morphogenesis.

The shape of fungal hyphae

To understand the mechanism of apical growth, it is important to have a precise knowledge of the geometry of the hypha, particularly the apical dome where most of the growth takes place. This is critical because apical growth involves a continuous transformation of the highly curved surface of the apical dome to the milder curvature of a near cylindrical surface. Reinhardt (1892) and subsequent researchers approximated the shape of a hypha by grafting a hemispherical (Bartnicki-Garcia, 1973) or a hemiellipsoidal dome (Trinci & Saunders, 1977) onto a cylindrical tube (Figs 5.1 A–E). These hybrid shapes were the bases of models of growth where the gradient of wall expansion was described by cosine function (spherical surface, Fig. 5.1 D), or partially approximated by a cotangent function (ellipsoidal surface, Fig. 5.1 E). In either case, the growth gradients terminated at the base of the dome. The dome region in these hybrid shapes is easily defined by the artificial boundary between the two shapes.

More recently, through a computer simulation of hyphal growth (see below), a new mathematical function was found which describes hyphal contour in its entirety. This function, named the hyphoid, describes not only the shape of the apical dome but also includes the near-cylindrical portion of the hyphal tube (Fig. 5.1 F). The hyphoid equation specifies a continuous, gradually decreasing change in curvature from the apical pole to the distal end of the hyphal tube. Although no boundary exists to mark an extension zone or dome region, for practical reasons and to retain a term that identifies the zone where most of the growth of a hypha takes place, we have agreed (C.E. Bracker & R. Lopez-Franco, pers comm.) to define the apical dome region in a hyphoid shape as the foremost portion lying between the apical pole and $2d$ (d is the distance of the VSC to the apical pole) (Fig. 5.1 F). Thus defined, the apical dome comprises the heart of the apical 'extension zone' since, theoretically, this dome region accounts for 64.8 % of the growth of a hypha.*

To avoid misunderstanding of the value and scope of the hyphoid equation (e.g. Koch, 1994), it is important to stress that the hyphoid, and any other proposed geometrical figure, represent idealized shapes. Given the intrinsic variation in the growth process, a perfect geometric shape would be extremely difficult to attain and maintain during growth. For a hypha to maintain a perfect hyphoid shape, its Spitzenkörper would have to

---

* In a previous article (Bartnicki-Garcia, 1990) the length of the apical dome was defined at $1d$; however, a dome twice as long seems more appropriate, cytologically, to define the region where most of the apical vesicular apparatus resides.

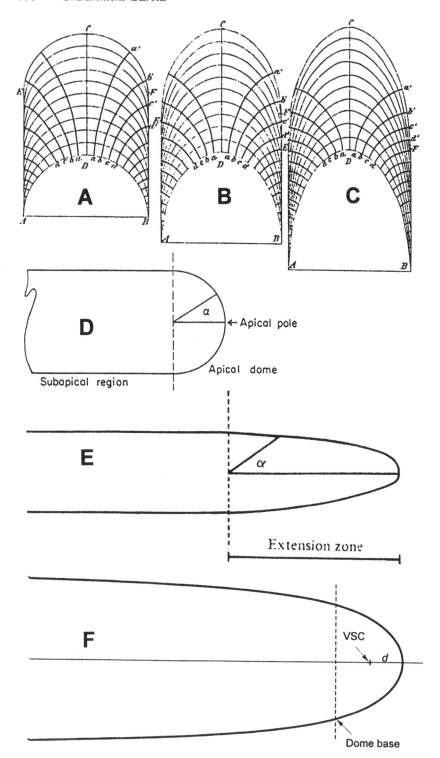

Fig. 5.1. Idealized shapes of fungal hyphae. A–C, Reinhardt's (1892) original drawings explaining the growth process for three differently shaped tips. **D**, Hybrid shape (hemisphere + cylinder) used by Bartnicki-Garcia (1973) (after Green, 1969) to develop model of wall growth. **E**, Hybrid shape (hemi-ellipsoid + cylinder) developed by Trinci & Saunders (1977) to describe the determination of tip shape (reproduced with permission from Gooday & Trinci, 1980). **F**, Hyphoid shape defined by the hyphoid equation $y = x \cot (xV/N)$ (Bartnicki-Garcia *et al.*, 1989).

follow a perfectly straight trajectory without oscillations or changes in its vesicle release rate during the growth sequence. This is contrary to observations on Spitzenkörper behaviour (Lopez-Franco, Bartnicki-Garcia & Bracker, 1994; Lopez-Franco, Howard & Bracker, 1995). The reported excellent correspondence (>95 %) between hyphoid and real hyphal shapes (Bartnicki-Garcia, Hergert & Gierz, 1989, 1990) was obtained for specimens that showed a normal undisturbed morphology and that had been ostensibly well preserved during handling for electron microscopy.

There are instances where the normal shape of the hyphal apex deviates significantly from the hyphoid shape. This is more common in lower fungi, particularly *Saprolegnia* or *Achlya*, where hyphae with long conical apices are often seen. These shapes and the underlying gradients of wall expansion can not be described by models based on cylindrical tubes, where expansion is *de facto* limited to the hemi-ellipsoidal, or hemispherical dome. The gradient of wall expansion specified by the hyphoid equation does not describe these highly tapered hyphal shapes either, but the hyphoid model can serve as the basis to deal quantitatively with the possible factors that cause some hyphae to deviate from the hyphoid shape and have a longer extension zone.*

Kinetics of hyphal growth

For nearly 3 decades, A.P.J. Trinci and his colleagues have amassed an impressive body of information on the growth kinetics of fungal hyphae and have given us new insights and concepts to understand and appreciate the physiology of hyphal growth. Trinci's studies established the basic exponential nature of hyphal growth (Trinci, 1969, 1971*a*, 1973, 1974), determined the dimensions of the apical growth zone (Collinge & Trinci,

---

* A new version of the model, with a modified VSC and a correspondingly different mathematical expression, has been developed. The modified VSC supports a gradient of wall expansion that tapers off more gradually in the apical region (G. Gierz & S. Bartnicki-Garcia, in prep.).

1974; Steele & Trinci, 1975), and explored the question of hyphal tip shape and growth (Trinci & Saunders, 1977; Saunders & Trinci, 1979). The concept of a *peripheral growth zone* (Trinci, 1971*b*) defined the extent to which a hyphal tip depends on the subapical region for its growth, and helped explain how hyphae of certain fungi can maintain their prodigious rates of elongation. Expanding on studies by Plomley (1959) and Katz, Goldstein & Rosenberger (1972), Caldwell & Trinci (1973) refined the concept of a *hyphal growth unit*, and defined it as the average length of hypha which supports the extension of each tip in the mycelium (Trinci, 1973). The hyphal growth unit was the basis for interpreting mycelial growth in terms of duplication cycles (Trinci, 1978). The latter is the physiological equivalent of the cell division cycle in unicellular organisms, and the means to compare cell growth and division between mycelial and single-cell fungi. Much of this work together with more recent studies on the kinetics and regulation of hyphal branching were reviewed by Trinci, Wiebe & Robson (1994).

Recently, a finer analysis of the kinetics of growth of fungi, made possible by video-enhanced contrast microscopy, led to the discovery that hyphal elongation does not proceed at a truly constant rate but occurs in brief pulses of alternating slow and fast growth with a somewhat regular periodicity (Lopez-Franco *et al.*, 1994). The cause of these pulses, detected in fungi from diverse taxonomic groups, has yet to be determined but there is the suspicion that they reflect a pulsed pattern of vesicle discharge in the growing apex. In two fungi, there was evidence correlating growth pulses with the merger of 'satellite' Spitzenkörper with the main Spitzenkörper (Lopez-Franco *et al.*, 1994).

Fungal morphogenesis equals cell wall morphogenesis

W.J. Nickerson first focused attention on the correlation between biochemical properties of the cell wall and the morphology of fungal cells (Nickerson & Falcone, 1956; Nickerson, 1963). He advocated viewing the cell wall as the primary form-giving structure of a fungal cell and elevated the rather simple changes in morphological development observed in fungi, e.g. mycelial–yeast dimorphism to the realm of morphogenesis (Nickerson, 1958). These conceptual refinements gave us a better framework for discussion and interpretation of experimental findings. Although the causal correlation between chemical composition of the cell wall and cell morphology sought by Nickerson did not materialize, the importance of wall structure and architecture in the genesis of cell shape became firmly established. This led to our current notion that the shape of the cell wall originates primarily from the manner in which wall polymers are deposited in space and not from the chemical nature of the polymers (Bartnicki-Garcia, 1969).

Vesicles and hyphal morphogenesis

Cytoplasmic vesicles have emerged as the centrepiece of the mechanism of cell wall growth in fungi. Vesicles deliver to the cell surface preformed wall polymers and enzymes needed to construct the cell wall. In fact, the most compelling case in biology for a causal correlation between vesicles and wall growth is found in the hyphal tips of fungi. The discovery of the importance of vesicles in apical growth came about in the 1960s when cytologists perfected the methodology for studying fungal cells in thin sections and a treasury of cytological knowledge began to mount. Several groups laboured diligently to elucidate the fine structure of growing hyphae and their efforts produced a bonanza of new knowledge. In 1968, McClure, Park & Robinson published the first electron micrographs showing that the Spitzenkörper of *Fusarium* spp. and *Aspergillus niger* was an aggregation of vesicles. Subsequent publications by Girbardt (1969), Grove & Bracker (1970) and Grove *et al.* (1970) became classical accounts of the cytological organization of the hyphal apex of fungi and the polarized distribution of organelles in hyphal cells. Other studies confirmed the presence of vesicles in hyphal tips of a wider range of fungi (Heath, Gay & Greenwood, 1971; Grove, 1978). By freeze substitution, a superior technique for organelle preservation, various conclusions made from images obtained by chemical fixation were confirmed and extended (Howard & Aist, 1979; Hoch & Howard, 1980; Howard, 1981; Roberson & Fuller, 1988; Vargas, Aronson & Roberson, 1993).

Hyphal wall construction: two secretory pathways

The vesicle concept provides an attractive foundation for the notion that cell wall formation is a discontinuous process; in other words, the growth of a cell wall is not a diffuse, continuous process but rather the sum total of numerous separate submicroscopic growth events or 'units of wall growth' (Bartnicki-Garcia, 1973, 1990). Each growth event results from the enzymes and materials discharged by vesicles.

A vesicle-mediated process of wall growth gives us a good handle to understand how the spatial regulation of wall synthesis could take place and, thus, it constitutes the foundation to explain the origin of cell shape (see below) (Bartnicki-Garcia, 1973; Bartnicki-Garcia et al., 1989).

Although much is known about the biosynthesis of individual cell wall polymers, especially chitin, chitosan, β-1,3-glucan, α-mannan, and a few others (see reviews by Ballou, 1976; Gooday & Trinci, 1980; Wessels, 1986; Cabib, 1987; Bartnicki-Garcia 1989; Ruiz-Herrera, 1992; Sentandreu, Mormeneo & Ruiz-Herrera, 1994), we still know very little about how these polymers become assembled into a cell wall (Selitrennikoff, 1983; Ruiz-Herrera, 1992). We do have evidence for a distinct division of labour during wall synthesis. Studies on the biosynthesis of chitin revealed the existence of a unique population of vesicles (chitosomes) responsible for delivery of chitin synthetase to the cell surface (Bracker, Ruiz-Herrera & Bartnicki-Garcia, 1976). This led to the realization that two distinct secretory pathways must be needed to elaborate the fungal cell wall (Bartnicki-Garcia, 1987b, 1990): 1) The chitin skeleton of the walls of most fungi is made by enzyme delivered as a zymogen in microvesicles (chitosomes) (Bracker et al., 1976; Bartnicki-Garcia et al., 1978; 1984; Hanseler, Nyhlen & Rast, 1983; Leal-Morales, Bracker & Bartnicki-Garcia, 1988); whereas, 2) The polymers that constitute the amorphous phase of the wall are preformed in the cytoplasm and secreted by larger vesicles (macrovesicles), which are typical secretory vesicles, commonly known in the mycological literature as apical vesicles (Girbardt, 1969; Grove & Bracker, 1970). These vesicles also secrete an assortment of enzymes needed for nutrient assimilation and also, presumably, for plasticizing the otherwise rigid wall to allow expansion. There is reason to believe, from cytochemical and immuno-cytochemical tests (Meyer, Parish & Hohl, 1976; Hardham, Suzaki & Perkin, 1986), that macrovesicles comprise a heterogeneous population but final biochemical characterization awaits the development of procedures for successful isolation of secretory vesicles in reasonably pure form.

The Spitzenkörper

Because of its location in the apical dome of growing hyphae, the Spitzenkörper has attracted much attention from those interested in understanding the mechanism of apical growth in fungi. Yet, not much notice seems to have been paid to Brunswik's (1924) original discovery of an iron-haematoxylin-staining body in the hyphal tips of Coprinus spp. (Fig. 5.2 A), until the 1950s when Girbardt's studies by phase contrast microscopy confirmed the existence of a Spitzenkörper in living hyphae.

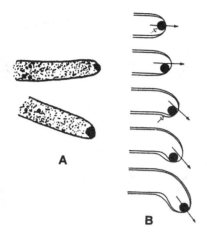

Fig. 5.2. The Spitzenkörper. **A**, First record of a Spitzenkörper found in the tips of *Coprinus narcoticus* (from Brunswik, 1924). **B**, Evidence that the position of the Spitzenkörper in the apical dome determines the direction of growth of a hypha of *Polystictus versicolor* (from Girbardt, 1973).

Girbardt (1957) correlated the appearance and disappearance of the Spitzenkörper with the growth of a hypha. He also noted that changes in the growth direction of hyphae were accompanied by changes in the position of the Spitzenkörper (Fig. 5.2 B).

The many forms of the Spitzenkörper

Since the first ultrastructural studies on the Spitzenkörper, its variation and complexity became evident. This gave rise to different interpretations as to whether or not the Spitzenkörper was only an aggregation of vesicles (Girbardt, 1969) or a more complicated structure with a central core of variable composition (Grove & Bracker, 1970). Subsequent studies have confirmed the complexity of the Spitzenkörper and the existence of cytoskeleton components, microtubules (Roos & Turian, 1977; Kwon, Hoch & Aist, 1991), and actin microfilaments (Bourett & Howard, 1991) in the Spitzenkörper region. Although electron microscopy had shown distinct differences in Spitzenkörper structure among the higher fungi, it was necessary to return to light microscopy to grasp more fully the morphological diversity and complex behaviour of the Spitzenkörper. In a video microscopic survey of more than 30 different fungal species, eight different patterns of Spitzenkörper organization were recognized (Fig. 5.3) and their highly complex dynamics and

114   S. Bartnicki-Garcia

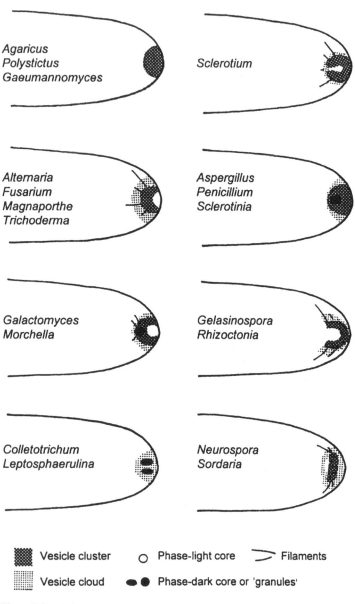

Fig. 5.3. Diagrammatic representation of the eight types of Spitzenkörper organization recognized by López-Franco & Bracker (1995) (adapted from López-Franco, 1992).

pleomorphism were recorded (Lopez-Franco, 1992; Lopez-Franco & Bracker, 1995).

The Spitzenkörper as a vesicle supply centre

A remarkable correlation emerged from a computer simulation of morphogenesis that led to the hyphoid model: the position of the vesicle supply centre (VSC) in the simulation coincided with the position of the Spitzenkörper in real hyphae (Bartnicki-Garcia *et al.*, 1989, 1990). This coincidence supports the notion that the Spitzenkörper functions as a centre for the final distribution of vesicles responsible for tip growth.

Although there is strong circumstantial evidence linking the Spitzenkörper with the growth of hyphae (Girbardt, 1957; McClure, Park & Robinson, 1968), the absence of a Spitzenkörper in hyphae of lower fungi (McClure *et al.*, 1968; Grove & Bracker, 1970) has been cause to question whether the Spitzenkörper plays an essential role in hyphal growth. Since the hyphoid equation describes hyphal morphology equally well for fungi having, or not having, conspicuous Spitzenkörper, we postulated (Bartnicki-Garcia *et al.*, 1989) that fungi lacking a Spitzenkörper (McClure *et al.*, 1968; Grove & Bracker, 1970) must have its functional equivalent, i.e. a site from which vesicles start on the final leg of their journey to the cell surface. Presumably, in these fungi such transient vesicles do not produce a localized agglomeration of sufficient density and/or refractivity to be visible by light microscopy. Recently, Vargas *et al.* (1993) have shown that a Spitzenkörper-like arrangement does exist in the hyphal tips of a class of lower fungi that was previously not believed to have a Spitzenkörper, the chytridiomycete *Allomyces macrogynus.*

Models of hyphal morphogenesis

The ultimate secret of hyphal morphogenesis lies in the events or processes that polarize growth to produce a tubular cell. Three basically different models have been proposed to explain how a hyphal tube is generated. Each model invokes different morphogenetic criteria.

1) Changes in the physical properties of the wall

Initially advocated by Robertson (1965), Park & Robinson (1966), and Saunders & Trinci (1979), this model was extensively refined by Wessels (1986, 1990) and Sietsma & Wessels (1994). Accordingly, the hyphal tube is manufactured by a steady-state process that transforms and expands the newly deposited plastic wall of the apical dome into a rigid cylindrical wall at the base of the dome. The newly added wall material is plastic due to the presence of individual polymer chains. This nascent wall becomes progressively cross-linked and thus develops resistance to turgor pressure.

Considerable experimental evidence has been gathered to support the notion that the cell wall is more plastic at the extreme apex and becomes more rigid at the base of the apical dome (Wessels & Sietsma, 1981; Vermeulen & Wessels, 1984; Sietsma, Sonnenberg & Wessels, 1985; Sonnenberg, Sietsma & Wessels, 1985). Although there is merit in considering that the overall physical properties of the apical wall may change significantly because of cross-links between β-1,3-glucan and chitin polymers, it is questionable whether any *additional** rigidification of the apical wall would regulate wall expansion. The fact that the apical wall maintains constant thickness during growth (Girbardt, 1969; Grove & Bracker, 1970; Trinci & Collinge, 1975) indicates that the spatially regulated secretion and construction of additional wall throughout the apical dome must be a more consequential event in apical morphogenesis. As formulated, the steady-state model assumes that the apical dome adopts a hemi-ellipsoidal shape (Wessels, 1986) but there is, as yet, no persuasive explanation of how a progressive increase in rigidity would give rise to a hemi-ellipsoidal dome wall.[†] Whether or not the extracytoplasmic process of wall rigidification could have an accessory role in hyphal morphogenesis remains to be seen.

2) The expanding cytoplasm moulds hyphal shape

In this model, favoured by Heath and co-workers (Heath, 1990, 1994) and by Money & Harold (Money & Harold, 1993; Money, 1994), the fungus is viewed primarily as a protozoon living inside a rigid tubular casing. Morphogenesis is thought to originate from the same elements that mould the shape of a wall-less protozoon, namely the underlying cytoskeleton. Specifically, a scaffolding of F-actin is believed to provide the shaping force that generates the hyphal tube. This model was developed for

---

* The wall of the apical dome, even at its most plastic point, must be strong enough to resist violent deformation by the high turgor pressure of the cytoplasm, otherwise the cell would explode under normal growth conditions (the counterclaim that an actin cap strengthens wall to prevent bursting is discussed on page 122).
† The cotangent relationship invoked to describe the gradient of wall expansion in the apical dome is not satisfactory since it specifies a gradient that begins with an *infinitely* high rate of expansion of the apical pole (alpha = 0°).

oomycetous fungi where there is clearly a dense array of filamentous actin (actin cap) associated with the hyphal apex (Heath, 1987; Temperli, Roos & Hohl, 1990; Harold & Harold, 1992). In other fungi, there is also a high concentration of actin at the apex but it is largely in the form of individual plaques (Yokoyama *et al.*, 1990; Roberson, 1992; Hasek & Bartnicki-Garcia, 1994). It is questionable whether the actin plaques would have the same presumed morphogenetic role. Barring other explanations reconciling the differences in the organization of apical actin (see Heath, 1994), the apparent absence of a well-defined array of filamentous actin in the hyphal tips of most fungi questions the general applicability of this model. Although the organization of the cytoskeleton may well have a major role in shape determination, especially in establishing the polarized distribution of organelles in a hypha, it is difficult to accept the notion that the weak forces generated by actin polymerization could have much of an impact on wall expansion, except under extraordinary circumstances where cultivation in extreme environments may have weakened the wall so that it could be deformed by the weak pressure generated by the cytoskeleton (Harold, Harold & Money, 1995).

Inhibitor studies do provide supporting evidence for a role of F-actin in hyphal morphogenesis. Cytochalasins, inhibitors of actin polymerization, cause the hyphal tips to stop elongation and swell (e.g. Betina, Micekova & Nemec, 1972; Grove & Sweigard, 1980). However, these and other related observations on interference with actin polymerization are only circumstantial proof for the morphogenetic role of F-actin proposed in this model (Heath, 1994). Alternatively, one could explain the inhibitor-induced distortions in hyphal growth as resulting from interference with cytoskeleton-assisted movement and organization of wall-building vesicles.

3) Patterns of vesicle migration: hyphoid model

In this model, the spatial distribution of wall-building vesicles plays a central role in shaping the wall and therefore determining cell shape (Bartnicki-Garcia, 1973; Bartnicki-Garcia *et al.*, 1989). The model assumes that extension growth can occur *anywhere* on the cell surface as a result of vesicle discharge and proposes that localized wall expansion is the result of three concomitant effects: deposition of new cell wall polymers, enzymic plasticizing action to loosen the rigid wall structure, and turgor to force wall expansion. Turgor is assumed to be an omnipresent force that expands the wall in those places where vesicle discharge creates a localized region of nascent wall. Turgor has no specific morphogenetic action; the site and amount of wall expansion would depend on the site and amount of enzyme discharged by vesicles. Accordingly, the key morphogen resides in the mechanism responsible for the polarized distribution of wall-building vesicles.

This model is supported experimentally by the finding of a sharp gradient of wall synthesis at the apex of *Mucor rouxii* hyphae (Bartnicki-Garcia & Lippman, 1969, 1977). Similar gradients have been found in various other fungi (Katz & Rosenberger, 1970; Gooday, 1971; Van der Valk & Wessels, 1977; Fèvre & Rougier, 1982). However, the postulated concomitant gradient of cell-wall loosening, to allow controlled intussusception of new wall components and expansion, remains to be experimentally demonstrated. Although there is suggestive evidence of a close linkage between a wall-building (chitin synthetase) and a wall-degrading (chitinase) activity (Gooday & Gow, 1990; Rast *et al.*, 1991), the crucial biochemical process that softens the fungal cell wall to allow expansion is yet to be identified. The fact that a gradient of chitin synthesis does exist in hyphal tips is strong evidence that this key enzymatic activity in wall construction is short-lived and reason to speculate that the presumed parallel wall-loosening activity would also be equally short-lived. Hence, it is not surprising that this elusive activity has been so difficult to demonstrate.

In its original qualitative formulation (Bartnicki-Garcia, 1973), the model assumed that the electric current flowing through growing hyphae provided the force and guidance to propel vesicles, electrophoretically, towards the apical pole and thus furnished the basis to understand polarity and the preferential discharge of vesicle at the fungal apex (Jaffe, 1968). But such an attractive idea, supported by Jaffe's ingenious experiments on electric current in apically growing cells, had to be abandoned when McGillviray & Gow (1987) and Schreurs & Harold (1988) showed that hyphal growth could take place just the same when the electric current was nil or when its polarity was reversed.

From the general assumptions made in the qualitative, vesicle-based model of fungal wall growth (Bartnicki-Garcia, 1973), G. Gierz and F. Hergert developed a computer simulation of fungal growth. The simulation led to the realization that morphogenesis in a vesicle-based system could be regulated by simply moving the internal source of vesicles: the so-called VSC or vesicle supply centre (Bartnicki-Garcia *et al.*, 1989). If the Spitzenkörper functions as a VSC (see below), then by the simple action of advancing in a straight line, while continuously releasing wall-destined vesicles, it would generate, automatically, a gradient of exocytosis needed to produce a hypha (next section).

The vesicle supply centre

In the hyphoid model (Bartnicki-Garcia *et al.*, 1989, 1990), vesicles are assumed to arise from an idealized point source (the VSC or vesicle supply centre); in real cells, the VSC may be viewed as the geometric centre of a complex system of the vesicle-producing or vesicle-releasing structures present in a fungal cell.

In theory, a single source of vesicles, i.e. a single Golgi apparatus, could generate a hypha provided that during the course of cell growth, it advanced linearly as it released its vesicles. But a single source of vesicles could not possibly account for the prodigious rate of hyphal extension so common in fungi (Grove & Bracker, 1970; Gooday & Trinci, 1980). Instead, fungi seem to have evolved an efficient mechanism to collect vesicles from the subapical cytoplasm (Grove & Bracker, 1970; Heath *et al.*, 1971; Barstow & Lovett, 1974; Trinci & Collinge, 1974; Howard & Aist, 1979) and translocate them to the tip. The model predicts that the collected vesicles would be first delivered to a distribution centre (VSC) from which they would be free to migrate in any random direction towards the cell surface.* Accordingly, the Spitzenkörper represents a terminal collection/distribution point for wall-destined vesicles.

The hyphoid equation: the shape of a perfect hypha

The vesicle-based simulation of hyphal growth led to a surprisingly simple geometric equation, named the *hyphoid*, which describes the shape of an ideal hypha in median, longitudinal cross-section (Fig. 5.4):

$$y = x \cot (x V/N).$$

This equation defines the shape and size of a fungal hypha by two *physiological* parameters: N is the amount of wall-destined vesicles released from the VSC per unit time; V is the rate of linear displacement of the VSC (see Bartnicki-Garcia *et al.*, 1989, 1990). The hyphoid equation was a totally

---

* The movement of vesicles to the plasma membrane is not likely to follow a Brownian walk but one facilitated by interaction of vesicles with the cytoskeleton. Since actin has been found in great abundance in the areas of wall formation, particularly at the growing tips (Hoch & Staples, 1985; Heath, 1987; Hasek & Bartnicki-Garcia, 1994), it seems reasonable to speculate that actin filaments may provide the tracks and/or the pulling force for the final displacement of vesicles from the Spitzenkörper to the plasma membrane.

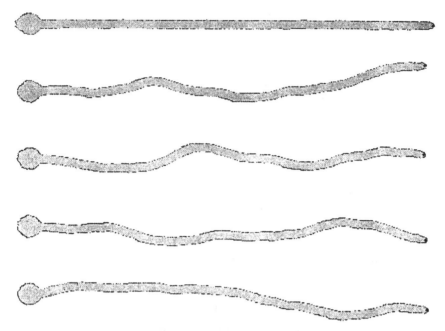

Fig. 5.4. Simulation of meandering in fungal hyphae. Hyphal shapes were programmed to grow according to the parameters of the hyphoid equation except that a transverse random oscillation of the VSC was introduced. The top shape was made with no VSC oscillation. The overall cell growth rate (N) and the rate of displacement of VSC (V) was the same in all cases. The different tortuous shapes result exclusively from the randomness programmed into the VSC oscillations. (From Bartnicki-Garcia *et al.*, 1995*a*).

unexpected but welcome result of describing a growth process based on a linearly advancing source of vesicles.* An excellent correspondence between the shape predicted by model and the shape of real hyphae has been illustrated for diverse fungi (Bartnicki-Garcia *et al.*, 1989, 1990).†

Modulation of hyphal growth

The availability of the hyphoid equation has opened the door to quantitate the morphological plasticity of fungi. For example, the diameter of a

---

* It should be noted (cf. Prosser, 1994), that the cotangent function in the hyphoid equation bears no conceptual relationship to the cotangent function used by Trinci & Saunders (1977) to describe the gradient of wall expansion in an hemi-ellipsoidal dome.
† The correspondence requires that the specimens selected for comparison pertain to normal, growing hyphae that had not been ostensibly distorted during manipulation for electron microscopy.

hypha would be governed by the *V/N* ratio, i.e. the interplay between the amount of vesicles released per unit time and the rate of displacement of the VSC: high *V/N* values would determine a hyphal morphology; low values, a yeast cell morphology. Manipulation of this ratio by computer graphics produces a full spectrum of morphologies seen in dimorphic fungi: from hyphal tubes to budding yeast cells (Bartnicki-Garcia & Gierz, 1993).

The morphology of a simulated hypha can be further modulated by factors that affect the trajectory of the VSC. Whereas the hyphoid equation describes the shape of a perfect hypha, in nature hyphae do not exhibit such perfect shapes; they usually meander, producing tubular shapes with various degrees of contortion. We believe lateral oscillations in the Spitzenkörper are responsible for the meandering appearance of ordinary hyphae (Bartnicki-Garcia & Gierz, 1993). Accordingly, the meandering behaviour of hyphae was simulated by incorporating a random oscillatory movement into the trajectory of the VSC (Fig. 5.4). And, by applying a sustained unidirectional bias on the VSC random oscillations, hyphal shapes with various degrees of curvature (e.g. the tropic bending of sporangiophores, loops of nematode-trapping fungi, or clamp connections) have been simulated (Bartnicki-Garcia, Bartnicki & Gierz, 1995*a*).

Hyphal morphogenesis models: general commentary

The three aforementioned models of hyphal morphogenesis embody processes that occur in living hyphae, and all need to be considered in any final comprehensive picture of hyphal development. There is no quarrel with the idea that hyphal growth may involve changes in physical properties of the wall (model 1), and that the cytoskeleton provides the dynamic scaffolding that gives spatial structure to the cytoplasm (model 2). But it seems more appropriate to view these phenomena as accessories to the main process in hyphal morphogenesis which is the construction of a cell wall by polarized exocytosis (model 3).* Ultimately, none of the current models defines the precise biochemical reaction(s) that confers polarity to a hypha.

---

* Recently, Koch (1994) declared the hyphoid model insufficient and proposed to complement the VSC concept with a 'soft spot' hypothesis. As presented, this hypothesis assumes that vesicles fuse directly with the wall and do so in greater numbers at the apical dome because of a presumed greater tendency of vesicles to fuse with the younger softer wall. Pending clarification of the largely undocumented challenge to the hyphoid equation, a chief problem with this amended version of the VSC model lies in the fact that vesicles cannot fuse directly with the wall because of the presence of an intervening plasma membrane.

In models 1 and 3, the cell wall plays a primary role in hyphal morphogenesis but a major conceptual difference separates these two models. In model 1, the apical wall is considered to be inherently plastic and rigidification is invoked to regulate the apex shaping process. In complete contrast, model 3 regards the existing apical wall as inherently rigid and localized enzymatic loosening is invoked to allow expansion.* Since there are no direct precise measurements on the relative strength of the wall at the apex to test these assumptions unequivocally, each model relies on circumstantial evidence to show the presence of enzymatic processes capable of rigidifying or lysing the cell wall.

There is a tendency among current proponents of models 1 or 2 (Sietsma & Wessels, 1994; Heath, 1994; Money, 1994) to minimize or dismiss the role of turgor as the main physical force in wall expansion and to adjudicate this function to the actin cytoskeleton present in abundance in the hyphal apex. Two different observations have been adduced in support of this view: i) hyphae of *Saprolegnia* or *Achlya* can grow in the near absence of turgor (Money & Harold, 1993), and ii) hyphal tip bursting occurs not at the apical pole but somewhere else in the apical dome. These observations, however, need to be carefully weighed. The conclusion that wall expansion can occur in the *near*[†] absence of turgor applies to a most unusual situation. It pertains to unique fungi that can not maintain turgor when cultivated in media of exceedingly high osmolarity. For ordinary fungi, the omnipresent high turgor (Adebayo, Harris & Gardner, 1971) provides a force far in excess of the calculated pressure exerted by an actin cap (Harold *et al.*, 1995) and is ostensibly more than what is needed to drive wall expansion. Claims that the site of bursting can be used to assess the relative strength of the wall in the apical region (Wessels, 1986; Jackson & Heath, 1990) run contrary to the main conclusion made in an earlier study of induced apical bursting (Bartnicki-Garcia & Lippman, 1972; see also Collinge & Markham, 1992). That study demonstrated that bursting was much more than a purely physical effect but one that required chemical (enzymatic) activity to weaken the wall. Accordingly, the results of bursting experiments need not disclose the weakest sites in the wall (before the shock) but,

---

* Although the mathematical equation that supports this model deals only with vesicle distribution, the ensuing localized process of wall growth that would follow vesicle discharge would require a spatially limited, coordinated interaction between wall synthesis and wall softening so that the wall would expand by a finite amount under the turgor pressure of the cytoplasm (Bartnicki-Garcia, 1973).

† In some of the writings, there is a tendency to express this condition as growth in the *complete* absence of turgor, yet there is no unequivocal proof that turgor was completely eliminated in these cells.

more likely, they show the sites where the greatest accidental discharge of lytic enzymes took place as a result of the shock treatment. Therefore, rather than having a physical role in wall expansion, the abundance of actin at the tips of growing fungi is probably an indicator of the involvement of actin in exocytosis. The possibility that the collective action of the actin microfilaments in the apex may be responsible for displacing the Spitzenkörper remains an intriguing prospect for consideration.

Presently, there is no quantitative formulation to correlate hyphal shape with a gradual change from plastic to rigid wall (model 1) or with the distribution of actin cytoskeleton at the tip (model 2). Model 3 does provide the mathematical means to deal quantitatively with the key cytological/physiological parameters of apical growth, namely vesicle dynamics. The predictions made by the hyphoid model about the role of the Spitzenkörper in hyphal morphogenesis are testable. Thus we have analysed instances where transient dislocations in the position of the Spitzenkörper produced permanent distortions in hyphal morphology. When the VSC was programmed to duplicate the movements of the Spitzenkörper before, during, and after a deformation, the resulting computer-generated shapes mimicked closely the observed deformations supporting the contention that the position and movement of the VSC determines the morphology of the fungal cell wall (Bartnicki-Garcia *et al.*, 1995*b*).

Hyphae and the boundaries of fungal taxonomy and phylogeny

Hyphae are not only a key to understanding the basic biology of individual fungi but also a key criterion for defining the frontiers of the fungal kingdom. Not surprisingly, given the universal impossibility of subdividing and stuffing any continuum of living organisms into neatly defined taxonomic boxes, the boundaries of the kingdom Fungi have been the subject of controversy. Textbooks typically define fungi by enumerating their most distinctive traits. But perhaps it is best, and simplest, to define fungi by invoking the two major characters that gave Whittaker (1969) the rationale to erect the fungal kindgom: organisms with a *mycelial* organization and a lifestyle devoted to decomposition and absorption of preformed organic matter. The eukaryotic mycelium provides an ideal anatomical basis for such a lifestyle.

Hyphae and the origin of fungi

I argued earlier that acquisition of the ability to assemble a microfibrillar cell wall was probably a major landmark in the evolution of primitive eukaryotes (Bartnicki-Garcia, 1984, 1987a). This event, believed to have occurred 1 to 1.5 billion years ago, was a preamble to the appearance of the fungal kingdom. When some of these primordial eukaryotic organisms developed the ability to grow long tubular cells by a process of apical growth, they set out the basis for the development of a mycelium and for the emergence of the fungal kingdom. Given the probable diphyletic origin of the fungi, it is reasonable to surmise that the feature which makes a fungus a fungus, the tip-growing hypha, was probably invented independently more than once, with somewhat different biochemical bricks used in each independent emergence (Bartnicki-Garcia, 1970, 1987a).

The Oomycetes, Chytridiomycetes and Hyphochytriomycetes are bona fide Fungi

Nobody questions that the higher fungi – *Ascomycetes* and *Basidiomycetes* – are fungi in the strictest sense, but other major classes of fungi display characters that distance them from the higher fungi to various degrees. With the exception of the *Zygomycetes*, other classes of lower fungi, *Chytridiomycetes*, *Oomycetes*, and *Hyphochytriomycetes*, have been the subject of relocation attempts outside the fungal kingdom (Cavalier-Smith, 1983, 1986; Margulis *et al.*, 1990).

If we accept that the central defining feature of a fungus is its ability to grow hyphae, all eukaryotic (non-photosynthetic) organisms with a mycelial habit should be regarded as bona fide members of the kingdom Fungi, and that of course would include organisms whose hyphal growing ability has become limited (e.g. yeasts and chytrids) but which retain close ties with hyphal species.

The *Oomycetes* are a particularly important case in point. These organisms have a long history of importance to basic and applied mycology, but recently their legitimacy among the fungi has been repeatedly questioned. That *Oomycetes* (plus the related *Hyphochytriomycetes*) are a unique group of fungi that differs significantly from the rest is not a new proposal. In fact, the distinctiveness of the *Oomycetes* had long been recognized on purely morphological grounds (see Gäumann & Dodge, 1928). The availability of biochemical markers, particularly the pathway of L-lysine biosynthesis (Vogel, 1964) and other properties such as cell wall chemistry (Bartnicki-Garcia, 1968) and tryptophan-pathway enzyme organization (Hütter & DeMoss, 1967) gave strong support to the claim that the *Oomycetes* are distinctly different from other aquatic and terrestrial fungi and must have evolved separately (Bartnicki-García, 1970).

Ultrastructural (Beakes, 1987) and molecular studies (Lovett & Haselby, 1971; Ohja, Dutta & Turian, 1975; Kwok, White & Taylor, 1986; Förster & Coffey, 1990; Förster, Oudemans & Coffey, 1990) have confirmed the distinctiveness of the *Oomycetes*.

There is no quarrel with interpreting such biochemical and morphological differences, as an indication that the *Oomycetes* and the higher fungi probably arose from different ancestors. The disagreement is with those who invoke these differences to dismember the kingdom Fungi by prescribing tests of phylogenetic purity while dismissing more compelling arguments that have kept the *Oomycetes* as an integral part of Mycology. Above all is the fact that *Oomycetes* share with the rest of the fungi the same key properties that characterize the fungal kingdom: mycelial morphology, and physiological and ecological properties suited for an existence devoted to external digestion and passive absorption of preformed organic matter.

The observed similarity of hyphal shape among taxonomically diverse fungi, including *Oomycetes* (*Pythium aphanidermatum*), *Ascomycetes* (*Fusarium culmorum, Aspergillus niger*), *Basidiomycetes* (*Armillaria mellea, Polystictus versicolor*) (Bartnicki-Garcia *et al.*, 1989), and *Zygomycetes* (*Gilbertella persicaria*) (unpubl.) suggests that the mechanism regulating hyphal morphogenesis is basically the same despite ostensible differences in the arrangement of vesicles at the apex.

The kingdom Fungi

The case for retaining a polyphyletic kingdom Fungi (Whittaker, 1969) is further strengthened by the fact that displaced fungi have been grouped with other organisms with which they share very little in common in overall physiological, morphological and ecological properties, e.g. the joining of *Oomycetes* with heterokont algae in the so-called *Chromista* kingdom (Cavalier-Smith, 1983) or the lumping together of all zoosporic fungi together with protozoa and algae in one gigantic kingdom: *Protoctista* (Margulis *et al.*, 1990). An admittedly diphyletic kingdom (Fig. 5.5) is definitely a more rational and practical solution to assemble and study the collection of organisms that we intuitively recognize as fungi.

> If it looks like a fungus,
> if it grows like a fungus,
> if it lives like a fungus,
> it has to be a fungus.*

---

* Modified from Bartnicki-Garcia. *The Oomycetes are bona fide fungi*. Abstracts of the 1994 meeting of the American Phytopathological Society, Albuquerque, New Mexico, August 6–10, 1994.

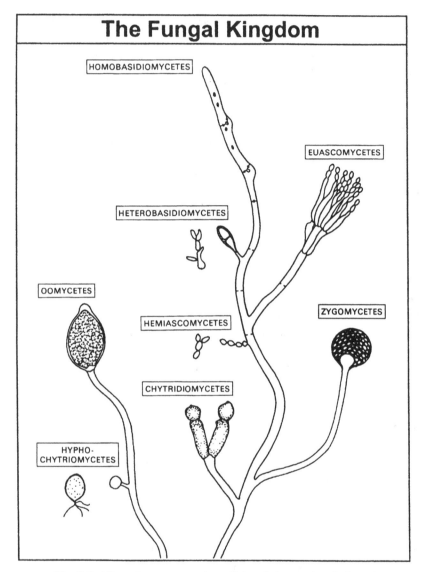

Fig. 5.5. Allegoric representation of the phylogenetic relationship of the major classes of fungi. The two trunk lines of cellulosic (left) and chitinous fungi (right) are believed to have evolved independently.

References

Adebayo, A.A., Harris, R.F. & Gardner, W.R. (1971). Turgor pressure of fungal mycelia. *Transactions of the British Mycological Society* **57**, 145–51.

Ballou, C. (1976). Structure and biosynthesis of the mannan component of the yeast cell envelope. *Advances in Microbial Physiology*, **14**, 93–157.

Barstow, W.E. & Lovett, J.S. (1974). Apical vesicles and microtubules in rhizoids of *Blastocladiella emersonii*: effects of actinomycin D and cycloheximide on development during germination. *Protoplasma*, **82**, 103–17.

Bartnicki-Garcia, S. (1968). Cell wall chemistry, morphogenesis, and taxonomy of fungi. *Annual Review of Microbiology*, **42**, 57–69.

Bartnicki-Garcia, S. (1969). Cell wall differentiation in the phycomycetes. *Phytopathology*, **59**, 1065–71.

Bartnicki-Garcia, S. (1970). Cell wall composition and other biochemical markers in fungal phylogeny. In *Phytochemical Phylogeny* (ed. J.B. Harborne), pp. 81–103. London: Academic Press.

Bartnicki-Garcia, S. (1973). Fundamental aspects of hyphal morphogenesis. In *Microbial Differentiation* (ed. J.M. Ashworth and J.E. Smith), pp. 245–267. Cambridge: Cambridge University Press.

Bartnicki-Garcia, S. (1984). Kingdoms with walls. In *Structure, Function, and Biosynthesis of Plant Cell Walls* (ed. W.M. Dugger, and S. Bartnicki-Garcia), pp. 1–18. Rockville, Maryland: Amer, Soc. Plant Physiology.

Bartnicki-Garcia, S. (1987a). The cell wall: a crucial structure in fungal evolution. In *Evolutionary Biology of the Fungi* (ed. A.D.M. Rayner, C.M. Brasier and D. Moore), pp. 389–403. Cambridge: Cambridge University Press.

Bartnicki-Garcia, S. (1987b). Chitosomes and chitin biogenesis. *Food Hydrocolloids*, **1**, 353–8.

Bartnicki-Garcia, S. (1989). The biochemical cytology of chitin and chitosan synthesis in fungi. In *Chitin and Chitosan* (ed. G. Skjak-Braek, T. Anthonsen and P.A. Sanford), pp. 23–35. London: Elsevier Applied Science.

Bartnicki-Garcia, S. (1990). Role of vesicles in apical growth and a new mathematical model of hyphal morphogenesis. In *Tip Growth in Plant and Fungal Cells* (ed. I.B. Heath), pp. 211–232. San Diego: Academic Press.

Bartnicki-Garcia, S. & Gierz, G. (1993). Mathematical analysis of the cellular basis of fungal dimorphism. In *Dimorphic Fungi in Biology and Medicine* (ed. H. Vanden Bossche, F.C. Odds and D. Kerridge), pp. 133–144. New York: Plenum Press.

Bartnicki-Garcia, S. & Lippman, E. (1969). Fungal morphogenesis: cell wall construction in *Mucor rouxii. Science*, **165**, 302–4.

Bartnicki-Garcia, S. & Lippman, E. (1972). The bursting tendency of hyphal tips of fungi: presumptive evidence for a delicate balance between wall synthesis and wall lysis in apical growth. *Journal of General Microbiology*, **73**, 487–500.

Bartnicki-Garcia, S. & Lippman, E. (1977). Polarization of cell wall synthesis during spore germination of *Mucor rouxii. Experimental Mycology*, **1**, 230–40.

Bartnicki-Garcia, S., Bracker, C.E., Reyes, E. & Ruiz-Herrera, J. (1978). Isolation of chitosomes from taxonomically diverse fungi and synthesis of chitin microfibrils *in vitro. Experimental Mycology*, **2**, 173–92.

Bartnicki-Garcia, S., Bracker, C.E., Lippman, E. & Ruiz-Herrera, J. (1984). Chitosomes from the wall-less 'slime' mutant of *Neurospora crassa. Archives of Microbiology*, **139**, 105–12.

Bartnicki-Garcia, S., Hergert, F. & Gierz, G. (1989). Computer simulation of fungal morphogenesis and the mathematical basis for hyphal (tip) growth. *Protoplasma*, **153**, 46–57.

Bartnicki-Garcia, S., Hergert, F. & Gierz, G. (1990). A novel computer model for generating cell shape: application to fungal morphogenesis. In *Biochemistry of Cell Walls and Membranes of Fungi* (ed. P.J. Kuhn, A.P.J. Trinci, M.J. Jung, M.W. Goosey and L.G. Copping), pp. 43–60. Berlin, Heidelberg: Springer-Verlag.

Bartnicki-Garcia, S., Bartnicki, D.D. & Gierz, G. (1995a). Determinants of fungal cell wall morphology: the vesicle supply centre. *Canadian Journal of Botany*, in press.

Bartnicki-Garcia, S., Bartnicki, D.D., Gierz, G., Lopez-Franco, R. & Bracker, C.E. (1995b). Evidence that Spitzenkörper behaviour determines the shape of a fungal hypha. Submitted for publication.

Beakes, G.W. (1987). Oomycete phylogeny: ultrastructural perspectives. In *Evolutionary Biology of the Fungi* (ed. A.D.M. Rayner, C.M. Brasier and D. Moore), pp. 405–421. Cambridge: Cambridge University Press.

Betina, V., Micekova, D. & Nemec, P. (1972). Antimicrobial properties of cytochalasins and their alteration of fungal morphology. *Journal of General Microbiology*, **71**, 343–9.

Bourett, T.M. & Howard, R.J. (1991). Ultrastructural immunolocalization of actin in a fungus. *Protoplasma*, **163**, 199–202.

Bracker, C.E. (1967). Ultrastructure of fungi. *Annual Review of Phytopathology*, **5**, 343–74.

Bracker, C.E., Ruiz-Herrera, J. & Bartnicki-Garcia, S. (1976). Structure and transformation of chitin synthetase particles (chitosomes) during microfibril synthesis *in vitro*. *Proceedings of the National Academy of Sciences, USA*, **73**, 4570–4.

Brunswik, H. (1924). Untersuchungen über Geschlechts und Kernverhältnisse bei der Hymenomyzetengattung *Coprinus*. In *Botanische Abhandlungen* (ed. K. Goebel), pp. 1–152. Jena: Gustav Fischer.

Cabib, E. (1987). The synthesis and degradation of chitin. In *Advances in Enzymology and Related Areas of Molecular Biology*, **59** (ed. A. Meister), pp. 59–101. New York: Wiley.

Caldwell, I.Y. & Trinci, A.P.J. (1973). The growth unit of the mould *Geotrichum candidum*. *Archiv für Mikrobiologie*, **88**, 1–10.

Castle, E.S. (1958). The topography of tip growth in a plant cell. *Journal of General Physiology*, **41**, 913–26.

Cavalier-Smith, T. (1983). A 6-kingdom classification and a unified phylogeny. In *Endocytobiology* II (ed. H.E.A. Schenk and W. Schwemmler), pp. 1027–1034. Berlin: deGruyter.

Cavalier-Smith, T. (1986). The kingdom Chromista: origin and systematics. In *Progress in Phycological Research* 4 (ed. F.E. Round and D.J. Chapman), pp. 309–347. Bristol: Biopress Ltd.

Collinge, A.J. & Markham, P. (1992). Ultrastructure of hyphal tip bursting in *Penicillium chrysogenum*. *FEMS Microbiology Letters*, **91**, 49–54.

Collinge, A.J. & Trinci, A.P.J. (1974). Hyphal tips of wild-type and spreading colonial mutants of *Neurospora crassa*. *Archives of Microbiology*, **99**, 353–68.

Fèvre, M. & Rougier, M. (1982). Autoradiographic study of hyphal cell wall synthesis of *Saprolegnia*. *Archives of Microbiology*, **131**, 212–15.

Förster, H. & Coffey, M.D. (1990). Sequence analysis of the small subunit ribosomal RNAs of three zoosporic fungi and implications for fungal evolution. *Mycologia*, **82**, 306–12.

Förster, H., Oudemans, P. & Coffey, M.D. (1990). Mitochondrial and nuclear DNA diversity within six species of *Phytophthora*. *Experimental Mycology*, **14**, 18–31.

Gäumann, E.A. & Dodge, C.W. (1928). *Comparative Morphology of Fungi*. New York: McGraw-Hill Book Co., Inc.

Girbardt, M. (1957). Der Spitzenkörper von *Polystictus versicolor*. *Planta*, **50**, 47–59.

Girbardt, M. (1969). Die Ultrastruktur der Apikalregion von Pilzhyphen. *Protoplasma*, **67**, 413–41.

Girbardt, M. (1973). Die Pilzzelle. In *Grundlagen der Cytologie* (ed. G.C. Hirsch, H. Ruska and P. Sitte), pp. 441–460. Jena: Fischer-Verlag.

Gooday, G.W. (1971). An autoradiographic study of hyphal growth of some fungi. *Journal of General Microbiology*, **67**, 125–33.

Gooday, G.W. & Gow, N.A.R. (1990). Enzymology of tip growth in fungi. In *Tip Growth in Plant and Fungal Cells* (ed. I.B. Heath), pp. 31–58. San Diego: Academic Press.

Gooday, G.W. & Trinci, A.P.J. (1980). Wall structure and biosynthesis in fungi. In *The Eukaryotic Microbial Cell. 30th Symposium of the Society for General Microbiology* (ed. G.W. Gooday, D. Lloyd and A.P.J. Trinci), pp. 207–251. Cambridge: Cambridge University Press.

Gow, N.A.R. (1994). Tip growth and polarity. In *The Growing Fungus* (ed. N.A.R. Gow and G.M. Gadd), pp. 277–297. London: Chapman & Hall.

Green, P.B. (1969). Cell morphogenesis. *Annual Review of Plant Physiology*, **20**, 365–94.

Gregory, P.H. (1984). The first benefactor's lecture. The fungal mycelium: An historical perspective. *Transactions of the British Mycological Society*, **82**, 1–11.

Grove, S.N. (1978). The cytology of hyphal tip growth. In *The Filamentous Fungi* (ed. J.E. Smith and D.R. Berry), pp. 28–50. London: Edward Arnold.

Grove, S.N. & Bracker, C.E. (1970). Protoplasmic organization of hyphal tips among fungi: vesicles and Spitzenkörper. *Journal of Bacteriology*, **104**, 989–1009.

Grove, S.N. & Sweigard, J.A. (1980). Cytochalasin A inhibits spore germination and hyphal tip growth in *Gilbertella persicaria*. *Experimental Mycology*, **4**, 239–50.

Grove, S.N., Bracker, C.E. & Morré, D.J. (1970). An ultrastructural basis for hyphal tip growth in *Pythium ultimum*. *American Journal of Botany*, **57**, 245–66.

Hanseler, E., Nyhlen, L.E. & Rast, D.M. (1983). Isolation and properties of chitin synthetase from *Agaricus bisporus* mycelium. *Experimental Mycology*, **7**, 17–30.

Hardham, A.R., Suzaki, E. & Perkin, J.L. (1986). Monoclonal antibodies to isolate-, species-, and genus-specific components on the surface of zoospores and cysts of the fungus *Phytophthora cinnamomi*. *Canadian Journal of Botany*, **63**, 311–21.

Harold, F.M. (1990). To shape a cell: and inquiry into the causes of morphogenesis of micro-organisms. *Microbiological Reviews*, **54**, 381–431.

Harold, R.L. & Harold, F.M. (1992). Configuration of actin microfilaments during sporangium development in *Achlya bisexualis*: comparison of two staining protocols. *Protoplasma*, **171**, 110–16.

Harold, F.M., Harold, R.L. & Money, N.P. (1995). What forces drive cell wall expansion? *Canadian Journal of Botany*, in press.

Hasek, J. & Bartnicki-Garcia, S. (1994). The arrangement of F-actin and microtubules during germination of *Mucor rouxii* sporangiospores. *Archives of Microbiology*, **161**, 363–9.

Hawksworth, D.L. (1991). The fungal dimensions of biodiversity: magnitude, significance, and conservation. *Mycological Research*, **95**, 641–55.

Heath, I.B. (1987). Preservation of a labile cortical array of actin filaments in growing hyphal tips of the fungus *Saprolegnia ferax*. *European Journal of Cell Biology*, **44**, 10–16.

Heath, I.B. (1990). The roles of actin in tip growth of fungi. *International Review of Cytology*, **123**, 95–127.

Heath, I.B. (1994). The cytoskeleton in hyphal growth, organelle movements, and mitosis. In *The Mycota* 1 (ed. J.G.H. Wessels and F. Meinhardt), pp. 43–65. Berlin: Springer-Verlag.

Heath, I.B., Gay, J.L. & Greenwood, A.D. (1971). Cell wall formation in the Saprolegniales: cytoplasmic vesicles underlying developing walls. *Journal of General Microbiology*, **65**, 225–32.

Hoch, H.C. & Howard, R.J. (1980). Ultrastructure of freeze-substituted hyphae of the basidiomycete *Laetisaria arvalis*. *Protoplasma*, **103**, 281–97.

Hoch, H.C. & Staples, R.C. (1985). The microtubule cytoskeleton in hyphae of *Uromyces phaseoli* germlings: its relationship to the region of nucleation and to the F-actin cytoskeleton. *Protoplasma*, **124**, 112–22.

Howard, R.J. (1981). Ultrastructural analysis of hyphal tip cell growth in fungi: Spitzenkörper, cytoskeleton and endomembranes after freeze-substitution. *Journal of Cell Science*, **48**, 89–103.

Howard, R.J. & Aist, J.R. (1979). Hyphal tip cell ultrastructure of the fungus *Fusarium*: improved preservation by freeze substitution. *Journal of Ultrastructural Research*, **66**, 224–34.

Hütter, R. & DeMoss, J.A. (1967). Organization of the tryptophan pathway: a phylogenetic study of the fungi. *Journal of Bacteriology*, **94**, 1896–907.

Jackson, S.L. & Heath, I.B. (1990). Evidence that actin reinforces the extensible hyphal apex of the oomycete *Saprolegnia ferax*. *Protoplasma*, **157**, 144–53.

Jaffe, L.F. (1968). Localization in the developing *Fucus* egg and the general role of localizing currents. *Advances in Morphogenesis*, **7**, 295–328.

Katz, D. & Rosenberger, R.F. (1970). The utilisation of galactose by an *Aspergillus nidulans* mutant lacking galactose phosphate-UDP glucose transferase and its relation to cell wall synthesis. *Archiv fur Mikrobiologie*, **74**, 41–51.

Katz, D., Goldstein, D. & Rosenberger, R.F. (1972). Model for branch initiation in *Aspergillus nidulans* based on measurements of growth parameters. *Journal of Bacteriology*, **109**, 1097–100.

Koch, A.L. (1994). The problem of hyphal growth in streptomycetes and fungi. *Journal of Theoretical Biology*, **171**, 137–50.

Kwok, S., White, T.J. & Taylor, J.W. (1986). Evolutionary relationships between fungi, red algae, and other simple eucaryotes inferred from total DNA hybridizations to a cloned basidiomycete ribosomal DNA. *Experimental Mycology*, **10**, 196–204.

Kwon, Y.H., Hoch, H.C. & Aist, J.R. (1991). Initiation of appressorium formation in *Uromyces appendiculatus*: organization of the apex, and the responses involving microtubules and apical vesicles. *Canadian Journal of Botany*, **69**, 2560–73.

Leal-Morales, C.A., Bracker, C.E. & Bartnicki-Garcia, S. (1988). Localization of chitin synthetase in cell-free homogenates of *Saccharomyces cerevisiae*: chitosomes and plasma membrane. *Proceedings of the National Academy of Sciences, USA*, **85**, 8516–20.

López-Franco, R. (1992). Organization and dynamics of the Spitzenkörper in growing hyphal tips. Doctoral Dissertation, Purdue University, W. Lafayette, Indiana, USA.

López-Franco, R. & Bracker, C.E. (1995). Variation and dynamics of the Spitzenkörper in growing hyphal tips. *Mycologia* (submitted).

López-Franco, R., Bartnicki-Garcia, S. & Bracker, C.E. (1994). Pulsed growth of fungal hyphal tips. *Proceedings of the National Academy of Sciences, USA*, **91**, 12228–32.

López-Franco, R., Howard, R.J. & Bracker, C.E. (1995). Satellite Spitzenkörper in growing hyphal tips. *Protoplasma* (in press).

Lovett, J.S. & Haselby, J.A. (1971). Molecular weights of the ribosomal ribonucleic acid of fungi. *Archiv für Microbiologie*, **80**, 191–204.

McClure, W.K., Park, D. & Robinson, P.M. (1968). Apical organization in the somatic hyphae of fungi. *Journal of General Microbiology*, **50**, 177–82.

McGillviray, A.M. & Gow, N.A.R. (1987). The transhyphal electrical current of *Neurospora crassa* is carried principally by protons. *Journal of General Microbiology*, **133**, 2875–81.

Marchant, R. & Smith, D.G. (1968). A serological investigation of hyphal growth in *Fusarium culmorum*. *Archiv für Mikrobiologie*, **63**, 85–94.

Margulis, L., Corliss, J.O., Melkonian, M. & Chapman, D.J. (ed.) (1990). *Handbook of Protoctista*. Boston: Jones & Bartlett.

Meyer, R., Parish, R.W. & Hohl, H.R. (1976). Hyphal tip growth in *Phytophthora*. Gradient distribution and ultrahistochemistry of enzymes. *Archiv für Mikrobiologie*, **110**, 215–24.

Money, N.P. (1994). Osmotic adjustment and the role of turgor in mycelial fungi. In *The Mycota* 1 (ed. J.G.H. Wessels and F. Meinhardt), pp. 67–88. Heidelberg: Springer-Verlag.

Money, N.P. & Harold, F.M. (1993). Two water moulds can grow without measurable turgor pressure. *Planta*, **190**, 426–430.

Moore, R.T. & McAlear, J.H. (1962). Fine structure of mycota. 7. Observations on septa of ascomycetes and basidiomycetes. *American Journal of Botany*, **49**, 86–94.

Nickerson, W.J. (1958). Biochemistry of morphogenesis. A report on Symposium VI. In *Proceedings of the Fourth International Congress of Biochemistry* (ed. O. Hoffmann-Ostenhof), pp. 191–209. London: Pergamon.

Nickerson, W.J. (1963). Symposium on biochemical bases of morphogenesis in fungi. IV. Molecular bases of form in yeasts. *Bacteriological Reviews*, **27**, 305–24.

Nickerson, W.J. & Falcone, G. (1956). Enzymatic reduction of disulfide bonds in cell wall protein of baker's yeast. *Science*, **124**, 318–19.

Ojha, M., Dutta, S.K. & Turian, G. (1975). DNA nucleotide sequence homologies between some zoosporic fungi. *Molecular and General Genetics*, **136**, 151–65.

Park, D. & Robinson, P.M. (1966). Aspects of hyphal morphogenesis in fungi. In *Trends in Plant Morphogenesis* (ed. E.G. Cutter), pp. 27–44. London: Longmans, Green & Co. Ltd.

Plomley, N.J.B. (1959). Formation of the colony in the fungus *Chaetomium*. *Australian Journal of Biological Sciences*, **12**, 53–64.

Prosser, J.I. (1994). Mathematical modelling of fungal growth. In *The Growing Fungus* (ed. N.A.R. Gow and G.M. Gadd), pp. 319–335. London: Chapman & Hall.

Rast, D.M., Horsch, M., Furter, R. & Gooday, G.W. (1991). A complex chitinolytic system in exponentially growing mycelium of *Mucor rouxii*: properties and function. *Journal of General Microbiology*, **137**, 2797–810.

Reinhardt, M.O. (1892). Das Wachsthum der Pilzhyphen. **23**, 479–566.

Roberson, R.W. (1992). The actin cytoskeleton in hyphal cells of *Sclerotium rolfsii*. *Mycologia*, **84**, 41–51.

Roberson, R.W. & Fuller, M.S. (1988). Ultrastructural aspects of the hyphal tip of *Sclerotium rolfsii* preserved by freeze substitution. *Protoplasma*, **146**, 143–9.

Robertson, N.F. (1965). Presidential address: the fungal hypha. *Transactions of the British Mycological Society*, **48**, 1–8.

Roos, U.-P. & Turian, G. (1977). Hyphal tip organization in *Allomyces arbuscula*. *Protoplasma*, **93**, 231–47.

Ruiz-Herrera, J. (1992). *Fungal Cell Wall: Structure, Synthesis, and Assembly*. Boca Raton: CRC Press.

Saunders, R.T. & Trinci, A.P.J. (1979). Determination of tip shape in fungal hyphae. *Journal of General Microbiology*, **110**, 469–73.

Schreurs, W.J.A. & Harold, F.M. (1988). Transcellular proton current in *Achlya bisexualis* hyphae: relationship to polarized growth. *Proceedings of the National Academy of Sciences USA*, **85**, 1534–8.

Selitrennikoff, C.P. (1983). Cell wall assembly of *Neurospora crassa*: lack of evidence for pre-existing cell wall acting as primer or template. *Developmental Biology*, **97**, 245–9.

Sentandreu, R., Mormeneo, S. & Ruiz-Herrera, J. (1994). Biogenesis of the Fungal Cell Wall. In *The Mycota* 1 (ed. J.G.H. Wessels and F. Meinhardt), pp. 111–124. Berlin: Springer-Verlag.

Sietsma, J.H. & Wessels, J.G.H. (1994). Apical wall biogenesis. In *The Mycota* (ed. J.G.H. Wessels & F. Meinhardt), **1**, pp. 126–141. Berlin: Springer-Verlag.

Sietsma, J.H., Sonnenberg, A.M.S. & Wessels, J.G.H. (1985). Localization by autoradiography of synthesis of (1->3)-β and (1->6)-β linkages in a wall glucan during hyphal growth of *Schizophyllum commune*. *Journal of General Microbiology*, **131**, 1331–7.

Smith, J.H. (1923). On the apical growth of fungal hyphae. *Annals of Botany*, **37**, 341–3.

Sonnenberg, A.S.M., Sietsma, J.H. & Wessels, J.G.H. (1985). Spatial and temporal differences in the synthesis of (1->3)-β and (1->6)-β linkages in a wall glucan of *Schizophyllum commune*. *Experimental Mycology*, **9**, 141–8.

Steele, G.C. & Trinci, A.P.J. (1975). The extension zone of mycelial hyphae. *New Phytologist*, **75**, 583–7.

Temperli, E., Roos, U. & Hohl, H.R. (1990). Actin and tubulin cytoskeletons in germlings of the oomycete fungus *Phytophthora infestans*. *European Journal of Cell Biology*, **53**, 75–88.

Trinci, A.P.J. (1969). A kinetic study of the growth of *Aspergillus nidulans* and other fungi. *Journal of General Microbiology*, **57**, 11–24.

Trinci, A.P.J. (1971*a*). Exponential growth of the germ tubes of fungal spores. *Journal of General Microbiology*, **67**, 345–8.

Trinci, A.P.J. (1971*b*). Influence of the width of the peripheral growth zone on the radial growth rate of fungal colonies on solid media. *Journal of General Microbiology*, **67**, 325–44.

Trinci, A.P.J. (1973). The hyphal growth unit of wild type and spreading colonial mutants of *Neurospora crassa. Archiv für Mikrobiologie*, **91**, 127–36.

Trinci, A.P.J. (1974). A study of the kinetics of hyphal extension and branch initiation of fungal mycelia. *Journal of General Microbiology*, **81**, 225–36.

Trinci, A.P.J. (1978). The duplication cycle and vegetative development in moulds. In *The Filamentous Fungi* (ed. J.E. Smith and D.R. Berry), pp. 132–160. London: Edward Arnold.

Trinci, A.P.J. & Collinge, A.J. (1975). Hyphal wall growth in *Neurospora crassa* and *Geotrichum candidum. Journal of General Microbiology*, **91**, 355–61.

Trinci, A.P.J. & Saunders, P.T. (1977). Tip growth of fungal hyphae. *Journal of General Microbiology*, **103**, 243–8.

Trinci, A.P.J., Wiebe, M.G. & Robson, G.D. (1994). The mycelium as an integrated entity. In *The Mycota* **1** (ed. J.G.H. Wessels and F. Meinhardt), pp. 175–193. Heidelberg: Springer-Verlag.

Van der Valk, P. & Wessels, J.G.H. (1977). Light and electron microscopic autoradiography of cell wall regeneration by *Schizophyllum commune* protoplasts. *Act Botanica Neerlandica*, **26**, 43–52.

Vargas, M.M., Aronson, J.M. & Roberson, R.W. (1993). The cytoplasmic organization of hyphal tip cells in the fungus *Allomyces macrogynus. Protoplasma*, **176**, 43–52.

Vermeulen, C.A. & Wessels, J.G.H. (1984). Ultrastructural differences between wall apices of growing and non-growing hyphae of *Schizophyllum commune. Protoplasma*, **120**, 123–31.

Vogel, H.J. (1964). Distribution of lysine pathways among fungi: Evolutionary implications. *American Naturalist*, **68**, 435–46.

Wessels, J.G.H. (1986). Cell wall synthesis in apical hyphal growth. *International Review of Cytology*, **104**, 37–79.

Wessels, J.G.H. (1990). Role of cell wall architecture in fungal tip growth generation. In *Tip Growth in Plant and Fungal Cells* (ed. I.B. Heath), pp. 1–29. San Diego: Academic Press.

Wessels, J.G.H. & Sietsma, J.H. (1981). Cell wall synthesis and hyphal morphogenesis: a new model for apical growth. In *Cell Walls '81* (ed. D.G. Robinson and H. Quader), pp. 135–142. Stuttgart: Wissen, Verlagsgesellschaft.

Whittaker, R.H. (1969). New concepts of kingdoms of organisms. *Science*, **163**, 150–60.

Yokoyama, K., Kaji, H., Nishimura, K. & Miyaji, M. (1990). The role of microfilaments and microtubules in apical growth and dimorphism of *Candida albicans. Journal of General Microbiology*, **136**, 1067–75.

6

Conidiogenesis, classification and correlation

B.C. SUTTON

Introduction

Conidiogenesis is understood by the majority of mycologists as the suite of physiological and morphological processes by which conidia are produced. Since the first taxonomic accounts of conidia-forming fungi appeared as early as 1729, these organisms have been included in general classifications for fungi. They have also been the subject of separate classifications designed solely to cope with their own particular problems of being an artificial group. These successive classificatory systems are the products of a continuous evolutionary process which has been discernible for many years and is still taking place. Some classifications have stood isolated as hypotheses which fairly quickly failed the tests when they were put into practical use. These were the deads ends. Others proved to be of some limited practical value at the time, or became established as templates from which some of the modern schemes now in use have developed. The fact is that the group of conidia-forming fungi is an artificial assemblage of organisms in which phylogenetic relationships are by and large unknown or not apparent. The result is that over all these initiatives in classification have hovered the spectres of basidiomycete, and especially ascomycete, taxonomy. This is because the conidial fungi are, or have evolved from, the mitotic (anamorphic) states of these meiotic (teleomorphic) groups. For various reasons, classifications of the ascomycetes and basidiomycetes have developed independently from those of the conidia-forming fungi, and apart from a few isolated attempts at organizational systems for conidia-forming fungi they have rarely taken account of relationships with ascomycetes and basidiomycetes. However, any systems which are aimed at producing phylogenetic classifications and are based on only part of the available evidence are of necessity flawed to some degree. Attempts to rationalize ascomycete and basidiomycete classifications with those of the conidial fungi based on conidiogenesis have only emerged in the last 40 years, since this aspect of the systematics of the group has been developed.

Not all classifications of conidia-forming fungi have been based on conidiogenesis. Concepts utilizing these features were first used for conidial fungi a little more than 100 years ago, but events leading to the dichotomy between classificatory systems in this group and those in the ascomycetes and basidiomycetes arose somewhat earlier. This chapter therefore has two themes. It traces the evolution of conidiogenesis as a taxonomic character which has contributed significantly to classifications or their equivalent for the conidial fungi. It also reviews the more recent trend which is to increasingly use conidiogenesis as an adjunct to ascomycete and basidiomycete systematics. Comments on molecular developments in relation to systematics of the group are also included.

Classifications prior to Fuckel and Saccardo based on conidiogenesis

To put into perspective the events of the last 100 years which have involved conidiogenesis, it is necessary to briefly review some earlier work. Prior to the 1860s, what we now call the mitosporic fungi (Sutton, 1993) (*syn.* conidia-forming fungi, conidial fungi, deuteromycetes, fungi imperfecti, asexual fungi) had been accommodated in the various general classifications of the fungi proposed up to that time. These attempts at placement have been reviewed in detail by Sutton (1973) and Kendrick (1981). With hindsight it seems most probable that the reasons such earlier efforts were unsuccessful was because they were made at a time when the significance of the events leading to the production of basidia and asci and the formation of conidia was not extensively appreciated. Without these characters being used in classificatory schemes, major separations and groupings were often based on features of fruiting structures or general growth habits observable with the low power microscope. This meant that members of the three groups were sometimes unrealistically juxtaposed. Moreover, the possibility that some mitosporic fungi were states of ascomycetes had not been generally realized, and comparable basidiomycete connections were poorly known and virtually undocumented.

This situation changed with the publication of the beautifully illustrated *Selecta Carpologia Fungorum* by Tulasne & Tulasne (1861–1865) where the authors synthesized the then limited amount of existing information on polymorphism in fungi and supplemented it by their own observations.

This established unequivocally that some mitosporic fungi were in fact stages or states in the life cycles of ascomycetes. Notwithstanding the occasional element of error and imagination in this early correlative work, it is the starting point for most later studies on the subject in relation to ascomycetes and their anamorphs. Somewhat later, Brefeld (1877, 1889) pioneered similar work on basidiomycetes and their anamorphs. However, it was Fuckel (1873) who codified the differences between meiotic (teleomorphic) and mitotic (anamorphic) states when he introduced the terms *Fungi Perfecti* and *Fungi Imperfecti*, respectively for them. It was not necessarily this step which established the dichotomy between classifications for the two groups, but more the fact that later, Saccardo (1873, 1880, 1884, 1886) effectively gave authority to the terminology by using it in his monumental *Sylloge Fungorum* which appeared between 1882 and 1931. At this time the primary criteria for classifying the mitosporic fungi were ones concerning conidiomatal structure. Certainly initially no information on conidiogenesis was incorporated and only limited observation on conidial morphology was used. Saccardo grafted characters of conidial shape, colour and septation on to conidiomatal structure as a means of separating subfamilies and genera. This Saccardoan system and the ones before it were referred to as morphological classifications by Sutton (1973) (Fig. 6.1). It formed the framework upon which most of the later major floristic and monographic works were based (Allescher, 1901, 1903; Bender, 1934; Clements & Shear, 1931; Diedicke, 1912–1915; Ferraris, 1910–1915; Grove, 1935, 1937; Lindau, 1904–1910; Migula, 1921, 1934; Vassiljevski & Karakulin, 1937, 1950). This system was also inextricably linked with the 'new host – new species' type of taxonomic approach in defining species which emerged during the earlier part of the nineteenth century. This was a trend holding sway in both the mitosporic fungi and ascomycetes at least to the middle of the twentieth century, and even persists in some work today. Taxonomic advances in many genera such as the *Cercospora.* Fresen. complex, *Septoria* Sacc., *Ascochyta* Lib. and *Phomopsis* (Sacc.) Bubàk are now severely hindered by this historical millstone around the mycological neck. It is only in genera like *Aspergillus* Micheli, *Penicillium* Link, *Fusarium* Link, *Bipolaris* Shoem. and related genera, *Phoma* Sacc. and *Colletotrichum* Corda where identification of species can now be accomplished without primary reference to the host substratum source of the fungus.

There were, however, undeniable practical advantages to this system, namely the comparative ease with which 'taxa' could be identified and the facility by which binomials could be introduced for what were at that time

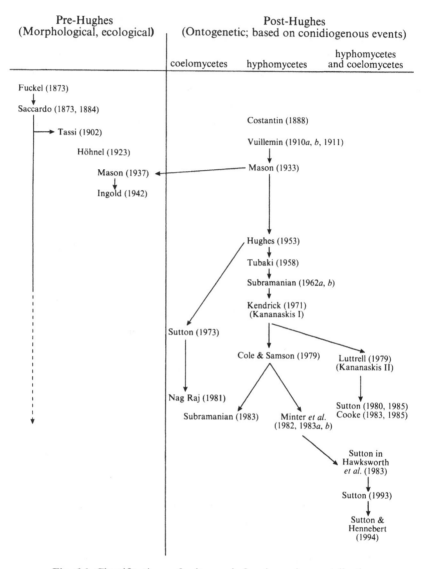

Fig. 6.1. Classifications of mitosporic fungi – major contributions.

considered to be 'new' taxa. At about the same time as Saccardo was con-
solidating this classificatory scheme and laying the ground rules for systematists for the next 75 years, Costantin (1888) suggested an alternative
means of classifying or dealing with some of the mitosporic fungi, in particular the hyphomycetes. This was the first introduction to hyphomycete
systematics of ideas and concepts about conidia and their relationships

with the structures on which they were formed. Such a scheme and the successors it spawned were termed ontogenetic classifications by Sutton (1973) (Fig. 6.1).

In the intervening 65 years between the work of Costantin (1888) and the major breakthrough made by Hughes (1953), the systematics of the mitosporic fungi was subject to a number of other hypotheses which did not achieve much degree of acceptance. Tassi (1902) took to extremes the Saccardoan system, to the extent of introducing many new generic names on the basis of a checker-board scheme composed of colour, septation and size of conidia and foliicolous or caulicolous habit of the conidiomata as co-ordinates. It resulted in a large number of newly described genera of coelomycetes, only a limited number of which are still accepted, the remainder either reduced to synonymy or regarded as of dubious application. Höhnel (1911, 1923) focused attention on pigmentation and structure of conidiomata in both the hyphomycetes and the coelomycetes and combined this with what he perceived as differences in the exogenous or endogenous production of conidia. The result was a comprehensive outline classification for the mitosporic fungi. However, this suffered from severe criticism by Petrak (1925) and was not taken up or developed by subsequent mycologists. Mason (1937) addressed the taxonomic significance of wet and dry conidia in relation to dispersal and their involvement as anamorphs of *Hypocreales*. Wakefield & Bisby (1941) adopted this as a framework for compiling the first British List of Hyphomycetes. Later, Ingold (1942) added a third biological type, the aquatic spore. These ideas seemed to have potential but as Hughes (1953) and later Kendrick (1981) commented, the taxonomic significance of biological spore types can be overstressed, inasmuch as they bear little relation to more stable developmental and morphological characters and must have evolved several times, even in closely related groups. As such the criteria were not to prove of lasting value in classification of the group.

A century of classifications using conidiogenesis

Costantin (1888) sought to classify 295 genera of hyphomycetes (which included some *Peronosporomycetes* and *Zygomycetes*) into 14 groups based on the varying ways in which conidia were attached to the structures on which they were borne. His work remained largely unquoted and it did not directly influence later developments. What can be stated unequivocally is that he had ideas which were far ahead of his time for he placed more emphasis on the attachment of conidia to conidiophores than their actual morphologies. In some ways he foreshadowed the work of Vuillemin (1910a,b, 1911) which almost suffered a similar fate. In fact, the latter remained unnoticed for more than 20 years until resuscitated by Mason (1933). Vuillemin was responsible for introducing some of the basic concepts concerning conidial development (thallic and blastic) that are still in use today and also for coining several new terms, including the much misused phialide. A concise summary of Vuillemin's ideas has been given by Kendrick (1981). The position of E.W. Mason was pivotal in the development of the role conidiogenesis was to play in the systematics of the mitosporic fungi for several reasons. In addition to reviving the principles explored by Vuillemin, extending them and coining additional terms, he was himself no mean philosopher and brought much lateral thinking to bear on systematic problems of the hyphomycetes and coelomycetes, especially in connection with the anamorphic relationships of the *Hypocreales*. The ideas he produced in print (Mason, 1933, 1937) and the more wide-ranging ones that must have arisen in the many informal stimulating discussions with staff during routine work at IMI in the late 1940s had their effect on S.J. Hughes. It was Hughes (Fig. 6.2) who subsequently produced his own innovative thoughts about these problems and provided the breakthrough that had been so long in coming.

In what is without question the most significant paper on systematics of mitosporic fungi this century, Hughes (1953) advanced the central thesis that there are only a limited number of ways in which conidia can be formed from other cells. It was postulated that these were of more taxonomic significance than the prevailing criteria then used in hyphomycete taxonomy, such as mononematous, synnematous or tuberculariaceous nature of conidiophores, the form, pigmentation and septation of mature conidia and the presence or absence of slime around them. He recognized just eight different methods of development in hyphomycetes (they were in fact a mixture of criteria, including conidial development and conidiogenous cell growth) and ascribed these to sections. He neither named the sections as formal taxa nor proposed any classification, even though classification was part of the title of his paper and classification was an important element in his discussion. He also indicated that the principles he introduced would be applicable to coelomycetes. This work was seemingly what many had been waiting for, whether they knew it or not. It was the turning point, for it quickly galvanized others into action and since that time there has appeared a steady stream of papers on the subject of conidiogenesis and associated events. These contributions have fallen into several categories. Periodically there have been attempts to codify the

Fig. 6.2. Dr S.J. Hughes DSc (Wales).

Hughesian principles into overall schemes of classification for the mitosporic fungi complete with formal nomenclatures above the rank of genus. Others have provided experimental evidence supporting and extending Hughes' ideas and in so doing have identified further combinations of conidiogenous events which have increased awareness of the theoretical and practical complexities involved. There have been a number of hypotheses concerned with organizing or refining the data. Especially there have been some misgivings over the morasse of terminology that afflicts the subject. Some have also attempted to extend the principles to the whole of the mitosporic fungi, rather than just the hyphomycetes. Goos (1956) was one of the first to review the events which had led up to the publication of the work by Hughes (1953). He was somewhat equivocal about the usefulness of conidial development in classification and identification, preferring to await the results of applying the ideas not only to cultures but also to material on natural substrata. In a later paper, Simmons (1966) thought that, even though only a few hundred fungi had then been characterized in the Hughesian scheme, it did illuminate the relationships of many groups whose natural proximity had been recognized in practice, in spite of their hitherto artificial and somewhat unrealistic separation in classical classificatory systems.

The first modification or development of the Hughesian ideas was by Tubaki (1958) who was primarily concerned with the correlation of Japanese leaf and stem hyphomycetes with their telomorphs. He accepted Hughes' eight sections and added a ninth, specifically to accommodate *Trichothecium roseum*. These views were reiterated and developed more formally in a later paper (Tubaki, 1962) where he used different types as the basis for six formally proposed divisions of hyphomycetes: *Blastosporae, Radulasporae, Aleuriosporae, Phialosporae, Porosporae* and *Arthrosporae*.

Meanwhile, Subramanian (1962*a*) worked with a wide range of Indian fungi on their natural substrata and was the first to use formation of conidia as a basis for family separation in the hyphomycetes. He recognized six different spore types: blastospore, gangliospore, phialospore, porospore, arthrospore and meristem-arthrospore which typified the *Torulaceae, Bactridiaceae, Tuberculariaceae, Helminthosporiaceae, Geotrichaceae* and *Coniosporaceae*, respectively. In another paper (Subramanian, 1962*b*) he elaborated on the reasoning behind these changes and recognized 25 sections distributed within these six families.

Barron (1968) was concerned with the identification of cultures of hyphomycetes from soil and took a somewhat less expansive approach than Subramanian (1962*b*). He developed a scheme, the origins of which were clearly derived from the six division approach by Tubaki (1962), but recognized ten series. These were again based on what have been interpreted as different types of conidia named according to the ways in which they were formed, but which actually are based on a mixture of conidiogenesis, growth of the conidiogenous cell and spatial arrangement of conidiogenous loci: *Blastosporae, Botryoblastosporae, Sympodulosporae, Aleuriosporae, Annellosporae, Phialosporae, Meristem Arthrosporae, Porosporae, Arthrosporae* and *Meristem Blastosporae*.

Ellis (1971, 1976) was even more conservative in approach and in treating 371 genera of dematiaceous hyphomycetes reverted to a six-group system: Non-meristematic, Meristematic, Basauxic, Holoblastic, Tretic and Phialidic. However, it differed from previous schemes because it took on board some of the recommendations on concepts and terminology agreed at the Kananaskis Conference in 1969 (Kendrick, 1971). However, the criteria used for distinguishing the groups were again mixed. The terms holoblastic, tretic and phialidic essentially describe ways in which conidia were thought to be formed whereas non-meristematic, meristematic and basauxic refer to conidiogenous cell or conidiophore growth. However, in terms of a framework on which to produce an identification manual it succeeded admirably. Moreover, it demonstrated once and for all the immense practical value and potential of the Hughesian criteria in identification and distinguishing between taxa, especially at the generic level.

Dong (1972) was presumably unaware of what had taken place in Kananaskis I (at least he did not cite the reference) so his classificatory scheme stands somewhat in isolation. It was based on the recognition of three types of conidium: arthroconidia (formed by articulation of hyphae), proconidia (formed by fertile hyphae and conidiogenous conidia, with conidia solitary or in acropetal chains), and euconidia (formed totally by fertile hyphae and not by conidia, with solitary conidia or ones formed in basipetal chains, with or without annellations). Three subgroups were proposed: *Arthrohyphomycetes*, *Blastohyphomycetes*, and *Euhyphomycetes*. Within these subgroups, eight sections were recognized: *Arthroconidiae*, *Blastoconidiae*, *Botryoblastoconidiae*, *Sympoduloconidiae*, *Poroconidiae*, *Aleurioconidiae*, *Meristem-aleurioconidiae*, and *Phialoconidiae*. These sections did not differ markedly from those accepted by earlier workers, but what the scheme did do was to impose a higher rank within which they were accommodated.

Subramanian (1983) summarized the evolution of a taxonomy for hyphomycetes and reviewed the use of conidiogenesis in classification of the group. Although regarded as tentative, he proposed an outline classification which brought together the significant and varied contributions that he had previously made to the subject. This excluded coelomycetes, as had so many efforts before, but did incorporate taxa with clamp connexions, i.e. of basidiomycete affinities. It comprised five orders and 27 families, of which eight were included in the *Moniliales*, one in the *Helminthosporiales*, 13 in the *Bactridiales*, four in the *Tuberculariales* and one in the *Geotrichaceae*. It has not achieved much degree of acceptance.

All these classificatory schemes were solely concerned with hyphomycete taxonomy, but while they were unfolding there were parallel efforts to apply Hughesian concepts to the coelomycetes. These were mostly initiated by Sutton (1961) and reviewed in later papers (Sutton, 1971, 1973; Nag Raj, 1981), culminating in the extensive accounts of the group of Sutton (1980) and Nag Raj (1993). The framework that Sutton (1980) used for *The Coelomycetes* was originally presented at the first conference on 'Taxonomy of Fungi' held in Madras in 1973, the text of which was published much later (Sutton, 1985). This 'suprageneric classification of the deuteromycetes' was designed to cope with not only hyphomycetes but also coelomycetes and as such was the first attempt since Höhnel (1923) to incorporate all mitosporic fungi in one scheme. It relied on differences in conidial development and wall relationships to separate taxa at the class, subclass and ordinal levels and thereafter used conidiomatal structure to define taxa at the subordinal level. Elements of this system were then incorporated in another comprehensive scheme outlined by Cooke (1983), to some extent based on an earlier system for hyphomycetes he proposed at the same Madras conference in 1973 and published later (Cooke, 1985). Meanwhile, Luttrell (1979) addressed the same problem of integrating hyphomycetes and coelomycetes and suggested a system based on conidiogenesis derived from the classification of Subramanian (1962*b*). One difference which presaged later developments was the separation at subclass level of ascomycete- and basidiomycete-related anamorphs. Some of these suprageneric ranks have been used sparingly in later works but universal acceptance, again, has not been achieved.

Since the early 1980s there have been no subsequent attempts to generate formal classificatory systems for the mitosporic fungi. The reasons for this are twofold. There have been significant theoretical considerations about conidiogenesis which indicate that the aforementioned compartmentalized approaches aimed at searching for the ideal phylogenetic classificatory system are neither practical nor really necessary. In fact Kendrick (1979) was firmly of the opinion that 'the dream of a natural classification of Fungi Imperfecti based exclusively on conidiogenesis has evaporated'. The second is that more attention has been focused on relating types of conidiogenous behaviour to ascomycete and basidiomycete systematics, perhaps in the belief that by accommodating all anamorphs and other mitosporic fungi within classificatory frameworks based on meiotic fungi, the group of 'imperfects' would 'disappear'.

Influence of terminology, concepts and ultrastructure

The systematics of the mitosporic fungi have been bedevilled by problems of terminology since studies on the group started and, no matter the era, these have stemmed from changes in concepts and the influx of new ideas. Long before the advent of conidiogenesis as a serious taxonomic tool, initiated by Costantin (1888) and given impetus by Vuillemin (1910*a,b*, 1911), the terms used for what we now know as conidia, conidiophores, conidiogenous cells and conidiomata were varied and sometimes singularly confused. There are many examples in the literature of conidia reported as sporidia, of conidiophores as basidia, of confusion between conidiogenous cells and conidiophores, and of conidiogenous cells described as obsolete, distinctly absent, obscure and not present! Now, many of these terms are regarded as largely redundant. Conidiogenesis and its associated events have been no exception to terminological inexactitude, for the problems had started immediately with Vuillemin (1910*a,b*, 1911) and thereafter were to some extent compounded by Mason (1933, 1937) and later workers. Such was the breadth of Vuillemin's work that in his three papers he touched upon almost every aspect that was later to cause difficulties in definition and terminology. In distinguishing between conidia vera and thallospores he introduced concepts that turned out to be fundamental. He also emphasized the importance of liberation and secession. With his four kinds of thallospore (arthrospores, blastospores, chlamydospores and aleuriospores) and the phialide, he introduced ideas, structures and terms which are not only at the very basis of the systematics of the group but ones which proved ripe for varied and confused interpretation later on. The acceptability of these terms and additional ones which were introduced by other workers and the different ways in which they were applied steered some mycologists towards proposing a succession of systems of classification. However, for those more interested in the processes and relationships rather than their application in this manner it was becoming increasingly obvious by the late 1960s that the sum of the parts of conidiogenesis was less than the total that many expected from it. There was debate over the acceptability of the aleuriospore and its relationship to the chlamydospore and gangliospore (Subramanian, 1962*b*; Barron, 1968; Carmichael, 1962). Madelin (1966) exposed the heterogeneity of what were regarded as phialides and some confusion reigned over the distinctions, if any, between phialides, proliferating phialides and annellides (Cole & Kendrick, 1969*a,b*; Sutton & Sandhu, 1969). The result was that conidiogenesis was not proving to be the universal taxonomic

Fig. 6.3. Dr W.B. Kendrick.

panacea for the group that had been hoped. The more mycologists delved into processes of conidium formation the more plastic those processes appeared to be (see also Madelin, 1979). No really workable classification had emerged to cope with all eventualities of structure and development, for there were always 'difficult' genera which failed to fit within the systems proposed. Coelomycetes had not been integrated with the hyphomycetes despite the increasing amount of accurate information on the group. Basic and secondary concepts were decidedly confused, and the terminology for describing them even more perplexing (Langeron, 1945; Luttrell, 1963; Subramanian, 1965; Barron, 1968; von Arx, 1970). At this time, the impact of transmission electron microscopy, scanning electron microscopy, time-lapse photomicrography, fluorescent staining and other techniques on these problems was just starting to be felt.

It was against this background that the next major advance took place and it occurred more by consensus than by any one individual. An international specialists' workshop was convened in 1969 by W.B. Kendrick (Fig. 6.3) at Kananaskis, Canada, and the proceedings were later edited by Kendrick (1971). The achievements of the conference were several. There was agreement on a clear separation between the processes leading to production of a conidium (conidiogenesis) and those which allow a conidiogenous cell to form a succession of conidia. One now considers these to be fundamental distinctions but Kananaskis I was the first time it had been spelled out clearly. Different terminologies were devised for each process and recommended lists of adjectives and nouns agreed. The corollary was a list of 30 terms rejected, replaced, or not recommended. This contained a surprising number which had hitherto enjoyed frequent use, such as aleuriospore, annellophore, arthrospore, blastoconidium, meristem arthrospore, phialospore, porospore, spore, sympodula, etc. A more logical and easily understood terminology was devised for describing the various aspects of conidiogenous events. It was also recognized that there are basic underlying similarities in wall layer involvement in conidial ontogenies which otherwise seem to be quite different, and it was agreed to recognize two basic modes of development. These are thallic (conidia are differentiated from a whole cell and if there is any enlargement of the conidial initial it occurs only after the initial has been delimited by a septum or septa) and blastic (conidia are differentiated from part of a cell and there is marked enlargement of a recognizable conidium initial before the initial is delimited by a septum or septa).

As the paper by Hughes (1953) had spawned a subsequent flurry of activity, in like manner but with different emphasis, so did the edited volume by Kendrick (1971). The Kananaskis I consensus was largely adopted by subsequent mycologists, mainly because most of them had actually agreed to the conclusions at the workshop. Thereafter one finds rather less emphasis placed on divining hypotheses and much more effort put into providing the practical evidence supporting and extending the Kananaskis I ideas, in other words there followed a period of consolidation, acquisition of data and experimentation. There was still room for general or specific theoretical contributions. Hughes (1971a,b) considered conidia in relation to gemmae of algae, bryophytes, and pteridophytes, and wrote about percurrent proliferations in fungi, algae and mosses, while Subramanian (1971, 1972a,b) discoursed on wall relationships and conducted general discussions on conidial ontogeny. His analysis of true and false conidial chains was particularly thought provoking. The experimental work carried out after Kananaskis I used techniques other than optical microscopy. Time lapse studies were initiated by Kendrick, Cole and others at the University of Waterloo, Canada, and transmission and scanning electron microscopy studies were carried out by several mycologists in many parts of the world, especially in Australia, Britain, Canada, France, Japan, Netherlands and the USA. In the 1970s alone ultrastructural and developmental data on conidiogenesis and related events in no less than 70 genera of hyphomycetes and coelomycetes had been published.

A useful summary of TEM work on conidiogenesis in hyphomycetes was provided by Mangenot and Reisinger (1976). This was later superseded by another milestone which was the synthesis of theory and experimental results made by Cole & Samson (1979). They sifted what by this time was a huge body of information on ultrastructure supplemented by time lapse studies, took the concepts agreed at Kananaskis I and confirmed, revised or extended them. These included a basic investigation of blastic (holoblastic and enteroblastic) and thallic (holothallic, holoarthric and enteroarthric) developments. Blastic conidia were the subject of a detailed analysis which demonstrated the many different ways in which successions of conidia are formed. Phialidic and annellidic developments were compared and arguments for their continued separation listed, including convergent types and intermediates such as proliferating phialides. Revised concepts of both phialidic and annellidic were given. Retrogressive and basauxic conidiogenous cells were shown to be widespread and heterogeneous. The different types of thallic development were also fully analysed. The conclusions of this synthesis were given not as a classification of mitosporic fungi but as a classification of developmental processes in which the origin from the conidiogenous cell, growth during early stages of conidial initiation, protoplasmic differentiation during early stages of conidial initiation and growth during maturation were emphasized. Summaries of the different arrangements of blastic and thallic conidia on conidiogenous cells, categories of development based on conidial arrangement and wall relations were given. Finally the variations in conidial and conidiogenous cell development were organized according to the origin of conidia (blastic, thallic), arrangement of conidia (solitary, botryose, catenulate), wall relations between conidia and conidiogenous cells (holoblastic, enteroblastic, holothallic, holoarthric, enteroarthric), conidiogenous cell growth during formation of successive conidia (determinate, proliferous, sympodial, percurrent, basauxic, retrogressive) and special modes of conidiogenous cell development (porogenous, phialidic, annellidic).

However, Cole & Samson (1979) left some problems unresolved and one of these was how phialides, especially proliferating phialides, and annellides differ, if indeed they differ at all. It was in addressing this particular conundrum, especially in the light of a conscious effort at maintaining some distance from imposing classifications for the group, that Minter, Kirk and Sutton (1982, 1983*a*) and Minter Sutton & Brady (1983*b*) evolved a quite different approach. They recognized three ways in which walls are laid down in hyphae: apical- , diffuse- and ring-wall building, ultrastructural evidence for which came indirectly and directly from Cole

& Samson (1979). A reconsideration of information already published on conidiogenesis led to identification of a basic number of processes surrounding conidiogenesis, such as conidial initiation, secession, and maturation, collarette production, and conidiogenous cell proliferation. They advised using precise words to describe the individual developmental stages, and, although the resulting statements are longer than the old use of a single term, they are unambiguous, accurate, scientific, and present an improvement on what had been used before. These conclusions were arrived at just as the page proofs of the 7th edition of Ainsworth & Bisby's *Dictionary of the Fungi* (Hawksworth, Sutton & Ainsworth, 1983) were being processed. Descriptive terminology for 13 fungi, including *Pseudallescheria*, *Cladosporium*, *Tritirachium*, *Trichoderma*, *Scopulariopsis*, *Spadicoides*, *Geotrichum*, *Pseudospiropes*, *Aspergillus*, *Trichothecium*, *Cladobotryum*, *Arthrinium* and *Basipetospora*, was rapidly organized and included at the last minute in *The Dictionary*. Thereafter, evidence accumulated supporting the concepts of wall-building and the terminology was applied in resolving various taxonomic problems (Descals, 1985; Glawe, 1984; Minter, 1985; Tiedt & Jooste, 1988; Wingfield, 1985; Wingfield, Van Wyk & Wingfield, 1987; Wingfield, Seifert & Webber, 1993; Van Wyk et al., 1987; Van Wyk, Wingfield & Marasas, 1988). The stimulus to extend descriptors to a wider range of hyphomycetes and coelomycetes was provided by input to the 8th edition of *The Dictionary*. In revising the entries for mitosporic fungi Sutton (1993) introduced a mosaic approach in identifying 34 different combinations of conidiogenous events. Shortly thereafter this number was expanded to 43 by Sutton & Hennebert (1994). Additional combinations of events have been recognized since that date, and it is likely that more will come to light in the future. These all contribute to the multiplex of conidiogenous events which is beginning to emerge. As such, the information will make description easier (see also the unitary parameters for conidiogenesis by Hennebert & Sutton, 1994), will cope with fungi which have not fitted comfortably into previous systems, will differentiate between closely related taxa, and will go a long way towards facilitating the rationalization of ascomycete and basidiomycete systematics with mitosporic fungi.

Correlations with teleomorphs involving conidiogenesis

Since the introduction of the names *Fungi Perfecti* and *Fungi Imperfecti* by Fuckel (1873) the classificatory systems for ascomycetes and basidiomycetes have proceeded in independent directions from those of their anamorphs and for other mitosporic fungi. Although details of conidial development had been used increasingly in systematic studies, descriptions and the identification of hyphomycetes and coelomycetes since Costantin (1888), these entrenched positions on separate classifications continued at least until the work of Hughes (1953). It took a comparatively short time for the new types of information popularized by Hughes to become assimilated into general studies on hyphomycetes and coelomycetes. Thereafter, the momentum increased and these types of data became included not only in descriptions and revisionary accounts but also in the various schemes of classification proposed for mitosporic fungi. Occasionally speculations on the use of conidiogenesis as another taxonomic criterion in the mainstream of ascomycete systematics were made (Fig. 6.4). The first of these attempts was by Tubaki (1958) who dealt solely with leaf and stem hyphomycetes and their correlations. Based on a phylogenetic classification of ascomycetes he sought similar relationships among hyphomycete anamorphs where he used the types of conidiophore and conidial development as the most stable criteria. Hughes' groups were modified at sectional level and superimposed on a number of orders of ascomycetes where correlations had been reported. In a later publication (Tubaki, 1981), which did not take account of the Kananaskis II meeting (Kendrick, 1979), he summarized these studies and extended them to take in many more examples, including basidiomycete relationships. It was shown that in the *Sphaeriales* all types of conidial development are represented and in the *Eurotiales* the picture is almost the same, with only the 'tretic' type not being present. On the other hand, the 'tretic' type predominates in the *Sphaeriales* and *Pleosporales* but is not found elsewhere. These are only a few examples of the sorts of conclusions which he made. The work gave a useful summary using such parameters. Müller (1971) felt that, in spite of the increase in knowledge and advancing concepts in the taxonomy of ascomycetes and imperfect forms, classification systems in neither group had achieved the necessary stability that enabled any synthesis to be achieved. He reiterated the point made by Tubaki (1958) that some kinds of conidial development are represented throughout the ascomycete system whereas others are limited to, or even typical of, one or a few orders. Similar thoughts were expressed by Subramanian (1971).

With seemingly little progress or initiative in these areas really taking place, W.B. Kendrick convened a second specialists' workshop at Kananaskis, the proceedings of which he later edited (Kendrick, 1979).

Major Contributions to Correlation of Ascomycetes and Basidiomycetes with Anamorphs

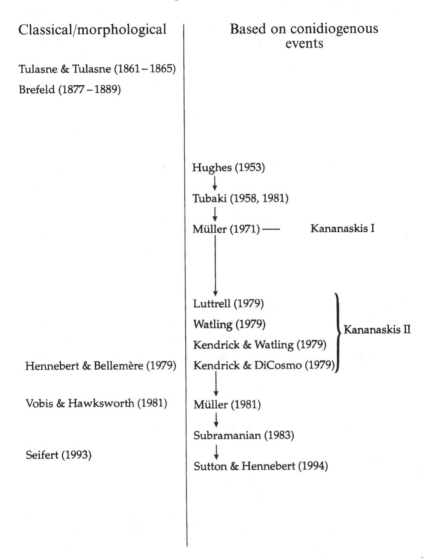

Classical/morphological

Based on conidiogenous events

Tulasne & Tulasne (1861–1865)

Brefeld (1877–1889)

Hughes (1953)

Tubaki (1958, 1981)

Müller (1971) —— Kananaskis I

Luttrell (1979)

Watling (1979)

Kendrick & Watling (1979)

Hennebert & Bellemère (1979)   Kendrick & DiCosmo (1979)

Kananaskis II

Vobis & Hawksworth (1981)   Müller (1981)

Subramanian (1983)

Seifert (1993)

Sutton & Hennebert (1994)

Fig. 6.4. Major contributions to correlation of Ascomycetes and Basidiomycetes with Anamorphs.

Kananaskis II addressed '*The Whole Fungus*', a title which is somewhat of a misnomer because the mycelium was not included (see Rayner, Chapter 8). One of the specific problems examined was how classifications of ascomycetous and basidiomycetous anamorphs and teleomorphs affect one another. There were some major inputs to this subject and a number of other contributions which although peripheral, were also relevant. Luttrell (1979) used septation to separate the mitosporic fungi into ascomycete- and basidiomycete-related groups at the subclass rank. He found that allying a classification of the ascomycetes to an arrangement of the anamorphs based on characters of conidiogenesis did not meet with success, and an integrated classification of hyphomycetes and coelomycetes actually raised questions concerning previously suggested correlations between teleomorphs and groupings of hyphomycetes based on types of conidiogenesis. Here was a positive indication that the value of conidiogenesis had been under-rated or under-utilized in ascomycete systematics. In three other chapters the known information on teleomorph and anamorph correlations in both ascomycetes and basidiomycetes was brought together and tabulated in various ways (Watling, 1979; Kendrick & Watling, 1979; Kendrick & DiCosmo, 1979). The value of any compilation lies in its comprehensive treatment or otherwise of the subject area it is intended to cover. In this respect the authors provided the most authoritative of lists hitherto assembled, with full reference to the work on which they were based. In addition, for the ascomycetes the correlations were rated for reliability in terms of evidence for affiliation, such as whether it was casual, circumstantial or experimental. This has been of invaluable help in assessing the value that can be placed on the correlations that have been reported and continue to be published. At about the same time, Hennebert & Bellemère (1979) produced a synthesis of anamorphs of discomycetes, including lichenized and non-lichenized genera. This demonstrated the by now expected varying degrees of homogeneity and heterogeneity between orders. Vobis & Hawksworth (1981) brought together the then known information on anamorphs of lichenized ascomycetes but found that too few had been accurately studied to draw even generalized conclusions. Müller (1979) returned to the subject he had addressed earlier (Müller, 1971), but arrived at much the same conclusion, i.e. that the systems are too divergent to permit a common base for integrating mitosporic fungi into the taxonomic systems for ascomycetes and basidiomycetes. However, he did stress that the use of anamorphs in this context is not simply helpful but absolutely indispensable. In an outline classification for hyphomycetes based on five types of conidiogenesis,

Subramanian (1983) included classes, orders and families of ascomycetes and basidiomycetes which had been correlated with members of these groups. However, little comment was given on the fitness or otherwise of the correlations, but he hoped that the presentation would stimulate the evolution of a better taxonomy. Seifert (1993) made some useful comparisons of and comment on correlations between teleomorphs and anamorphs but did not emphasize conidiogenesis in this context. He drew attention to examples of exclusive anamorph–teleomorph connections at the generic level, examples of anamorph genera associated with many teleomorph genera, and teleomorph genera with many associated anamorph genera. He warned of the assumption that teleomorph taxonomy is necessarily correct, and implied the corollary that comparison of anamorphs in separate teleomorph families does not naturally mean a correlation of like with like. By the 1990s the multiplex approach derived from the work of Minter *et al.* (1982, 1983*a,b*) for arranging conidiogenous events had been suggested (Sutton, 1993). Sutton & Hennebert (1994) used this framework in analysing the information on correlations compiled by Kendrick & DiCosmo (1979), but with the ascomycete system updated by Eriksson & Hawksworth (1993). Rather than using anamorph generic names, they assessed data on conidiogenous events known for various orders and families of non-lichenized ascomycetes. This allowed a subjective appraisal of the homogeneous or heterogeneous status of the correlations reported. Approximately half the correlations for ascomycete families were homogeneous or possibly homogenous and half heterogeneous. In the *Trichocomaceae*, *Hypocreaceae*, *Xylariaceae* and *Venturiaceae*, the use of anamorphs has contributed enormously to the systematics of these families (see refs in Reynolds & Taylor, 1993; Hawksworth, 1994). However, in the *Sordariales* and large families with many taxa such as the *Amphisphaeriaceae*, *Pleosporaceae*, *Asterinaceae*, *Dothioraceae* and others, the evidence from conidiogenous events is contradictory, confused, unreliable or very sparse, indicating that these areas are ones in which future effort must be focused. Correlations in two contrasting groups of ascomycetes were discussed to show how this type of synthesis can be of value. Connections in the *Sclerotiniaceae* and *Pezizaceae* showed the nature and dimensions of the problems. At these extremes there is uniformity of correlations and conidiogenous events in the former and mostly heterogeneity in the latter. The other example taken was the genus *Mycosphaerella* which has had up to 27 different anamorphic states correlated with various species. By organizing the anamorphs according to conidiogenous events it was possible to draw two conclusions. First, anamorphs do little to support subgeneric and sectional organizations that have previously been proposed for *Mycosphaerella*, which suggests that the subgeneric taxonomy of the genus is suspect or unresolved. Second, the groupings according to conidiogenous events combined with conidiomatal structure offer the prospect of a more rational means of separation of taxa within *Mycosphaerella*, which could have merit at subgeneric level. This analysis remains to be tested.

Conclusions

Whereas it is possible to identify some genera, families, classes and orders of ascomycetes where the known anamorph correlations are homogeneous and have therefore contributed significantly to the systematics of the group, there are far more in which the correlations are non-existent or comparatively rare, poorly or questionably documented, unconfirmed or grossly heterogeneous. It is in these areas where major effort is required in future. A century of conidiogenesis and associated events has introduced new and valuable criteria which have refined the taxonomy and identification of mitosporic fungi, has overcome the successive impulses to utilize conidiogenesis as the basis for systems of classification for the mitosporic fungi alone, has participated in the wide use of experimental techniques to explore and understand morphological and developmental variation of mitosporic fungi, and finally has ushered in the dawn of a belated recognition that many of the mitosporic fungi are integral/occasional components of complex life cycles. As such, the integration of mitosporic fungi into classificatory systems for meiotic fungi needs to be accomplished very much more effectively than heretofore.

However, although conidiogenesis is the latest in a succession of morphologically based means of resolving systematic problems it will certainly not be the sole answer in future. It has been often stated that, as more and more correlations are made, the mitosporic fungi will be gradually assimilated into the ascomycete and basidiomycete classifications so that at some time in the longer term future there will be no separate group for the mitosporic fungi. This denies the fact that mitosporic fungi can evolve (even though Müller's ratchet is an acknowledged constraint), and presupposes huge investment in solving what for many species of fungi is an academic problem. Neither is the case. Mitosporic evolution is a possible and partial explanation for the comparatively small numbers of confirmed correlations, even more minuscule if so far only 5% of fungi have been recognized, according to Hawksworth (1991). Molecular characters offer very much more potential for not only closing the gap between the two classifications but also for determining relationships between mitosporic fungi which have never been correlated with teleomorph states (Taylor, 1994) and might never be so because they have evolved into different taxa. What can be said for conidiogenesis is that it has pointed mycologists in a more phylogenetic direction, and any rationalization with meiosporic classifications and/or the application of molecular techniques will be so much the better placed than before.

References

Allescher, A. (1901). *Rabenhorst's Kryptogamen Flora* I. Die Pilze. VI. Fungi Imperfecti: Hyalin-sporige Sphaerioideen. Leipzig: E. Kummer.
Allescher, A. (1903). *Rabenhorst's Kryptogamen Flora* I. Die Pilze. VII. Fungi Imperfecti: Gefarbt-sporige Sphaerioideen sowie Nectrioideen, Leptostromaceen, Excipulaceen und der Ordnung der Melanconieen. Leipzig: E. Kummer.
Barron, G.L. (1968). *The Genera of Hyphomycetes from Soil*. Baltimore: Williams & Wilkins Co.
Bender, H.B. (1934). *The Fungi Imperfecti: Order Sphaeropsidales*. New Haven, Connecticut: Tuttle, Morehouse & Taylor Co.
Brefeld, O. (1877). *Botanische Untersuchungen über Schimmelpilze*, 3, 1–226. Leipzig.
Brefeld, O. (1889). *Untersuchungen aus dem Gesammtgebeite der Mykologie*, 8, 1–305. Leipzig.
Carmichael, J.W. (1962). *Chrysosporium* and some other aleuriosporic hyphomycetes. *Canadian Journal of Botany*, 40, 1137–73.
Clements, F.E. & Shear, C.L. (1931). *The Genera of Fungi*. New York: Hafner.
Cole, G.T. & Kendrick, W.B. (1969a). Conidium ontogeny in hyphomycetes. The phialides of *Phialophora, Penicillium*, and *Ceratocystis. Canadian Journal of Botany*, 47, 779–89.
Cole, G.T. & Kendrick, W.B. (1969b). Conidium ontogeny in hyphomycetes.The annellophores of *Scopulariopsis brevicaulis. Canadian Journal of Botany*, 47, 925–9.
Cole, G.T. & Samson, R.A. (1979). *Patterns of Development in Conidial Fungi*. London, San Francisco, Melbourne, Pitman.
Cooke, W.B. (1983). Toward a system for the Fungi imperfecti. *Revista de Biologia*, 12, 279–96.
Cooke, W.B. (1985). A proposed artificial hierarchical system of classification for the Moniliales (Fungi Imperfecti). In *Taxonomy of Fungi* Pt 2. Proceedings of the International Symposium on Taxonomy of Fungi, University of Madras, 1973 (ed. C.V. Subramanian), pp. 387–396.
Costantin, J. (1888). *Les Mucédinés simples: Matériaux pour l'histoire des Champignons*. Paris: Klincksieck.
Descals, E. (1985). Conidia as modified hyphae. *Proceedings of the Indian Academy of Sciences (Plant Sciences)*, 94, 209–27.
Diedicke, H. (1912–1915). *KryptogamenFlora Mark Brandenburg*. IX. Pilze. VII. Sphaeropsideae, Melanconieae. Leipzig: Borntraeger Bros.

Dong, B.X. (1972). Contributions à l'étude taxonomique des Hyphomycetes (Deuteromycetes). I. Esquisse d'un nouvelle classification. *Ceskà Mykologie*, **26**, 155–66.

Ellis, M.B. (1971). *Dematiaceous Hyphomycetes*. Kew: Commonwealth Mycological Institute.

Ellis, M.B. (1976). *More Dematiaceous Hyphomycetes*. Kew: Commonwealth Mycological Institute.

Eriksson, O.E. & Hawksworth, D.L. (1993). Outline of the ascomycetes – 1993. *Systema Ascomycetum*, **12**, 51–257.

Ferraris, T. (1910–1915). *Flora Italia Cryptogama* Pars I: Fungi. Hyphales. Tuberculariaceae – Stilbaceae. fasc. **1**, 1–198; fasc. **8**, 199–534; fasc. **10**, 535–846; fasc. **13**, 847–979. Firenze.

Fuckel, L. (1873). Symbolae mycologicae, Beiträge zur Kenntniss der Rheinischen Pilze. Zoreiter Nächtrag. *Jahrbüch Nassauschen Vereins Naturkunde*, **27–28**, 1–99.

Glawe, D.A. (1984). *Bloxamia truncata* in artificial culture. *Mycologia*, **76**, 741–5.

Goos, R.D. (1956). Classification of the Fungi Imperfecti. *Proceedings of the Iowa Academy of Science*, **63**, 311–20.

Grove, W.B. (1935). *British Stem- and Leaf-Fungi* **1**. Cambridge: Cambridge University Press.

Grove, W.B. (1937). *British Stem- and Leaf-Fungi* **2**. Cambridge: Cambridge University Press.

Hawksworth, D.L. (1991). The fungi dimension of biodiversity: magnitude, significance, and conservation. *Mycological Research*, **95**, 641–55.

Hawksworth, D.L. (ed.) (1994). *Ascomycete Systematics: problems and perspectives in the nineties*. NATO ASI Series. Series A: Life Sciences **269**.

Hawksworth, D.L., Sutton, B.C. & Ainsworth, G.C. (1983). *Ainsworth & Bisby's Dictionary of the Fungi*. Kew: Commonwealth Mycological Institute.

Hennebert, G.L. & Bellemère, A. (1979). Les formes conidiennes des Discomycetes. Essai taxonomique. *Revue de Mycologie*, **43**, 259–315.

Hennebert, G.L. & Sutton, B.C. (1994). Unitary parameters in conidiogenesis. In *Ascomycete Systematics: problems and perspectives in the nineties*. NATO ASI Series. Series A: Life Sciences **269** (ed. D.L. Hawksworth), pp. 65–76.

Höhnel, F. (1911). Zur Systematik der Sphaeropsideen und Melanconieen. *Annales Mycologici*, **9**, 258–65.

Höhnel, F. (1923). System der fungi imperfecti Fuckel. *Mykologisches Untersuchungen*, **1**, 301–69.

Hughes, S.J. (1953). Conidiophores, conidia and classification. *Canadian Journal of Botany*, **31**, 577–659.

Hughes, S.J. (1971a). Percurrent proliferations in fungi, algae and mosses. *Canadian Journal of Botany*, **49**, 215–31.

Hughes, S.J. (1971b). On conidia of fungi and gemmae of algae, bryophytes and pteridophytes. *Canadian Journal of Botany*, **49**, 1319–39.

Ingold, C.T. (1942). Aquatic Hyphomycetes of decaying alder leaves. *Transactions of the British Mycological Society*, **25**, 339–417.

Kendrick, B. (1971). *Taxonomy of Fungi Imperfecti*. University of Toronto Press.

Kendrick, B. (1979). Introduction. In *The Whole Fungus* **1** (ed. B. Kendrick), pp. 11–15. Ottawa: National Museum of Natural Science, National Museums of Canada.

Kendrick, B. (1981). The history of conidial fungi. In *Biology of Conidial Fungi* **1** (ed. G.T. Cole & B. Kendrick), pp. 3–18. Academic Press.

Kendrick, B. & DiCosmo, F. (1979). Teleomorph–anamorph connections in Ascomycetes. In *The Whole Fungus* 1 (ed. B. Kendrick), pp. 283–410. Ottawa: National Museum of Natural Science, National Museums of Canada.

Kendrick, B. & Watling, R. (1979). Mitospores in Basidiomycetes. In *The Whole Fungus* 2 (ed. B. Kendrick), pp. 473–545. Ottawa: National Museum of Natural Science, National Museums of Canada.

Langeron, M. (1945). *Précis de Mycologie.* Masson, France.

Lindau, G. (1904–1907). *Rabenhorst's Kryptogamen-Flora von Deutschland, Osterreich, und der Schweiz* 1 Band. Pilze. 8 Abt. Fungi imperfecti, Hyphomycetes (1 hälfte), Mucedinaceae, Dematiaceae (Phaeosporae und Phaeodidymae). Leipzig, E. Kummer.

Lindau, G. (1907–1910). *Rabenhorst's Kryptogamen-Flora von Deutschland, Osterreich, und der Schweiz* 1 Band. Pilze. 9 Abt. Fungi imperfecti, Hyphomycetes (2 hälfte), Dematiaceae (Phaeophragmiae bis Phaeostaurosporae), Stilbaceae, Tuberculariaceae, etc. Leipzig: E. Kummer.

Luttrell, E.S. (1963). Taxonomic criteria in *Helminthosporium. Mycologia*, **55**, 643–74.

Luttrell, E.S. (1979). Deuteromycetes and their relationships. In *The Whole Fungus* 1 (ed. B. Kendrick), pp. 241–264. National Museum of Natural Sciences, Canada and The Kananaskis Foundation.

Madelin, M.F. (1966). The genesis of spores in higher fungi. In *The Fungus Spore* (ed. M.F. Madelin), pp. 15–36. London: Butterworth.

Madelin, M.F. (1979). An appraisal of the taxonomic significance of some different modes of producing blastic conidia. In *The Whole Fungus* 1 (ed. B. Kendrick), pp. 63–80. National Museum of Natural Sciences, Canada and The Kananaskis Foundation.

Mangenot, F. & Reisinger, O. (1976). Form and function of conidia as related to their development. In *The Fungal Spore: form and function* (ed. D.J. Weber & W.M. Hess), pp. 79–846.

Mason, E.W. (1933). Annotated account of fungi received at the Imperial Mycological Institute, List II. (Fascicle 2). *Mycological Papers*, **3**, 1–67.

Mason, E.W. (1937). Annotated account of fungi received at the Imperial Mycological Institute, List II. (Fascicle 3 – General Part). *Mycological Papers*, **4**, 69–99.

Migula, W. (1921). *Kryptogamen-Flora von Deutschland, Deutsch-Osterreich und der Schweiz.* III. Pilze. 4 Teil. 1 Abt. Fungi imperfecti: Sphaeropsidales, Melanconiales. Berlin, Hugo Bernmühler Verlag.

Migula, W. (1934). *Kryptogamen-Flora von Deutschland, Deutsch-Osterreich und der Schweiz.* III. Pilze. 4 Teil. 2 Abt. Fungi imperfecti: Hyphomycetes. Leipzig, Akademische Verlagsgesellschaft M.B.H.

Minter, D.W. (1985). A re-appraisal of the relationships between *Arthrinium* and other hyphomycetes. *Proceedings of the Indian Academy of Sciences (Plant Sciences)*, **94**, 281–308.

Minter, D.W., Kirk, P.M. & Sutton, B.C. (1982). Holoblastic phialides. *Transactions of the British Mycological Society*, **79**, 75–93.

Minter, D.W., Kirk, P.M. & Sutton, B.C. (1983a). Thallic phialides. *Transactions of the British Mycological Society*, **80**, 39–66.

Minter, D.W., Sutton, B.C. & Brady, B.L. (1983b). What are phialides anyway? *Transactions of the British Mycological Society*, **81**, 109–120.

Müller, E. (1971). Imperfect-Perfect Connections in Ascomycetes. In *Taxonomy of Fungi Imperfecti* (ed. B. Kendrick), pp. 184–201. University of Toronto Press.

Müller, E. (1979). Factors inducing asexual and sexual sporulation in fungi (mainly ascomycetes). In *The Whole Fungus*, 1 (ed. B. Kendrick), pp. 265–82. Ottawa: National Museum of National Science, National Museums of Canada.

Müller, E. (1981). Relations between conidial anamorphs and their teleomorphs. In *Biology of the Conidial Fungi* 1 (ed. G.T. Cole & B. Kendrick), pp. 145–69. New York: Academic Press.

Nag Raj, T.R. (1981). Coelomycete systematics. In *Biology of Conidial Fungi* 1 (ed. G.T. Cole & B. Kendrick), pp. 43–84. New York: Academic Press.

Nag Raj, T.R. (1993). *Coelomycetous Anamorphs with Appendage-bearing Conidia.* Waterloo, Canada: Mycologue Publications.

Nag Raj, T.R. & Kendrick, W.B. (1971). Genera Coelomycetarum I. *Urohendersonia. Canadian Journal of Botany*, **49**, 1853–62.

Petrak, F. (1925). Mykologische Notizen VIII. *Annales Mycologici*, **23**, 1–143.

Reynolds, D.R. & Taylor, J.W. (ed.) (1993). *The Fungal Holomorph: mitotic, meiotic and pleomorphic speciation in fungal systematics.* Wallingford: CAB International.

Saccardo, P.A. (1873). *Mycologiae Venetae Specimen.* Padua: P. Prosperini.

Saccardo, P.A. (1880). Conspectus generum fungorum italiae inferiorum nempe ad Sphaeropsideas, Melanconieas et Hyphomyceteas pertinentium systemate sporologico dispositorum. *Michelia*, **2**, 1–39.

Saccardo, P.A. (1884). *Sylloge Fungorum* 3. Padua.

Saccardo, P.A. (1886). *Sylloge Fungorum* 4. Padua.

Seifert, K.A. (1993). Integrating anamorphic fungi into the fungal system. In *The Fungal Holomorph: mitotic, meiotic and pleomorphic speciation in fungal systematics* (ed. D.R. Reynolds and J.W. Taylor), pp. 79–85. Wallingford: CAB International.

Simmons, E.G. (1966). The theoretical bases for classification of the Fungi Imperfecti. *Quarterly Review of Biology*, **41**, 113–23.

Subramanian, C.V. (1962*a*). A classification of the hyphomycetes. *Current Science*, **31**, 409–11.

Subramanian, C.V. (1962*b*). A classification of the hyphomycetes. *Bulletin of the Botanical Survey of India*, **4**, 249–59.

Subramanian, C.V. (1965). Spore types in the classification of the Hyphomycetes. *Mycopathologia et Mycologia Applicata*, **26**, 373–84.

Subramanian, C.V. (1971). Conidial ontogeny in fungi imperfecti with special reference to cell wall relationships. *Journal of the Indian Botanical Society. Golden Jubilee Volume*, **50A**, 51–9.

Subramanian, C.V. (1972*a*). Conidial chains, their nature and significance in the taxonomy of hyphomycetes. *Current Science*, **41**, 43–9.

Subramanian, C.V. (1972*b*). Conidium ontogeny. *Current Science*, **41**, 619–24.

Subramanian, C.V. (1983). *Hyphomycetes: Taxonomy and Biology.* New York: Academic Press.

Sutton, B.C. (1961). Coelomycetes I. *Mycological Papers*, **80**, 1–16.

Sutton, B.C. (1971). Conidial ontogeny in pycnidial and acervular fungi. In *Taxonomy of Fungi Imperfecti* (ed. B. Kendrick), pp. 263–78. University of Toronto Press.

Sutton, B.C. (1973). Coelomycetes. In *The Fungi: an advanced treatise* **IVA** (ed. G.C. Ainsworth, F.K. Sparrow and A.S. Sussman), pp. 513–82. New York: Academic Press.

Sutton, B.C. (1980). *The Coelomycetes.* Kew: Commonwealth Mycological Institute.

Sutton, B.C. (1985). Suprageneric classification of the Deuteromycotina. In *Taxonomy of Fungi Pt 2.* Proceedings of the International Symposium on Taxonomy of Fungi, University of Madras, 1973 (ed. C.V. Subramanian), pp. 379–86.

Sutton, B.C. (1993). Mitosporic Fungi (Deuteromycetes) in the *Dictionary of the Fungi.* In *The Fungal Holomorph: mitotic, meiotic and pleomorphic speciation in fungal systematics* (ed. D.R. Reynolds and J.W. Taylor), pp. 27–55. Wallingford: CAB International.

Sutton, B.C. & Hennebert, G.L. (1994). Interconnections amongst anamorphs and their possible contribution to ascomycete systematics. In *Ascomycete Systematics: problems and perspectives in the nineties.* NATO ASI Series. Series A: Life Sciences **269** (ed. D.L. Hawksworth), pp. 77–100.

Sutton, B.C. & Sandhu, D.K. (1969). Electron microscopy of conidium development and secession in *Cryptosporiopsis* sp., *Phoma fumosa*, *Melanconium bicolor*, and *M.apiocarpum. Canadian Journal of Botany*, **47**, 745–9.

Tassi, F. (1902). I generi *Phyllosticta* Pers., *Phoma* Fr., *Macrophoma* (Sacc.) Berl. et Vogl. e i loro generi analoghi, giusta la legge d'analogia. *Bullettino Laboratorio Orto Botanico Siena*, **5**, 1–72.

Taylor, J.W. (1994). A contemporary view of the holomorph: nucleic acid sequence and computer databases are changing fungal classification. In *The Fungal Holomorph: mitotic, meiotic and pleomorphic speciation in fungal systematics* (ed. D.R. Reynolds and J.W. Taylor), pp. 3–13. Wallingford: CAB International.

Tiedt, L.R. & Jooste, W.J. (1988). Ultrastructure of collarette formation in *Fusarium* sect. *Liseola* and some taxonomic implications. *Transactions of the British Mycological Society*, **90**, 531–6.

Tubaki, K. (1958). Studies on the Japanese hyphomycetes. V. Leaf and stem group with a discussion of the classification of hyphomycetes and their perfect stages. *Journal of the Hattori Botanical Laboratory*, **20**, 142–244.

Tubaki, K. (1962). Taxonomic study of hyphomycetes. *Annual Report of the Institute of Fermentation, Osaka*, **1**, 25–54.

Tubaki, K. (1981). *Hyphomycetes: their perfect–imperfect connections.* J. Cramer, Vaduz.

Tulasne, L.R. & Tulasne, C. (1861–65). *Selecta Carpologia Fungorum:* **1–3.** Imperial Press, Paris.

Van Wyk, P.S., Venter, E., Wingfield, M.J. & Marasas, W.F.O. (1987). Development of macroconidia in *Fusarium. Transactions of the British Mycological Society*, **88**, 347–53.

Van Wyk, P.S., Wingfield, M.J. & Marasas, W.F.O. (1988). Differences in synchronization of stages of conidial development in *Leptographium* species. *Transactions of the British Mycological Society*, **90**, 451–6.

Vassiljevski, N.I. & Karakulin, B.P. (1937). *Fungi Imperfecti Parasitici* Pars I Hyphomycetes. Moscow & Leningrad: Academiae Scientiarum URSS.

Vassiljevski, N.I. & Karakulin, B.P. (1950). *Fungi Imperfecti Parasitici* Pars II Melanconiales. Moscow & Leningrad: Academiae Scientiarum URSS.

Vobis, G. & Hawksworth, D.L. (1981). Conidial lichen-forming fungi. In *Biology of Conidial Fungi* 1 (ed. G.T. Cole & B. Kendrick), pp. 245–72. New York: Academic Press.

von Arx, J.A. (1970). On the ontogeny of fungus spores. *Netherlands Journal of Plant Pathology*, **76**, 147–51.

Vuillemin, P. (1910*a*). Matériaux pour une classification rationelle des Fungi imperfecti. *Compte Rendu Hebdomadaire Séances de l'Academie des sciences Paris*, **150**, 882–4.

Vuillemin, P. (1910*b*). Les conidiosporés. *Bulletin de la Société des sciences de Nancy*, **2**, 129–72.

Vuillemin, P. (1911). Les aleuriosporés. *Bulletin de la Société des sciences de Nancy*, **3**, 151–72.

Wakefield, E.M. & Bisby, G.R. (1941). List of hyphomycetes recorded for Britain. *Transactions of the British Mycological Society*, **25**, 49–126.

Watling, R. (1979). The morphology, variation and ecological significance of anamorphs in the Agaricales. In *The Whole Fungus* **2** (ed. B. Kendrick), pp. 453–72. National Museum of Natural Science, National Museums of Canada, Ottawa.

Wingfield, M.J. (1985). Reclassification of *Verticicladiella* based on conidial development. *Transactions of the British Mycological Society*, **85**, 81–93.

Wingfield, M.J., Seifert, K.A. & Webber, J.N. (eds). (1993). Ceratocystis *and* Ophiostoma: *taxonomy, ecology and pathology*. St Paul, Minn: American Phytopathological Society.

Wingfield, M.J., Van Wyk, P.S. & Wingfield, B.D. (1987). Reclassification of *Phialocephala* based on conidial development. *Transactions of the British Mycological Society*, **89**, 509–20.

7

The flagellated fungal spore

MELVIN S. FULLER

Introduction

In 1963, when I made a conscious effort to devote a portion of my research
to the study of the flagellated fungal spore, the best summary of previous
work on the biology of fungal motile cells was an address entitled 'The
Zoospore' given by one of the British Mycological Society's earlier presidents (Waterhouse, 1962). Shortly thereafter, I had an opportunity to meet
and discuss my early ultrastructural work on fungal zoospores with one of
your most famous plant cell biologists, Dr Irene Manton. This pioneering
student of motile cells in algae and green plants had, with her co-workers,
included fungi (Manton *et al.*, 1952) in studies that were responsible for
the major discovery that all eukaryotic flagella shared a 9 + 2 organization.

Before discussing flagellated fungal cells, I will provide two definitions
of fungi that I use routinely, the more restricted for those organisms that
are members of the clade that constitutes the Kingdom Fungi, and the
broader definition for those non-photosynthetic groups of organisms that
have absorptive nutrition, walls surrounding their cells and that exhibit
what I call the 'fungus way of life'. Barr (1992) has suggested that the
broader, polyphyletic group be called a 'Union of Fungi'. Bowman *et al.*
(1992) established beyond doubt that the flagellated *Chytridiomycetes* are
members of the same clade as the *Ascomycetes* and *Basidiomycetes*, a
conclusion suggested by the earlier study of Förster *et al.* (1990). The
Oomycetes and *Hyphochytriomycetes*, however much they may live and
look like fungi, are members of a group (Kingdom *Cromista* of Cavalier-
Smith, 1993) that I will call straminipiles (Liepe *et al.*, 1994). Thus, being
fully aware of the phylogenetic diversity of the groups generally studied by
mycologists, I will discuss the flagellated spores of the zoospore-producing
Chytridiomycetes, Oomycetes, Hyphochytriomycetes, Labyrinthulomycetes

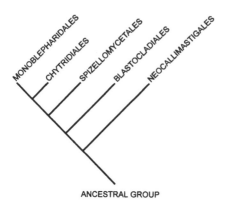

Fig. 7.1. Orders in Kingdom Fungi that have zoospores.

and *Plasmodiophoromycetes* in this paper. The last class of these fungus-like organisms is still studied by mycologists but almost certainly (Barr, 1992) represents a third kingdom that is separate from Fungi and straminipiles. I will, until more molecular data (Liepe *et al.*, 1994) become available, use the class *Labyrinthulomycetes* for the labyrinthulids and thraustochytrids. The labyrinthulids are deeply rooted in the straminipiles (Leipe *et al.*, 1994) and their closeness to the thraustochytrids still needs to be determined. The fungus-like organisms that produce flagellate cells were once put in the Class *Phycomycetes* which is now considered an unnatural group and has been replaced by the five classes discussed above. When this paper capitalizes the word Fungi it refers to organisms in what we know as the Kingdom Fungi whereas fungi and zoosporic fungi are used for all organisms having the fungus way of life. Figures 7.1 and 7.2 represent, respectively, the groups of Fungi and straminipiles treated herein; the *Plasmodiophoromycetes* are not included in either figure.

Flagellated fungus spores (zoospores), except for gametes, are uninucleate reproductive packages or cells whose primary function is as 'homing' or 'site selection agents' (Deacon, 1988; Deacon & Donaldson, 1993). They do not increase their numbers by division during the motile period and their metabolic processes are directed toward producing the energy necessary to arrive at a new substratum. The evidence available (Whittaker, Shattock, & Shaw, 1992; Olson & Fuller, 1971) indicates that fungal zoospores are in the G_1 phase of the cell cycle. Several workers since Waterhouse (1962) have reviewed the fungal zoospore, including Hickman and Ho (1966), Fuller (1977), Lange and Olson (1979), Carlile (1986) and Deacon and Donaldson (1993). I will try herein to treat those aspects of flagellated fungal cells that have been covered more briefly in other reviews

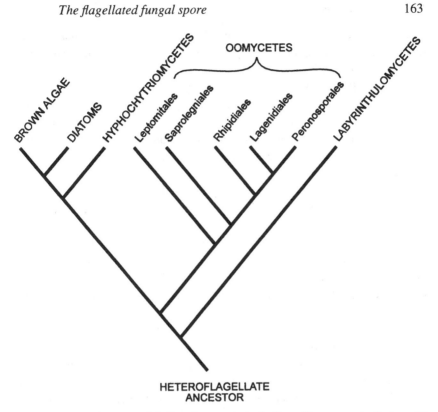

Fig. 7.2. Straminipiles (in part) including Classes and Orders of zoosporic fungi in group.

and lightly treat some aspects I believe have been thoroughly covered elsewhere. Because of their importance as pathogens of commercially significant vascular plants, the depth of data is greater for the *Peronosporales*. Where I think further study of members of the *Chytridiomycetes* and *Plasmodiophoromycetes* might yield important comparative data, it will be indicated. I have restricted myself to the motile cells, even though there is much exciting biology related to zoospore development (Hyde *et al.*, 1991*b*), release (Money & Webster, 1989) and encystment (Deacon & Donaldson, 1993).

Structure of Fungal Zoospores

Very early (Scherffel, 1925) in the modern study of zoosporic fungi, emphasis was placed on the importance of number of flagella and their placement on swimming motile cells. Scherffel (1925) recognized three main lines with respect to zoospores: the chytrid line, the *Monoblepharis* line and the *Saprolegnia-Peronospora* line. Weston (1935) suggested that the chytrid-*Monoblepharis* line was a single evolutionary group and Sparrow (1935) adopted this idea in his paper on the inter-relationships of the *Chytridiales*. Couch (1938) and Vlk (1938) both considered that whether the flagellum was whiplash or tinsel, i.e. type of flagellum, was also important. Interestingly, Couch (1941) stated: '. . . that the aquatic Phycomycetes as constituted at present contain several (at least four) distinct classes and probably are not a monophyletic series'. The groups were: *Chytridiales*, with a single, posteriorly directed whiplash flagellum; what we now call the *Hyphochytriales*, with a single, anteriorly directed tinsel (with tripartite hairs) flagellum; a biflagellate line with spores having one tinsel (with tripartite hairs) and one whiplash flagellum; and the *Plasmodiophorales*, also biflagellate, but with two whiplash flagella of unequal length. It was Sparrow (1935, 1943, 1960, 1973), who firmly established the importance of flagellar number, placement and type in modern treatments of zoosporic fungi. These observations still hold for separation of classes of fungi with motile cells and subsequent structural discussion will use classes and orders as set forth in Figs 7.1 and 7.2.

Koch (1961) summarized his light microscopy observations on the internal structure of the major types of motile cells in the *Chytridiomycetes*. It was becoming clear that electron microscopy observations of shadowed preparations of zoospores including his own (Koch, 1956) and those of others, e.g. Manton *et al.* (1952), on biflagellate cells would not be sufficient to reveal much about the internal structure of zoospores. Turian and Kellenberger (1956), observing motile gametes of *Allomyces*, and Cantino *et al.* (1963) looking at the zoospores of *Blastocladiella emersonii*, would be the first to demonstrate the merits of thin sections for studying fungal motile cells. At a symposium entitled *The Fungus Spore*, held at the University of Bristol, Fuller (1966) treated observations on the ultrastructure of uniflagellate spores and Colhoun (1966) did the same for biflagellate spores. It was already apparent that the ultrastructure of zoospores was variable in the different taxonomic groups of 'fungi' that had motile cells. By 1980 (Barr, 1980), enough was known about the ultrastructure of zoospores in the *Chytridiales* to allow Barr to use these characteristics as the basis for removing a new order, the *Spizellomycetales*, from the *Chytridiales*. Today, the ultrastructure of zoospores is used in parallel with other morphological data and molecular sequences to assess relationships among the zoosporic 'fungi'.

What kinds of organelles are in the package we call a zoospore and how does the arrangement of the organelles vary? In some ways zoospores are like model cells; they contain all of the organelles you would package in a typical cell whose function is to propagate an organism and efficiently move it from one substratum to another. The included organelles are the nucleus, ribosomes, one or more mitochondria, one or more microbodies, lipid, a Golgi or Golgi equivalent, vesicles or vacuoles, inclusions, and the flagellum with its associated structures. While I will attempt to discuss the highlights of zoospore structure, it is impossible to review such a vast literature in the space allotted for this paper. The papers cited will allow one to access details for most organisms that have been studied.

Motile cells of zoosporic fungi are uninucleate with variation being observed in the shape of the nucleus as well as the position of the nucleus relative to other organelles, particularly the kinetosome. Most reports of zoospores with more than one nucleus (e.g. Desjardins, Zentmyer & Reynolds, 1970) are probably the result of sporangia developing under conditions unfavourable for cleavage of the spores. Frequently, the nucleus is closely associated with the kinetosome and it is likely that there are physical connections between the nucleus and the kinetosome. Microtubules (Lingle & Barstow, 1983; Barr & Désaulniers, 1992) that originate near the kinetosome could also be responsible for maintaining close proximity between the nucleus and the kinetosomes in zoospores of some *Spizellomycetales*, the *Blastocladiales* and the *Oomycetes*. The nucleus in the zoospores of the *Hyphochytriomycetes* is also closely associated with the kinetosome (Fuller & Reichle, 1965; Cooney, Barr, & Barstow, 1985) but, since only two microtubules pass from the kinetosome to the nucleus, it seems likely that other structures are responsible for positioning the nucleus. Ribosomes are variously distributed in fungal zoospores. They may be evenly distributed throughout the cytoplasm of the zoospore (all biflagellates and most, if not all, *Spizellomycetales*), arranged in loosely grouped aggregates or helical arrays (the *Neocallimastigales*), arranged in groups of ribosomes that are partially delimited and traversed by endoplasmic reticulum (*Hyphochytriomycetes* and *Monoblepharidales*) or may be grouped and distinctly separated from the rest of the cytoplasm by modified endoplasmic reticulum (most *Chytridiales* and the *Blastocladiales*). The separated ribosomes are most apparent in the *Blastocladiales* where the structure containing them is very distinct and has an appearance that has resulted in its being called a nuclear cap.

Mitochondria, lipids and microbodies are often closely associated with one another in fungal zoospores as was first pointed out by Powell (1978).

This association will be discussed briefly after describing the individual components of what was termed (Powell, 1978) a microbody–lipid globule complex (MLC). With the exceptions of the obligately anaerobic *Neocallimastigales* and some of the wholly fermentative zoospore-producing forms (Lingle & Barstow, 1983; Held *et al.*, 1969) that have vestiges of mitochondria that do not participate in respiration, fungal zoospores have mitochondria with cristae. The cristae of the *Plasmodiophoromycetes* (Miller & Dylewski, 1983) and the three classes that are in the straminipiles are tubular; those zoospore-producing organisms in the Kingdom Fungi have mitochondria with flattened cristae. Most zoospores contain several mitochondria (e.g. Powell & Roychoudhury, 1992) and, although *Blastocladiella emersonii* with a single mitochondrion is the best known exception, one must always examine serial sections since reconstructions may (Beakes, Canter & Jaworski, 1988) demonstrate that what appear to be sections of individual mitochondria are actually profiles of a single mitochondrion. Zoospores, with the possible exception of those in the anaerobic, gut-inhabiting *Neocallimastigales* where lipids are not apparent in electron micrographs (Heath, Bauchop & Skipp, 1983), contain one or more lipid bodies. The order *Chytridiales*, as treated by Sparrow (1960), had many species with a single prominent oil globule. Barr (1980), in extracting the *Spizellomycetales* from the *Chytridiales*, left those species that had a single large lipid body associated with microbodies and/or mitochondria in the *Chytridiales*. Most other groups treated here have zoospores that characteristically contain many lipid bodies. Microbodies are present in the zoospores of all groups and they may be called, based upon the physiological functions that have been associated with them, glyoxosomes, peroxisomes or hydrogenosomes. Associated with the microbodies and lipids in the *Chytridiales* and *Monoblepharidales* is a structure that I (Fuller, 1966), in the best Latin I could summon at the 1966 Colston Symposium in Bristol, termed a rumposome because of its position in the zoospores of *Monoblepharella*. No one has been able to attribute a function to this striking structure that consists of a honeycomb-like array of interconnected tubules; and, thus, it has never been given a more respectable name that would better describe its role in the motile cells where it is found. The rumposome is nearly always found in close association with one or more microbodies and is often appressed to a lipid globule as in, e.g. *Entophlyctis luteolus* (Longcore, 1995).

   Powell (1978) defined four classes of MLCs that paralleled the four orders of *Chytridiomycetes* existing at the time. Each of the four types of MLCs had one or more subtypes. Later, Powell and a co-worker (Powell &

Roychoudhury, 1992) described a fifth type of MLC that also had several subtypes. Longcore (1993) described what is a fifth subtype of the type 5 MLC in *Lacustromyces hiemalis*. It seems certain that more types will be described as the ultrastructure of more fungal zoospores is examined. Powell, Lehnen & Bortnick (1985) cautioned against using the term MLC for microbodies associated with lipid in zoospores of fungal groups other than the *Chytridiomycetes* (Lange & Olson, 1979; Olson *et al.*, 1984). Several workers have described paracrystalline bodies (Lange & Olson, 1979; Taylor & Fuller, 1981; Longcore, 1995) or striated inclusions (Longcore, 1992) in members of the *Chytridiales* and at least one member of the *Blastocladiales* (Lacarotti & Federici, 1984) but no function has been attributed to these inclusions and their systematic utility, if any, remains to be determined.

Some of the many vesicles found in zoospores have been functionally characterized while others await further study. One of the most exciting programmes where the function of vesicles in fungal zoospores has been studied is that of Hardham (1989, 1992) with *Phytophthora cinnamomi*. Using glutaraldehyde and formaldehyde fixed zoospores (the latter fixative often rupturing the cells and exposing internal vesicles and their contents), Hardham (1989) was able to immunize mice and select monoclonal antibodies to surface antigens and to contents of the vesicles of ruptured spores. Using several of these antibodies, Gubler and Hardham (1990, 1991) employed colloidal gold of different sizes and beautifully demonstrated that one class of surface vesicles, located in the dorsal region of zoospores encysting on a root, produced a glycoprotein coat over most of the surface of the cyst while another class known as ventral vesicles was secreted at the site of attachment. These two vesicle types appear to be associated with the adhesion of encysting cells whereas the large peripheral vesicles that others (Sing & Bartnicki-Garcia, 1975*a*; Estrada-Garcia *et al.*, 1989; Estrada-Garcia, Callow & Green, 1990) have implicated as being involved in adhesion in related organisms, remained at the periphery of encysted cells and later migrated away from the cyst surface. These conflicting observations suggest that morphological sorting of vesicle types may lead to confusion, with structures that are functionally different being given the same name in related organisms. A fourth antibody that was generated to cyst walls labelled the peripheral cisternae (Hemmes, 1983; Cho & Fuller, 1989*a*) characteristic of *Phytophthora* zoospores. These cisternae break down during encystment and elements derived from them can be labelled in the new cyst wall. Lehnen and Powell (1989) concluded that vesicles known as K bodies in zoospores of the *Saprolegniales* contained carbohydrate and were involved in the adhesion of encysting zoospores. Structures called U-bodies in *Phytophthora palmivora* were found (Powell & Bracker, 1986) to be a type of microbody. Powell, Lehnen and Bortnick (1985) clarified the terminology of K bodies (= U bodies) and justified their reasons for using K_2 in many but not all (Lehnen & Powell, 1989) subsequent papers. Many *Oomycetes*, especially *Phytophthora* spp., contain vesicles that have been called 'fingerprint' vesicles or vacuoles and that Lunney and Bland (1976) called phospholipid vesicles on the basis of their composition. Kazama (1973) localized acid phosphatase in several vesicle types in the zoospores of *Thraustochytrium* sp.

Fungal zoospores lack cell walls and are surrounded by a cell membrane, or at most a membrane plus a cell coat (Taylor & Fuller, 1981; Dorward & Powell, 1983; Powell, 1994; Longcore, 1995). Powell (1994) discusses instances where extracellular coats also extend around the plasma membrane or flagellar sheath surrounding the flagellum. Interestingly, the cell coat is absent from a region of the sheath opposite an electron dense core in the transition region between the kinetosome and flagellum. *Monoblepharella* zoospores, in addition to a coat (Powell, 1994), also have (Fuller & Reichle, 1968) an extracellular extension of unknown composition at their anterior ends as they swim.

Whether cell coats are present or absent, fungal zoospores have to survive in aqueous media of relatively high water potentials when compared with their cytoplasm. To survive, they need a means of pumping water out, at least as fast as it enters, or risk bursting due to osmotic stress. The vesicles concerned with removal of water from zoospores have been called contractile vacuoles (Hoch & Mitchell, 1972) but I am not aware that anyone has found associated cytoskeletal elements that might mediate contraction of the vacuoles. Several workers (e.g. Crump & Branton, 1966; Hoch & Mitchell, 1972; Grove & Bracker, 1978; Cho & Fuller, 1989b) have shown that these vesicles open and close at 4–6 second intervals. Until contractile, cytoskeletal elements are shown to be associated with closing of the vesicles, I prefer the name water exclusion vacuole (WEV; Grove & Bracker, 1978). The WEV consists of several vesicles that are larger than any of the other vesicles in zoospores and usually have rough or coated, cup-shaped vesicles associated with them (Hoch & Mitchell, 1972; Grove & Bracker, 1978; Cho & Fuller, 1989b). The most complete study of the WEV is that of Cho and Fuller (1989b) who found the large, coated and cup-shaped vesicles as well as what structurally appeared to be clathrin-coated vesicles in the area of the water exclusion apparatus of

Phytophthora palmivora. Shields (pers. comm.) has used an antibody to clathrin and has been able to localize the molecule in the region of the WEV and in areas of the plasma membrane of *Monoblepharella* spp. zoospores. Most of what has been published about WEVs comes from the study of oomycetous zoospores. Powell (1981), who described vacuoles that pulsated at 10-second intervals in the zoospores of the chytridiomycete *Caulochytrium protostelioides*, has the most convincing published evidence for WEVs in that group. The pulsating vacuoles, when examined ultrastructurally, had large bristle-coated vesicles associated with them. Much structural work remains to be done on the vacuoles associated with water exclusion, especially in those groups of zoosporic fungi that are not in the *Oomycetes*.

Most fungal zoospores that have been studied extensively with the electron microscope have a Golgi (Taylor & Fuller, 1981) or Golgi equivalent (Feeney & Treimer, 1979) and, where the typical cisternae with vesicles are not observable, one should look for Golgi marker enzymes such as thiamine pyrophosphatase and inosine 5′-diphosphatase. Comparing the Golgi observed in the swimming zoospore with those in, e.g. sporangia developing zoospores (Taylor & Fuller, 1981), suggests that the Golgi apparatus is likely to be inactive during the motile period.

The flagellum of the zoospore and structures that are related to it (see Fig. 7.3) are difficult to summarize and I treat them last. Associated with the flagellum, or flagella, are the kinetosomes (basal bodies), flagellar roots and the transition zone that lies between the flagellum and kinetosome. The number and types of flagella, i.e. tinsel and/or whiplash, and their placement in the groups of zoosporic fungi were detailed earlier. All fungal flagella have the familiar 9 + 2 appearance in cross section and detailed structural and functional asepcts of flagella of eukaryotes are treated elsewhere (Alberts *et al.*, 1994). Tinsel flagella have axonemes that bear tripartite hairs that presumably will be found to be associated with the peripheral axonemal microtubules as in the case of the algal straminipile, *Ochromonas danica* (Markey & Bouck, 1977). These hairs, in addition to being on the flagellum, have been seen in intracellular packets in zoospores of *Rhizidiomyces* (Fuller & Reichle, 1965), in zoospores of the *Saprolegniales* (Heath, Greenwood, & Griffiths, 1970) and in hyphae and zoospores of *Phytophthora cinnamomi* (Cope & Hardham, 1994). The study of Cope and Hardham (1994), using three monoclonal antibodies to flagellar antigens, identifies domains on flagella that are not present on the membrane around the body of the zoospore. Their work also indicates that our understanding of tripartite hairs in zoosporic fungi is likely to

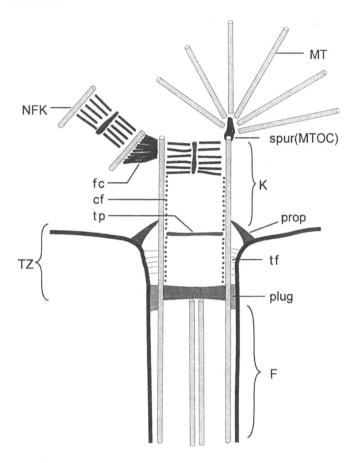

Fig. 7.3 Representative flagellum to illustrate associated structures discussed in text. **K**, kinetosome; **NFK**, non-flagellate kinetosome; **TZ**, transitional zone; **F**, flagellum; **MT**, microtubule; **fc**, fibrous connection; **cf**, concentric fibre; **tp**, transitional plate; **tf**, transitional fibres.

increase; one of the antibodies was to the mastigonemes and allowed immunolocalization of the epitope in hyphae as well as zoospores. With this antibody, it should be possible in future studies to examine the early development of mastigonemes and to determine whether their addition to and organization on the flagellum is similar to that in related algal groups.

Kinetosomes share a structure characteristic of centrioles and, indeed, they are transformed centrioles with sections through their basal regions having the 'cartwheel' structure typical of centrioles (Alberts *et al.*, 1994), i.e. there are nine sets of triplet microtubules. Kinetosomes are often longer than the centrioles of vegetative cells and little work has been done on changes that occur prior to formation of axonemal microtubules or changes that are associated with kinetosomes reverting to centrioles. Olson and Fuller (1968) examined sections through the kinetosome of several *Phlyctochytrium* spp. (now in *Spizellomyces*) and documented the transition from nine sets of three microtubules at the base to nine sets of two microtubules at the end where the flagellum forms. Centrioles generally occur in pairs in vegetative cells and this is true, with very few exceptions (Lucarotti & Federici, 1984; Martin, 1971), of kinetosomes in motile cells. One needs to serial section several zoospores before concluding that a second kinetosome or centriole is not associated with the flagellum-bearing centriole. Fuller and Reichle (1968) concluded that the second centriole was absent in *Monoblepharella* but Reichle (1972) looked at many more sections and found that it was present.

This chapter adopts the terminology recommended by Andersen *et al.* (1991) for the flagellar/ciliary apparatus in protists. In a uniflagellate group such as the *Hyphochytriomycetes*, what we know about evolutionary history and relationships suggests that the nonflagellate centriole should be called a 'barren' kinetosome because it might have given rise to a flagellum in ancestral organisms and does in other straminipiles. In the case of the *Chytridiomycetes* where we do not know the closely related groups among flagellate or centric organisms, or whether the second centriole has ever been a kinetosome or can become one in subsequent generations, the recommendation (Andersen *et al.*, 1991) is that the second centriole be called a 'nonflagellate' kinetosome. If the choanoflagellates (Mollicone & Longcore, 1994) turn out to be the most closely related group of centric organisms, it would be appropriate to use barren kinetosome for the *Chytridiomycetes*. The angle with respect to one another of the two kinetosomes in fungal motile cells is variable. They may be parallel, as in most members of the *Chytridiales* and *Monoblepharidales*, or, as in the *Spizellomycetales* and the rest of the zoospore-producing groups, angled with respect to one another but at an angle of less than 180°. The relationship of the kinetosomes to one another in representatives of *Oomycetes*, *Labyrinthulomycetes* and *Hyphochytriomycetes* is clearly documented and discussed by Barr and Allan (1985). There may be concentric fibres (Fig. 7.3) in the cores of the kinetosomes of *Chytridiomycetes* and *Oomycetes* (Barr, pers. comm.). The concentric fibres are generally more conspicuous in the transition zone between the kinetosome and flagellum where they appear as broken circles.

Renaud and Smith (1964) made the first and some of the most complete observations on flagellum development and the cleavage of zoospores in the *Chytridiomycetes.* Hyde and co-workers (Hyde, Gubler & Hardham, 1991*a*; Hyde *et al.*, 1991*b*) have made extensive observations on zoo-sporogenesis in freeze-substituted sporangia of *Phytophthora* spp. In all cases where the process of flagellum formation has been studied exten-sively, a transitional (terminal) plate (Fig. 7.3) usually forms at the flagel-lum end of the kinetosome and flagellum growth is initiated by extension of the outer nine pairs of microtubules into what has been called the primary cleavage vesicle. The actual mechanism of cleavage and later stages of flagellum delimitation are still being studied (Hyde *et al.*, 1991*b*). Because there are no studies of freeze substituted, cleaving sporangia of *Chytridiomycetes* to date, one only has chemically fixed material to compare with the observations of Hyde *et al.* (1991*b*).

The transition zone of the flagellar apparatus is the region that begins where the shift from triplet to doublet organization occurs in the kineto-some and ends at the point where the central pair of microtubules begins in the flagellum. The transition zone may include transitional fibres (Fig. 7.3) that are thin electron opaque strands running from points between the A- and B-tubules to the plasma and/or flagellum membrane. Concentric fibres, as mentioned in the discussion of kinetosomes, may also be present in the transition zone. Members of the *Chytridiomycetes*, except for the *Neocallimastigales*, have props (Olson & Fuller, 1968) that are fibrillar extensions of the C-tubules that extend to the plasma membrane and ter-minate near the point where the flagellum begins (Fig. 7.3). The hypothesis that these props may function in maintaining stability at the juncture between the flagellum sheath and the zoospore plasma membrane is sup-ported by the observations in *Olpidium* spp. of props (Lange & Olson, 1978) or partial props (Barr & Hadland-Hartman, 1977) at this juncture in a fungus in which more than half of the flagellar axoneme is inside the zoospore. Many of the zoospores of the *Chytridiales* have an electron opaque 'plug' (Fig. 7.3) just before the point where the two central micro-tubules begin. This opaque material assumes various shapes and may be inside as well as outside the ring of nine doublets that compose the flagel-lum at this point. Some members of the straminipiles have electron opaque rings in the transition zone and these rings have been described (Barr & Allan, 1985) as 'concertina like' for the *Oomycetes* and as 'bell-shaped' in *Thraustochytrium*, a representative of the *Labyrinthulomycetes*. The transition zones of the flagella in the *Plasmodiophoromycetes* (Barr & Allan, 1982) are very different from those of the biflagellate straminipiles and much simpler than those of any of the other fungal groups, again sup-porting the hypothesis that these organisms belong in a third group.

Flagellar roots and connecting fibres are the last category of structures related to flagella and kinetosomes that will be discussed. Flagellar roots may be fibrous, microtubular or amorphous. Barr and Désaulniers (1988) have, through the use of serial sections beginning in the flagellum and going through the kinetosome, numbered the tubules of the kinetosome to correspond with the numbering of tubules in the flagellum, thus allowing for precise comparisons of locations for kinetosome-associated roots and fibres. This numbering will be useful in future systematic comparisons of these structures in zoosporic fungi and was used by Longcore (1995) in a recent study of zoospores of *E. luteolus*. Fibrous roots and connecting fibrils are often striated and are sometimes referred to as rhizoplasts. In most cases there is a fibrous connection (Fig. 7.3) between the kinetosome and the nonflagellate kinetosome or between the two flagellum-bearing kinetosomes in biflagellate spores. The connection between the two kinetosomes may be more fibrillar with striations and have an overall fan shape as in *Phytophthora infestans* (Barr & Désaulniers, 1990). Electron opaque areas that may be microtubule organizing centres are often associated with the kinetosomes. These areas can be electron-opaque bars (Barr & Désaulniers, 1988), plate-like bodies (Barr & Désaulniers, 1988) and electron opaque areas of various shapes and complexity called spurs (Barr, 1981). Some members of the *Spizellomycetales* have extensive, striated rhizoplasts (Barr, 1981) that extend from the kinetosome base to the nucleus. Zoospores of all members of the *Monoblepharidales* that have been examined (see Mollicone & Longcore, 1994 for most recent discussion) have a disc-shaped rhizoplast that is striated and from which microtubules radiate (Fuller & Reichle, 1968). Frequently, rhizoplasts and amorphous regions, attached to or close to kinetosomes, have microtubules associated with them (Fig. 7.3) and, although they have been referred to as microtubule organizing centres (Barr, 1981), I am unaware of any experimental evidence from zoosporic fungi that these structures can serve as nucleation sites for microtubules in an *in vitro* system. Although several fungal zoospores have been described as lacking microtubules, most zoospores have one or more sets of microtubules associated with the kinetosomes or the kinetosome-associated structures. Most of the *Chytridiomycetes* have a single set of microtubules originating at the kinetosome or from kinetosome-associated structures and radiating forward or laterally in the zoospore. In zoospores that have a rumposome, the microtubules most often travel from the kinetosome region to the rumposome. Barr and Désaulniers (1988) believe that this, among other things, suggests that the rumposome's function is related to the control of zoospore movement.

The microtubules in the zoospores of the *Blastocladiales* (Fuller & Calhoun, 1968) are highly organized and, with few exceptions, characteristically occur in nine sets of three microtubules each, precisely the same number of microtubules as in the kinetosome base. This striking correspondence suggests that the same control is operating on cellular as on kinetosomal microtubules. Microtubule arrays become much more numerous and complex in the groups that are allied with the straminipiles. While there are some different conclusions as to the numbers of arrays of microtubules (Barr, 1992; Hardham, 1992) that relate to the kinetosome regions, there are multiple groups with varying numbers of microtubules. One should consult the many cited papers by Barr and his co-workers for detailed information on microtubular rootlets in the straminipiles.

Early studies on the comparative ultrastructure of the fungal motile cells (Fuller, 1966) indicated that ultrastructural characteristics would become as useful to systematists as flagellar type and number had been earlier to Sparrow (1960) and others in assessing relationships among zoosporic fungi. The distinct 'nuclear cap' of the *Blastocladiales*, already apparent in 1966, is not seen in any other group of zoospore-producing organisms and, when encountered, makes it possible to assign these fungi to the appropriate order. In 1966, we certainly did not appreciate how important ultrastructural characteristics would become as a result of work by Barr, Olson and Lange, Powell, Longcore and many others cited in this paper. Before we knew how to amplify DNA and compare sequences, the ultrastructural characteristics of the flagella and associated organelles in biflagellate fungal spores allowed Barr (1981) and others to present compelling evidence for the *Oomycetes* and related groups being aligned with the heterokont algae as championed earlier by Bessey (1942), among others. Powell (1978), in recognizing the ubiquity of the microbody-lipid globule complex in zoosporic fungi, and Barr (1980), in separating the *Spizellomycetales* from the *Chytridiales* on the basis of zoospore ultrastructure, led in the modern use of ultrastructural characteristics in the systematics of zoosporic fungi. Today, zoospore characteristics are used at all levels in the systematics of the fungi that have them.

Recently, Li, Heath and Packer (1993) used structural characteristics, mostly of zoospores, to do a cladistic analysis of the *Chytridiomycetes* and Longcore (1995) has an excellent discussion of the application of ultrastructural data to the systematics of the *Chytridiomycetes*. Barr (1992) also summarizes the importance of zoospore ultrastructure in understanding the evolutionary relationships of fungi. Some morphological characteristics of zoospores such as the transitional zone and the presence of tripartite hairs are conserved (Barr & Désaulniers, 1992) and will be useful at the kingdom and class levels. Other characteristics such as flagellar roots and the many structures related to kinetosomes are highly variable and will be useful at the family, the genus and even the species levels (Barr & Désaulniers, 1992). I look forward to watching the development of a new systematics for zoosporic fungi that is based upon ultrastructural and molecular sequence data.

Biology of fungal zoospores

Zoospore composition

As indicated by Waterhouse (1962), and as clearly demonstrated throughout the edited work by Fuller and Jaworski (1987), large quantities of zoospores are readily obtained from most zoosporic fungi. Thus, studies of zoospore composition could easily be made. Nevertheless, quantitative studies of zoospore composition are few in number with the most complete data having been obtained by Bimpong (1975) for *Phytophthora palmivora* and by Suberkropp & Cantino (1972, 1973) for *Blastocladiella emersonii*. Other interesting but less complete studies of molecules that have been found in zoospores will be noted.

For *B. emersonii* (Suberkropp & Cantino, 1973), the average dry weight of freshly harvested spores was 47.0 pg with a partial breakdown as follows: 54–57% protein, 18–23% nucleic acid, 7–9% lipid and 3–5% carbohydrate. Bimpong's (1975) figures for *P. palmivora* can be converted to percentages and would be as follows: 43.07±14.49% protein, nucleic acids not measured, 24.82±5% lipid and 11.8±1.7% carbohydrate. Lovett (1963) examined preparations of *B. emersonii* nuclear caps and found them to consist of 60% protein and 40% ribonucleic acid; the cap represented 18% of the dry weight of the zoospore and contained 69% of the total RNA in the spores. Gubler *et al.* (1990) localized calmodulin in both flagella of *Phytophthora cinnamomi* with there being a calmodulin-rich region at the base of the tinsel flagellum. While Gubler *et al.* (1990) suggested that calmodulin might be involved in mediating calcium-regulated motility, they did not do experiments to show this. Li and Heath (1994) and others have demonstrated the presence of actin in fungal zoospores.

The presence of carbohydrates and what we now refer to as glycoconjugates on the surfaces of fungal zoospores was pointed out early by Sing and Bartnicki-Garcia (1975*b*) in their studies of *P. palmivora*. The laboratories of Clarke (Hardham, Ralton & Clarke, 1982) and Hardham (see

Hardham, 1992, for review) have looked extensively at surface glycoconjugates in *P. cinnamomi* and Hardham's group continues to study the function of these zoospore surface molecules along with other spore surface antigens. Binding of concanavalin A to glycoproteins containing α-mannans on the plasma membranes or surfaces of fungal zoospores often results in lysis of the spores (Sing & Bartnicki Garcia, 1975b; Jen & Haug, 1979). Hopefully, we will soon learn much more about the structure and function of molecules on the surfaces of zoospores. Peroxidase has also been associated with the surfaces of fungal zoospores (Powell & Bracker, 1986; Berbee, & Kerwin, 1993).

Zoospore metabolism

Barash, Klisiewicz and Kosuge (1965) indicated that swimming spores of *Phytophthora drechsleri* can metabolize a variety of sugars, amino acids, organic acids and fatty acids at very low rates. However, these compounds were metabolized more rapidly when zoospores had been swimming for several hours or after encystment occurred, suggesting that the zoospores were not dependent upon an exogenous energy source. More recent studies (Penington *et al.*, 1989; Suberkropp & Cantino, 1973) suggest, in contrast, that zoospores are unable to take up and metabolize sugars, amino acids and other organic compounds. Certainly, metabolism in the fungal zoospore is greatly reduced, controlled and primarily directed toward producing energy for flagellar movement and the maintenance of osmotic balance with the external medium. The molecular work, summarized in Lovett (1975) indicates that syntheses of macromolecules are mostly turned off in swimming zoospores of *B. emersonii*. A more recent study of *B. emersonii* (Jaworski & Wilson, 1989) indicates that the messenger RNAs for protein synthesis that must occur before germ tube formation from encysted spores can take place are present in the zoospores but probably do not function in the spores because inhibitory proteins associate with the mRNAs and prevent their translation. Suberkropp and Cantino (1973) looked at the composition of *B. emersonii* spores after they had been swimming for five hours and concluded that they were metabolizing, in order of importance, protein, lipid and polysaccharide; no evidence was obtained for nucleic acid metabolism by swimming spores. Using a Warburg respirometer, the respiratory quotient for the swimming spores of *B. emersonii* was determined to be 0.92 which Suberkropp and Cantino (1973) interpreted as supporting their conclusion 'that roughly equal quantities of lipid and polysaccharide plus about triple that amount of protein were being metabolized . . .'. Suberkropp and Cantino (1973) also presented ultrastructual evidence that lipids were being broken down during the five-hour period of motility. In *P. palmivora*, Bimpong (1975) concluded that the major source of energy was lipids with lesser amounts of protein and carbohydrate being metabolized. Lipids with ~9 kcal g^{-1} are a more efficient way of storing energy than, e.g. mycolaminarin with ~4 kcal g^{-1}. Bimpong (1975) found evidence for most of the Krebs cycle enzymes and two glyoxylate cycle enzymes, malate synthase and isocitrate lyase, in zoospores. She concluded that the glyoxylate cycle was probably very active in zoospores by providing metabolites for the Krebs cycle. Mills and Cantino (1975) presented biochemical evidence that glyoxylate cycle enzymes are in the microbodies of *B. emersonii*. Several workers, e.g. Powell (1976), have presented cytochemical evidence that catalase and glyoxylate cycle enzymes are in the microbodies of fungal zoospores. Zoospores of the *Neocallimastigales* require an anaerobic environment for survival and must generate the energy for movement and cell maintenance during the motile period by using anaerobic metabolism associated, in part, with hydrogenosomes (Li & Heath, 1993). Similarly, zoospores of other zoosporic fungi that are indifferent to oxygen (Held *et al.*, 1969; Robertson & Emerson, 1982) must, although I know of no studies of the metabolism of these zoospores, derive their energy from fermentative metabolism.

Swimming patterns

Couch (1941), using dark field microscopy and an improvised stroboscopic light made some of the first and most complete observations on the movement of zoospores and the role of the flagella in their movement. He observed flagellation and swimming patterns of representative posteriorly uniflagellate (whiplash type) spores, anteriorly uniflagellate (tinsel type) spores, anteriorly biflagellate (one flagellum of each type) oomycete spores and laterally biflagellate oomycete spores. Posteriorly uniflagellate spores may swim in a wide circular or straight path without rotating on their axes. They may also rotate on their axes and swim in a straight line or along a spiral path. Waves that move the spore forward are generated at the spore and travel to the flagellum tip. Couch (1941) sometimes observed the image of the flagellum as a straight line while at other times he saw two broad arcs that appeared to be anchored at the ends. The image observed was a function of the plane in which the waves were being propagated relative to the observer and could vary, particularly if the zoospore was rotating as it moved across the field. Many later workers have made similar observations on rotating spores as they traversed a helical path. Chytrid zoospores were often described (Sparrow, 1960) as hopping and darting about and having periods where they displayed short, convulsive springs. However, Couch (1941) felt this type of movement had been over-emphasized in describing the movement of chytrid zoospores. Miles and Holwill (1969), in one of the early studies of movement in fungal zoo-spores that employed high-speed cine-photographic techniques, studied turning movements by the zoospores of *B. emersonii*. They concluded that turning was the result of asymmetric flagellar movement relative to the orientation of the spore.

Although they dealt mainly with gametes and zygotes, the analyses of motility in *Allomyces* done by Pommerville (1978) provided an excellent means for studying the patterns of zoospore movement under different conditions. Pommerville used dark-field illumination in a manner very similar to that used earlier by Ho and Hickman (1967), with time expo-sures of known but varying lengths that allowed him to record the path a motile spore traversed and to determine the speed of swimming under different conditions. Pommerville (1978) was able to show that male gametes, female gametes and zygotes swam at respective rates of 100, 60 and 140 μm s⁻¹. Both male and female gametes had smooth swimming runs along small arcs with the smooth runs being interrupted by short, jerking motions and changes of direction. Female gametes showed many more jerking movements and, as a consequence, travelled shorter dis-tances. When male gametes were responding to a female-produced hormone, sirenin, they travelled longer distances in broader arcs with fewer changes in direction of swimming; those changes that did occur facilitated homing in on the attractant. Biflagellate zygotes of *Allomyces* exhibited a smooth swimming pattern along a helical path and their move-ment was interrupted by fewer changes in direction.

Couch (1941) found that the anteriorly uniflagellate, tinsel-bearing zoo-spores of *Rhizidiomyces* swam with great rapidity in wide spiral paths that led straight across the field of view. The swimming spores rotated on their axes as they travelled across an area in a broad spiral. Although Couch (1941) felt that the waves that propelled the zoospores of *Rhizidiomyces* forward originated at the flagellum tip, it seems more likely (Fawcett, 1961) that they originated at the body of the spore. Jahn, Landman and Fonseca (1964) offer a reasonable explanation as to how a flagellum bearing mastigonemes could, with a sine wave generated at its base, pull a spore forward in a liquid medium. Holwill (1982), based on observations of biflagellate *P. palmivora* zoospores, also felt that the forward movement was mediated by the mastigonemes on the anterior flagellum. Much still needs to be learned about the forward movement of anteriorly uniflagellate zoospores and our knowledge could be greatly increased if mutants with no, or reduced numbers of, mastigonemes were available.

Couch's (1941) observations on primary and secondary zoospores in the *Saprolegniales* led him to conclude that primary, anteriorly biflagellate spores were very poor swimmers. The two flagella on primary spores were attached at the anterior end with the tinsel flagellum directed forward and the whiplash to the side or backward. Couch concluded that both flagella contributed to the movement of primary spores and that the anterior flagellum was the more active. After short periods of swimming slowly and awkwardly in broad circles with rotating and frequent turns, the primary zoospores encysted to re-emerge as secondary spores at a later time. Secondary-type zoospores in the *Oomycetes* are laterally biflagellate but still swim with the tinsel flagellum directed forward and the whiplash to the back. Although Couch (1941) concluded that both flagella were contributing equally to the movement of secondary type zoospores, most authors before and since the work of Couch, as summarized in Carlile (1983), have concluded that the anterior flagellum is more important in spore movement. According to Carlile (1983) the tinsel flagellum is more responsible for the forward movement of the *Phytophthora* zoospore and provides 10 times the thrust of the whiplash flagellum which acts like a rudder and is responsible for directional movement of the zoospore. The secondary type zoospore rotates on its axis and swims (Allen & Newhook, 1973) in a helical path. The same authors (Allen & Newhook, 1974) subsequently determined that ethanol suppresses the spontaneous turning activity of zoospores of *Phytophthora cinnamomi*. More recent observations using tracings of video recordings (Thomas & Butler, 1989; Donaldson & Deacon, 1993) of swimming secondary type spores of *Oomycetes* have allowed, not only for more precise recording of the spiral path and its amplitude, but also for determining the effects of ions and various drugs on flagellum movements and the paths travelled by swimming spores. Thomas and Butler (1989) showed that the paths travelled by zoospores of *Achlya heterosexualis* were greatly influenced by ions, e.g. 10^{-3} M KCl caused the spores to move more slowly and to change direction frequently, resulting in their swimming in circles whereas 1 mM Ca^{2+} caused the spore to swim in almost a straight line. Donaldson & Deacon (1993) were able to maintain zoospores of *Pythium aphanidermatum* in five (normal plus four perturbed) swimming modes by varying Ca^{2+} ion concentrations using a chelator (EGTA), calmodulin antagonists (dibucaine, trifluoperazine), a Ca^{2+} ionophore (A23187), a membrane ion flux inhibitor (amiloride) and Ca^{2+} channel-blockers (La^{3+}, verpamil). The calmodulin antagonists significantly reduced the speed of swimming, suggesting that they were affecting the tinsel flagellum that has more calmodulin (Gubler et al., 1990) and the other antagonists more likely affected the rudder function of the whiplash flagellum. The paper by Donaldson and Deacon (1993) should be consulted for some of the most complete data on directional changes, linear speed and velocity; velocity takes into account the amplitude of the helical swimming pattern.

Merz (1992) cites unpublished work of Diriwaechter on the primary zoospores of Spongospora subterranea and describes their motion as consisting of slow and steady movements. Barr and Allan (1982) briefly described the movement of secondary zoospores of Polymyxa graminis where both flagella tended to be directed posteriorly, even though the shorter emerges from the forward end of the swimming spore. Using light microscopy and video equipment, Merz (1992) observed the movement of secondary zoospores of the plasmodiophorid S. subterranea in which the kinetosomes are arranged at approximately 180° to one another. Freshly emerged zoospores showed characteristic swimming patterns, moving in straight lines followed by sudden changes to circular movements before returning to straight-line movement. The longer of the two flagella was the posteriorly directed one and it generally trailed behind the swimming spores; it was used as a whip during times of circular movements.

Zoospores are extremely sensitive to environmental conditions and can quickly encyst when they encounter an unfavourable environment. Periods of motility can be as short as a few minutes or as long as two or three days (Hickman & Ho, 1966). Carlile (1986) estimates the average swimming time for most zoospores to be about 10 hours and calculates that, if the average spore swims at a speed of 160 μm s^{-1}, a zoospore might swim 576 mm in one hour. Hickman and Ho (1966) give other speeds for swimming zoospores and discuss the effects of environmental parameters on swimming. The ionic concentration of the medium in which zoospores are swimming is extremely important and has been studied extensively by Bruce Grant and his students. Griffith, Iser and Grant (1988) found, as have others (Thomas & Butler, 1989; Donaldson & Deacon, 1993), that Ca^{2+} was extremely important to zoospore motility. Between 0.2 and 1 μM free Ca^{2+}, swimming of zoospores could be maintained for several hours. When the concentration of free Ca^{2+} fell below 0.1 μM, zoospores slowly lysed. Teakle and Gold (1964) reported that 0.05–0.1 M solutions of glycine, L-proline, L-histidine, D-isoleucine and L-isoleucine enhanced the motility of zoospores of the chytrid *Olpidium brassicae* as did 0.5–5% rabbit serum, 0.2–2% bovine serum albumin and 0.01 M phosphate buffer. *Lagenidium giganteum* zoospores remained motile longer (Lord & Roberts, 1985) when placed in solutions of 0.2 and 0.05 g/litre Bacto-peptone. Several authors in Fuller and Jaworski (1987) give fomulations for media that they indicate are best for maintenance of motility of zoospores in the fungi treated in the individual contributions.

As was first pointed out by Weston (1919) with *Dictyuchus*, many zoospores, especially in the *Saprolegniaceae* of the *Oomycetes*, have the capacity to encyst and swim repeatedly. Subsequently, Salvin (1940) demonstrated five successive swimming periods as a result of repeated emergence of secondary zoospores in a species of *Achlya*. Cerenius and Söderhäll (1984) demonstrated repeated zoospore emergence from cysts of the crayfish parasite *Aphanomyces astaci*. Isolated cysts of *A. astaci*, produced by vortexing or adding growth medium, could be washed and resuspended in 1 mM Ca^{2+} where three consecutive generations of motile secondary zoospores could be achieved. If 50 mM Ca^{2+} was added to cysts within 15 min of encystment, the cysts always germinated to form a vegetative mycelium. More recently (Dieguez-Uribeondo, Cerenius & Söderhäll, 1994) a new record of six periods of motility was demonstrated in *Saprolegnia parasitica* when cysts were washed repeatedly in sterile lake water.

Several treatments are known to immobilize fungal zoospores. Érsek, Hölker & Höfer (1991) found that 100 mM lithium (Li^+) caused an irreversible but non-lethal immobilization of the zoospores of *Phytophthora infestans*. The spores retained their flagella and pyriform shape but had lost the ability to encyst. They concluded that the Li^+ affected the cellular levels of Ca^{2+} and K^+ required for encystment and the paralysed spores only encysted when supplemental amounts of these two essential cations were added to the external medium. Dill and Fuller (1971) reported on the effects of a 0.2 M mixture of L-leucine and L-lysine on the motile cells of representatives of the *Chytridiomycetes* and *Hyphochytriomycetes* as well as two algae and one ciliate. The treatment, still unexplained, with the two amino acids (Dill & Fuller, 1971) resulted in an immobilization of the flagella of all the organisms tested. Reversibility of the immobilization was not tested. The authors (Dill & Fuller, 1971) do report that washing the immobilized spores and resuspending them in growth medium resulted in synchronous encystment and germination. Olson and Fuller (1971) refined the methods of Dill and Fuller (1971) to produce synchronously growing germlings of *Allomyces neomoniliformis*.

Physical confinement or space constraints can cause premature encystment of zoospores of *Phytophthora* spp. (Benjamin & Newhook, 1982). Artificial systems with glass beads that simulated the spaces between soil particles, decreased the motile period of zoospores with some species being better able to remain motile in the liquid between the beads. The behaviour of zoospores, especially those of *Phytophthora* spp., in soils has been discussed in detail by Duniway (1979) and Duniway and Gordon (1986). The concentration of zoospores in a suspension can also affect their behaviour and motility (Thomas & Peterson, 1990).

Handling water

How the zoospores of these fungi handle life in a medium with high water potential is a most intriguing problem and must be approached with a diversity of methodologies. Almost certainly, water is constantly moving from the external medium into the zoospores and, if the spores are not to burst, a considerable amount of membrane has to be recycled when the water is excluded by collapse or contraction of water-containing vacuoles. Cho and Fuller (1989*b*) presented a model of how they believed membranes and associated vesicles were being recycled in *P. palmivora* but were unable to internalize any label in the coated vesicles that were being produced at the cell surface or at the collapsing water exclusion vacuole. Cerenius, Rennie and Fowke (1988) succeeded in getting cationized ferritin into coated vesicles of zoospores of *Aphanomyces euteiches* but did not follow the fate of the ferritin to the extent necessary to relate it to the model of Cho and Fuller (1989*b*). Hardham's laboratory (Hardham, Suzaki & Perkin, 1985) has a monoclonal antibody to the WEV in *P. cinnamomi* that could be useful in determining movement of membranes. The small size of the zoospores prevented Cho and Fuller (1989*b*) from localizing actin around the WEV but improving methodologies will mean future students of fungal zoospores can determine whether exclusion of water involves contractile elements of the cytoskeleton.

Tactic movements

Tactic movements of flagellate fungal cells have been reviewed recently by Deacon & Donaldson (1993) and I believe the reader would be best served by consulting their excellent review that, although it emphasizes observations on species of *Phytophthora* and *Pythium*, does not neglect other groups of zoosporic fungi. The reader might also wish to consult the papers of Muehlstein and Amon (1987a,b) and a recent study of electro-taxis by Morris and Gow (1993). Fungal zoospores respond tactically to light, chemicals, electrical fields, ionic fields, gravity, streams in liquids and to one another. Progress is being made in understanding the mechanisms of tactic movements by fungal zoospores but I consider identification of receptors (see Deacon & Donaldson, 1993 for a discussion) as one of the major challenges that lies ahead with respect to the biology of fungal motile cells. Cahill and Hardham (1994) have, by taking advantage of chemotaxis and electrotaxis of *Phytophthora cinnamomi* zoospores and by using species specific monoclonal antibodies, developed an immunoassay specific for propagules of *P. cinnamomi* in the field.

Virus transmission

Zoosporic fungi that are parasitic on roots of annual and perennial vascu-lar plants can transmit plant viruses (Hiruki & Teakle, 1987; Adams, 1991) from plant to plant by carrying the virus on the zoospore when it moves to a new host root. *Olpidium* spp. were the first zoosporic fungi to be impli-cated in the transmission of vascular plant viruses (Grogan *et al.*, 1958; Teakle, 1962). Viruses that have been associated with *Olpidium* spp. include cucumber necrosis virus (CNV) (McLean *et al.*, 1994), tobacco necrosis virus (TNV) (Teakle, 1962), tobacco stunt virus (Hiruki & Teakle, 1987) and lettuce big-vein virus (Grogan *et al.*, 1958; Tomlinson & Garret, 1964). Another zoosporic fungus, *Polymyxa graminis* (a member of the *Plasmodiophoromycetes*), is involved in the transmission of barley yellow mosaic virus (Adams, Swaby, & Jones, 1988) and of wheat spindle streak mosaic virus (Slykhuis & Barr, 1978). *Spongospora subterranea*, also a plasmodiophorid, transmits potato mop-top virus (Merz, 1992). It seems likely that other root-inhabiting, zoosporic fungi will be implicated as vectors of plant viruses. The best available evidence (Temmink, Campbell & Smith, 1970; Stobbs, Cross & Monocha, 1982) indicates that these viruses attach or bind to the plasma membranes surrounding the flagella and zoospores when they are released from sporangia and before they leave the host cells. The virus particles are internalized when the zoospores retract their flagella and encyst. Stobbs *et al.* (1982) also demonstrated endocytosis of virus particles and obtained evidence that transmission of the virus required an intact protein coat. It is not clear how the virus gets from the fungal cytoplasm to the cells of the vascular plant, although Stobbs *et al.* (1982) suggest that some detachment of viral particles could occur during zoospore encystment. Some virus particles are known

(Hiruki & Teakle, 1987) to persist for extended periods in resting structures of *Olpidium* and *Spongospora*. Recent studies (McLean *et al.*, 1994) also indicate that virus coat protein is involved in the specificity of virus transmission by *O. bornovanus*. When the virus coat protein gene of CNV was replaced with a tomato bushy stunt (a virus not transmitted by *O. bornovanus*) protein coat gene, the virus was no longer transmitted by the fungus. Zoospores that can transmit viral diseases of roots also have a potential to serve as vectors of foreign genes in experiments directed toward transforming the host plant. Zhang *et al.* (1994) used a plasmid carrying the chloramphenicol acetyl transferase (CAT) gene as a reporter in conjunction with dissociated capsid protein of TNV to demonstrate transient expression of the CAT gene in wheat roots. Some unrelated work (Bryngelsson *et al.*, 1988) with *Plasmodiophora brassicae* indicates that this 'fungal' parasite is capable of taking host DNA into its cells although no evidence is presented to indicate that this DNA is incorporated into the fungal genome.

Conclusions

Considerable information has been added to our knowledge of the fungal zoospore since the address by Waterhouse (1962). Many zoospores have been examined with the transmission electron microscope and a new systematics, based in great part on the ultrastructure of motile cells, is emerging for the zoosporic fungi. We are slowly addressing biological problems related to fungal motile cells and this knowledge will enhance our understanding of and ability to control these organisms in nature. Many exciting problems on the biology of fungal zoospores have been looked at only superficially and the future is exciting for students of these fungi. The free-wheeling opportunists who combine their fascination for fungal motile cells with a minimum of constraints relative to technique will move our knowledge of these cells to incomprehensible levels during the next 100 years of the British Mycological Society.

References

Adams, M.J. (1991). Transmission of plant viruses by fungi. *Annals of Applied Biology*, **118**, 479–92.
Adams, M.J., Swaby, A.G. & Jones, P. (1988). Confirmation of the transmission of barley yellow mosaic virus (BaYMV) by the fungus *Polymyxa graminis*. *Annals of Applied Biology*, **112**, 133–41.
Alberts, B., Bray, D., Lewis, J., Raff, M., Roberts, K. & Watsonn, J.D. (1994). *Molecular Biology of the Cell*. Garland, New York.

Allen, R.N. & Newhook, F.J. (1973). Chemotaxis of zoospores of *Phytophthora cinnamomi* to ethanol in capillaries of soil pore dimensions. *Transactions of the British Mycological Society*, **61**, 287–302.

Allen, R.N. & Newhook, F.J. (1974). Suppression by ethanol of spontaneous turning activity in zoospores of *Phytophthora cinnamomi*. *Transactions of the British Mycological Society*, **63**, 383–5.

Andersen, R.A., Barr, D.J.S., Lynn, D.H., Melkonian, M., Moestrum, Ø. & Sleigh, M.A. (1991). Terminology and nomenclature of the cytoskeletal elements associated with the flagellar/ciliary apparatus in protists. *Protoplasma*, **164**, 1–8.

Barash, I., Klisiewicz, J.M. & Kosuge, T. (1965). Utilisation of carbon compounds by zoospores of *Phytophthora drechsleri* and their effect on motility and germination. *Phytopathology*, **55**, 1257–61.

Barr, D.J.S. (1980). An outline for the reclassification of the Chytridiales, and for a new order, the Spizellomycetales. *Canadian Journal of Botany*, **58**, 2380–94.

Barr, D.J.S. (1981). The phylogenetic and taxonomic implications of flagellar rootlet morphology among zoosporic fungi. *BioSystems*, **14**, 359–70.

Barr, D.J.S. (1992). Evolution and kingdoms of organisms from the perspective of a mycologist. *Mycologia*, **84**, 1–11.

Barr, D.J.S. & Allan, P.M.E. (1982). Zoospore ultrastructure of *Polymyxa graminis* (Plasmodiophoromycetes). *Canadian Journal of Botany*, **60**, 2496–504.

Barr, D.J.S. & Allan, P.M.E. (1985). A comparison of the flagellar apparatus in *Phytophthora, Saprolegnia, Thraustochytrium* and *Rhizidiomyces*. *Canadian Journal of Botany*, **63**, 138–54.

Barr, D.J.S. & Désaulniers, N.L. (1988). Precise configuration of the chytrid zoospore. *Canadian Journal of Botany*, **66**, 869–76.

Barr, D.J.S. & Désaulniers, N.L. (1990). The flagellar apparatus in the *Phytophthora infestans* zoospore. *Canadian Journal of Botany*, **68**, 2112–18.

Barr, D.J.S. & Désaulniers, N.L. (1992). The flagellar apparatus in zoospores of *Phytophthora, Pythium*, and *Halophytophthora*. *Canadian Journal of Botany*, **70**, 2163–9.

Barr, D.J.S. & Hadland-Hartman, V.E. (1977). Zoospore ultrastructure of *Olpidium cucurbitacearum* (Chytridiales). *Canadian Journal of Botany*, **55**, 3063–74.

Beakes, G.W., Canter, H.M. & Jaworski, G.H.M. (1988). Zoospore ultrastructure of *Zygorhizidium affluens* and *Z. planktonicum*, two chytrids parasitizing the diatom *Asterionella formosa*. *Canadian Journal of Botany*, **66**, 1054–67.

Benjamin, M. & Newhook, F.J. (1982). Effect of glass microbeads on *Phytophthora* zoospore motility. *Transactions of the British Mycological Society*, **78**, 43–6.

Berbee, M.L. & Kerwin, J.L. (1993). Ultrastructural and light microscopic localization of carbohydrates and peroxidase/catalases in *Lagenidium giganteum* zoospores. *Mycologia*, **85**, 734–43.

Bessey, E.A. (1942). Some problems in fungus phylogeny. *Mycologia*, **34**, 355–79.

Bimpong, C.E. (1975). Changes in metabolic reserves and enzyme activities during zoospore motility and cyst germination in *Phytophthora palmivora*. *Canadian Journal of Botany*, **53**, 1411–16.

Bowman, B.H., Taylor, J.W., Brownlee, A.G., Lee, J., Lu, S. & White, T.J. (1992). Molecular evolution of the fungi: relationship of the Basidiomycetes, Ascomycetes, and Chytridiomycetes. *Molecular Biology and Evolution*, **9**, 285–96.

Bryngelsson, T., Gustafsson, M., Green, B. & Lind, C. (1988). Uptake of host DNA by the parasitic fungus *Plasmodiophora brassicae*. *Physiological and Molecular Plant Pathology*, **33**, 163–71.

Cahill, D.M. & Hardham, A.R. (1994). Exploitation of zoospore taxis in the development of a novel dipstick immunoassay for the specific detection of *Phytophthora cinnamomi*. *Phytopathology*, **84**, 193–200.

Cantino, E.C., Lovett, J.S., Leak, L.V. & Lythgoe, J. (1963). The single mitochondrion, fine structure, and germination of the spore of *Blastocladiella emersonii*. *Journal of General Microbiology*, **31**, 393–404.

Carlile, M.J. (1983). Motility, taxis, and tropism in *Phytophthora*. In *Phytophthora: Its Biology, Taxonomy, Ecology and Pathology* (ed. D.C. Erwin, S. Bartnicki-Garcia, & P.H. Tsao), pp. 95–107. American Phytopathological Society.

Carlile, M.J. (1986). The zoospore and its problems. In *Water, Fungi and Plants*, British Mycological Society Symposium 11 (ed. P.G. Ayres & L. Boddy), pp. 105–118. Cambridge: Cambridge University Press.

Cavalier-Smith, T. (1993). Kingdom Protozoa and its 18 phyla. *Microbiological Reviews*, **57**, 953–94.

Cerenius, L., Rennie, P. & Fowke, L.C. (1988). Endocytosis of cationized ferritin by zoospores of the fungus *Aphanomyces euteiches*. *Protoplasma*, **144**, 119–24.

Cerenius, L. & Söderhäll, K. (1984). Repeated zoospore emergence from isolated spore cysts of *Aphanomyces astaci*. *Experimental Mycology*, **8**, 370–7.

Cho, C. & Fuller, M.S. (1989*a*). Ultrastructural organization of freeze-substituted zoospores of *Phytophthora palmivora*. *Canadian Journal of Botany*, **67**, 1493–9.

Cho, C. & Fuller, M.S. (1989*b*). Observations of the water expulsion vacuole of *Phytophthora palmivora*. *Protoplasma*, **149**, 47–55.

Colhoun, J. (1966). The biflagellate zoospore of aquatic Phycomycetes with particular reference to *Phytophthora* spp. In *The Fungus Spore*, Colston Papers Number 18 (ed. M.F. Madelin), pp. 85–92. London: Butterworths.

Cooney, E.W., Barr, D.J.S. & Barstow, W.E. (1985). The ultrastructure of the zoospore of *Hyphochytrium catenoides*. *Canadian Journal of Botany*, **63**, 497–505.

Cope, M. & Hardham, A.R. (1994). Synthesis and assembly of flagellar surface antigens during zoosporogenesis in *Phytophthora cinnamomi*. *Protoplasma*, **180**, 158–68.

Couch, J.N. (1938). Observations on cilia of aquatic Phycomycetes. *Science*, **88**, 476.

Couch, J.N. (1941). The structure and action of the cilia in some aquatic phycomycetes. *American Journal of Botany*, **28**, 704–13.

Crump, E. & Branton, D. (1966). Behavior of primary and secondary zoospores of *Saprolegnia* sp. *Canadian Journal of Botany*, **44**, 1393–400.

Deacon, J.W. (1988). Behavioural responses of fungal zoospores. *Microbiological Sciences*, **5**, 249–52.

Deacon, J.W. & Donaldson, S.P. (1993). Molecular recognition in the homing responses of zoosporic fungi, with special reference to *Pythium* and *Phytophthora*. *Mycological Research*, **97**, 1153–71.

Desjardins, P.R., Zentmyer, G.A. & Reynolds, D.A. (1970). On the binucleate condition of the quadriflagellated zoospores of *Phytophthora palmivora*. *Mycologia*, **62**, 421–7.

Dieguez-Uribeondo, J., Cerenius, L. & Söderhäll, K. (1994). Repeated zoospore emergence in *Saprolegnia parasitica*. *Mycological Research*, **98**, 810–15.

Dill, B.C. & Fuller, M.S. (1971). Amino acid immobilization of fungal motile cells. *Archives of Microbiology*, **78**, 92–8.

Donaldson, S.P. & Deacon, J.W. (1993). Changes in motility of *Pythium* zoospores induced by calcium and calcium modulating drugs. *Mycological Research*, **97**, 877–83.

Dorward, D.W. & Powell, M.J. (1983). Cytochemical detection of polysaccharides and the ultrastructure of the cell coat of zoospores of *Chytriomyces aureus* and *Chytriomyces hyalinus*. *Mycologia*, **75**, 209–20.

Duniway, J.M. (1979). Water relations of water molds. *Annual Review of Phytopathology*, **17**, 431–60.

Duniway, J.M. & Gordon, T.R. (1986). Water relations and pathogen activity. In *Water, Fungi and Plants*, British Mycological Society Symposium 11 (ed. P.G. Ayres & L. Boddy), pp. 119–37. Cambridge: Cambridge University Press.

Érsek, T., Hölker, U. & Höfer, M. (1991). Non-lethal immobilization of zoospores of *Phytophthora infestans* by Li$^+$. *Mycological Research*, **95**, 970–2.

Estrada-Garcia, T., Green, J.R., Booth, J.M., White, J.G. & Callow, J.A. (1989). Monoclonal antibodies to cell surface components of zoospores and cysts of the fungus *Pythium aphanidermatum* reveal species-specific antigens. *Experimental Mycology*, **13**, 348–55.

Estrada-Garcia, T., Callow, J.A. & Green, J.R. (1990). Monoclonal antibodies to the adhesive cell coat secreted by *Pythium aphanidermatum* zoospores recognize 200×10^3 M$_r$ glycoproteins stored within large peripheral vesicles. *Journal of Cell Science*, **95**, 199–206.

Fawcett, D. (1961). Cilia and flagella. In *The Cell* **II**, Chap. 4 (ed. J. Brachet & A.E. Mirsky), pp. 217–97. New York: Academic Press.

Feeney, D.M. & Treimer, R.E. (1979). Cytochemical localization of Golgi marker enzymes in *Allomyces macrogynus*. *Experimental Mycology*, **3**, 157–63.

Förster, H., Coffey, M.D., Elwood, H. & Sogin, M.L. (1990). Sequence analysis of the small subunit ribosomal RNAs of three zoosporic fungi and implications for fungal evolution. *Mycologia*, **82**, 306–12.

Fuller, M.S. (1966). Structure of the uniflagellate zoospores of aquatic Phycomycetes. In *The Fungus Spore*, Colston Papers Number 18 (ed. M.F. Madelin), pp. 67–84. London: Butterworths.

Fuller, M.S. (1977). The zoospore, hallmark of the aquatic fungi. *Mycologia*, **69**, 1–20.

Fuller, M.S. & Calhoun, S.A. (1968). Microtubule–kinetosome relationships in the motile cells of the Blastocladiales. *Zeitschrift für Zellforschung*, **84**, 526–33.

Fuller, M.S. & Jaworski, A.J. (1987). *Zoosporic Fungi in Teaching and Research*. Athens, GA: Southeastern Publishing Corporation.

Fuller, M.S. & Reichle, R.E. (1965). The zoospore and early development of *Rhizidiomyces apophysatus*. *Mycologia*, **57**, 946–61.

Fuller, M.S. & Reichle, R.E. (1968). The fine structure of *Monoblepharella* sp. zoospores. *Canadian Journal of Botany*, **46**, 279–83.

Griffith, J.M., Iser, J.R. & Grant, B.R. (1988). Calcium control of differentiation in *Phytophthora palmivora*. *Archives of Microbiology*, **149**, 565–71.

Grogan, R.G., Zink, F.W., Hewitt, W.B. & Kimble, K.A. (1958). The association of *Olpidium* with the big-vein disease in lettuce. *Phytopathology*, **48**, 292–7.

Grove, S.N. & Bracker, C.E. (1978). Protoplasmic changes during zoospore encystment and cyst germination of *Pythium aphanidermatum*. *Experimental Mycology*, **2**, 51–98.

Gubler, F., Jablonsky, P.P., Duniec, J. & Hardham, A.R. (1990). Localization of calmodulin in flagella of zoospores of *Phytophthora cinnamomi*. *Protoplasma*, **155**, 233–8.

Gubler, F. & Hardham, A.R. (1990). Protein storage in large peripheral vesicles in *Phytophthora* zoospores and its breakdown after cyst germination. *Experimental Mycology*, **14**, 393–404.

Gubler, F. & Hardham, A.R. (1991). The fate of peripheral vesicles in zoospores of *Phytophthora cinnamomi* during infection of plants. In *Electron Microscopy of Plant Pathogens* (ed. K. Mendgen & D.-E. Lesemann), pp. 197–210. Springer-Verlag.

Hardham, A.R. (1989). Lectin and antibody labelling of surface components of spores of *Phytophthora cinnamomi*. *Australian Journal of Plant Physiology*, **16**, 19–32.

Hardham, A.R. (1992). Cell biology of pathogenesis. *Annual Review of Plant Physiology and Plant Molecular Biology*, **43**, 491–526.

Hardham, A.R., Ralton, J.E. & Clarke, A.E. (1982). Cell surface characteristics of *Phytophthora cinnamomi* zoospores. *Micron*, **13**, 387–8.

Hardham, A.R., Suzaki, E. & Perkin, J.L. (1985). The detection of monoclonal antibodies specific for surface components on zoospores and cysts of *Phytophthora cinnamomi*. *Experimental Mycology*, **9**, 264–8.

Heath, I.B., Bauchop, T. & Skipp, R.A. (1983). Assignment of the rumen anaerobe *Neocallimastix frontalis* to the Spizellomycetales (Chytridiomycetes) on the basis of its polyflagellate zoospore. *Canadian Journal of Botany*, **61**, 295–307.

Heath, I.B., Greenwood, A.D. & Griffiths, H.B. (1970). The origin of flimmer in *Saprolegnia, Dictyuchus, Synura* and *Cryptomonas*. *Journal of Cell Science*, **7**, 445–61.

Held, A.E., Emerson, R., Fuller, M.S. & Gleason, F.H. (1969). *Blastocladia* and *Aqualinderella*: fermentative water molds with high carbon dioxide optima. *Science*, **165**, 706–9.

Hemmes, D.E. (1983). Cytology of *Phytophthora*. In *Phytophthora: Its Biology, Taxonomy, Ecology and Pathology* (ed. D.C. Erwin, S. Bartnicki-Garcia and P.H. Tsao), pp. 9–40. American Phytopathological Society.

Hickman, C.J. & Ho, H.H. (1966). Behaviour of zoospores in plant pathogenic phycomycetes. *Annual Review of Phytopathology*, **4**, 195–220.

Hiruki, C. & Teakle, D.S. (1987). Soil-borne viruses of plants. In *Current Topics in Vector Research* **3** (ed. F.K. Harris), pp. 177–215. New York: Springer-Verlag.

Ho, H.H. & Hickman, C.J. (1967). Asexual reproduction and behavior of zoospores of *Phytophthora megasperma* var. *sojae*. *Canadian Journal of Botany*, **45**, 1963–81.

Hoch, H.C. & Mitchell, J.E. (1972). The ultrastructure of zoospores of *Aphanomyces euteiches* and of their encystment and subsequent germination. *Protoplasma*, **75**, 113–38.

Holwill, M.E.J. (1982). Dynamics of eukaryotic flagellar movement. In *Eukaryotic and Prokaryotic Flagella*. Symposium of the Society for Experimental Biology 35 (ed. E.B. Amos & J.G. Duckett), pp. 289–312. Cambridge: Cambridge University Press.

Hyde, G.J., Gubler, F. & Hardham, A.R. (1991a). Ultrastructure of zoosporogenesis in *Phytophthora cinnamomi*. *Mycological Research*, **95**, 577–91.

Hyde, G.J., Lancelle, S.A., Hepler, P.K. & Hardham, A.R. (1991b). Freeze substitution reveals a new model for sporangial cleavage in *Phytophthora*, a result with implications for cytokinesis in other eukaryotes. *Journal of Cell Science*, **100**, 735–46.

Jahn, T.L., Landman, M.D. & Fonseca, J.R. (1964). The mechanism of locomotion of flagellates. II. Function of the mastigonemes of *Ochromonas*. *Journal of Protozoology*, **11**, 291–6.

Jaworski, A.J. & Wilson, J.B. (1989). Decreased activity of *Blastocladiella emersonii* zoospore ribosomes: correlation with developmental changes in ribosome-associated proteins. *Developmental Biology*, **135**, 340–8.

Jen, C.J. & Haug, A. (1979). Concanavalin A-induced lysis of zoospores of *Blastocladiella emersonii*. *Experimental Cell Research*, **120**, 425–8.

Kazama, F.Y. (1973). Ultrastructure of *Thraustochytrium* sp. zoospores. III. Cytolysomes and acid phosphatase distribution. *Archives of Microbiology*, **89**, 95–104.

Koch, W.J. (1956). Studies of the motile cells of chytrids. I. Electron microscope observations of the flagellum, blepharoplast and rhizoplast. *American Journal of Botany*, **43**, 811–19.

Koch, W.J. (1961). Studies of the motile cells of chytrids. III. Major types. *American Journal of Botany*, **48**, 786–8.

Lange, L. & Olson, L.W. (1978). The zoospore of *Olpidium radicale*. *Transactions of the British Mycological Society*, **71**, 43–55.

Lange, L. & Olson, L.W. (1979). The uniflagellate phycomycete zoospore. *Dansk Botanisk Arkiv*, **33**, 7–95.

Lehnen, L.P. & Powell, M.J. (1989). The role of kinetosome-associated organelles in the attachment of encysting secondary zoospores of *Saprolegnia ferax* to substrates. *Protoplasma*, **149**, 163–74.

Leipe, D.D., Wainright, P.O., Gunderson, J.H., Porter, D., Patterson, D.J., Valois, F., Himmerich, S. & Sogin, M.L. (1994). The stramenopiles from a molecular perspective: 16S-like rRNA sequences from *Labyrinthuloides minuta* and *Cafeteria roenbergensis*. *Phycologia*, **33**, 369–77.

Li, J. & Heath, I.B. (1993). Chytridiomycetous gut fungi, oft overlooked contributors to herbivore digestion. *Canadian Journal of Microbiology*, **39**, 1003–13.

Li, J. & Heath, I.B. (1994). The behavior of F-actin during the zoosporic phases of the Chytridiomycete gut fungi *Neocallimastix* and *Orpinomyces*. *Experimental Mycology*, **18**, 5769.

Li, J., Heath, I.B. & Packer, L. (1993). The phylogenetic relationships of the anaerobic chytridiomycetous gut fungi (Neocallimasticaceae) and the Chytridiomycota. II. Cladistic analysis of structural data and description of Neocallimasticales ord. nov. *Canadian Journal of Botany*, **71**, 393–407.

Lingle, W.L. & Barstow, W.E. (1983). Ultrastructure of the zoospore of *Blastocladia ramosa* (Blastocladiales). *Canadian Journal of Botany*, **61**, 3502–13.

Longcore, J.E. (1992). Morphology, occurrence, and zoospore ultrastucture of *Podochytrium dentatum* sp. nov. (Chytridiales). *Mycologia*, **84**, 183–92.

Longcore, J.E. (1993). Morphology and zoospore ultrastructure of *Lacustromyces hiemalis* gen. et sp. nov. (Chytridiales). *Canadian Journal of Botany*, **71**, 414–25.

Longcore, J.E. (1995). Morphology and zoospore ultrastructure of *Entophlyctis luteolus* sp. nov. (Chytridiales): Implications for chytrid taxonomy. *Mycologia*, **87**, 25–33.

Lord, J.C. & Roberts, D.W. (1985). Solute effects on *Lagenidium giganteum*: zoospore motility and bioassay reproducibility. *Journal of Invertebrate Pathology*, **46**, 160–5.

Lovett, J.S. (1963). Chemical and physical characterization of 'nuclear caps' isolated from *Blastocladiella* zoospores. *Journal of Bacteriology*, **85**, 1235–46.

Lovett, J.S. (1975). Growth and differentiation of the water mold *Blastocladiella emersonii*: cytodifferentiation and the role of ribonucleic acid and protein synthesis. *Bacteriological Reviews*, **39**, 345–404.

Lucarotti, C.J. & Federici, B.A. (1984). Ultrastructure of the gametes of *Coelomomyces dodgei* Couch (Blastocladiales, Chytridiomycetes). *Protoplasma*, **121**, 77–86.

Lunney, C.Z. & Bland, C.E. (1976). Ultrastructural observations of mature and encysting zoospores of *Pythium proliferum* de Bary. *Protoplasma*, **90**, 119–37.

Manton, I., Clarke, B., Greenwood, A.D. & Flint, E.A. (1952). Further observations on the structure of plant cilia, by a combination of visual and electron microscopy. *Journal of Experimental Botany*, **8**, 204–15.

Markey, D.R. & Bouck, G.B. (1977). Mastigoneme attachment in *Ochromonas*. *Journal of Ultrastructural Research*, **59**, 173–7.

Martin, W.W. (1971). The ultrastructure of *Coelomomyces punctatus* zoospores. *Journal of the Elisha Mitchell Scientific Society*, **87**, 209–21.

McLean, M.A., Campbell, R.N., Hamilton, R.I. & Rochon, D.M. (1994). Involvement of the cucumber necrosis virus coat protein in the specificity of fungus transmission by *Olpidium bornovanus*. *Virology*, **204**, 840–2.

Merz, U. (1992). Observations on swimming pattern and morphology of secondary zoospores of *Spongospora subterranea*. *Plant Pathology*, **41**, 490–4.

Miles, C.A. & Holwill, M.E.J. (1969). Asymmetric flagellar movement in relation to the orientation of the spore of *Blastocladiella emersonii*. *Journal of Experimental Biology*, **50**, 683–7.

Miller, C.E. & Dylewski, D.P. (1983). Zoosporic fungal pathogens of lower plants. What can be learned from the likes of *Woronina*? In *Zoosporic Plant Pathogens* (ed. S.T. Buczacki), pp. 249–283. Academic Press.

Mills, G.L. & Cantino, E.C. (1975). The single microbody in the zoospore of *Blastocladiella emersonii* is a 'symphyomicrobody'. *Cell Differentiation*, **4**, 35–43.

Mollicone, M.R.N. & Longcore, J.E. (1994). Zoospore ultrastructure of *Monoblepharis polymorpha*. *Mycologia*, **86**, 615–25.

Money, N.P. & Webster, J. (1989). Mechanism of sporangial emptying in *Saprolegnia*. *Mycological Research*, **92**, 45–9.

Morris, B.M. & Gow, N.A.R. (1993). Mechanism of electrotaxis of zoospores of phytopathogenic fungi. *Phytopathology*, **83**, 877–82.

Muehlstein, L.K. & Amon, J.P. (1987a). Chemotaxis in zoosporic fungi. In *Zoosporic Fungi in Teaching and Research* (ed. M.S. Fuller & A.J. Jaworski), pp. 284–5. Southeastern Publishing Corporation.

Muehlstein, L.K. & Amon, J.P. (1987b). Examining the photoresponse in zoosporic fungi. In *Zoosporic Fungi in Teaching and Research* (ed. M.S. Fuller & A.J. Jaworski), pp. 286–7. Southeastern Publishing Corporation.

Olson, L.W., Cerenius, L., Lange, L. & Söderhäll, K. (1984). The primary and secondary spore cyst of *Aphanomyces* (Oomycetes, Saprolegniales). *Nordic Journal of Botany*, **4**, 681–96.

Olson, L.W. & Fuller, M.S. (1968). Ultrastructural evidence for the biflagellate origin of the uniflagellate fungal zoospore. *Archives of Microbiology*, **63**, 237–50.

Olson, L.W. & Fuller, M.S. (1971). Leucine–lysine synchronization of *Allomyces* germlings. *Archives of Microbiology*, **78**, 76–91.

Penington, C.J., Iser, J.R., Grant, B.R. & Gayler, K.R. (1989). Role of RNA and protein synthesis in stimulated germination of zoospores of the pathogenic fungus. *Phytophthora palmivora. Experimental Mycology*, **13**, 158–68.

Pommerville, J. (1978). Analysis of gamete and zygote motility in *Allomyces. Experimental Cell Research*, **113**, 161–72.

Powell, M.J. (1976). Ultrastructure and isolation of glyoxysomes (microbodies) in zoospores of the fungus *Entophlyctis* sp. *Protoplasma*, **89**, 1–27.

Powell, M.J. (1978). Phylogenetic implications of the microbody–lipid globule complex in zoosporic fungi. *BioSystems*, **10**, 167–80.

Powell, M.J. (1981). Zoospore structure of the mycoparasitic chytrid *Caulochytrium protostelioides* Olive. *American Journal of Botany*, **68**, 1074–89.

Powell, M.J. (1994). Production and modifications of extracellular structures during development of chytridiomycetes. *Protoplasma*, **181**, 123–41.

Powell, M.J. & Bracker, C.E. (1986). Distribution of diaminobenzidine reaction products in zoospores of *Phytophthora palmivora. Mycologia*, **78**, 892–900.

Powell, M.J., Lehnen, L.P. & Bortnick, R.N. (1985). Microbody-like organelles as taxonomic markers among Oomycetes. *BioSystems*, **18**, 321–34.

Powell, M.J. & Roychoudhury, S. (1992). Ultrastructural organization of *Rhizophlyctis harderi* zoospores and redefinition of the type 1 microbody–lipid globule complex. *Canadian Journal of Botany*, **70**, 750–61.

Reichle, R.E. (1972). Fine structure of *Oedogoniomyces* zoospores, with comparative observations on *Monoblepharella* zoospores. *Canadian Journal of Botany*, **50**, 819–24.

Renaud, F.L. & Smith, H. (1964). The development of basal bodies and flagella in *Allomyces arbusculus. Journal of Cell Biology*, **23**, 338–54.

Robertson, J.A. & Emerson, R. (1982). Two members of the Blastocladiaceae II. Morphogenetic responses to O_2 and CO_2. *American Journal of Botany*, **69**, 812–17.

Salvin, S.B. (1940). The occurrence of five successive swarming stages in a non-sexual *Achlya. Mycologia*, **32**, 148–54.

Scherffel, A. (1925). Endophytische Phycomycete-Parasiten der Bacillariaceen un einige neu Monadinen. *Archives für Protistenkunde*, **52**, 1–141.

Sing, V.O. & Bartnicki-Garcia, S. (1975a). Adhesion of *Phytophthora palmivora* zoospores: electron microscopy of cell attachment and cyst wall fibril formation. *Journal of Cell Science*, **18**, 123–32.

Sing, V.O. & Bartnicki-Garcia, S. (1975b). Lysis of zoospores of *Phytophthora palmivora* induced by concanavalin A. *Experientia*, **31**, 643–4.

Slykhuis, J.T. & Barr, D.J.S. (1978). Confirmation of *Polymyxa graminis* as a vector of wheat spindle streak mosaic virus. *Phytopathology*, **68**, 639–43.

Sparrow, F.K. (1935). The interrelationships of the Chytridiales. *Proceedings of the Sixth International Botanical Congress, Amsterdam*. **2**, p. 181.

Sparrow, F.K. (1943). *The Aquatic Phycomycetes, Exclusive of the Saprolegniaceae and Pythium*. Ann Arbor, University of Michigan Press.

Sparrow, F.K. (1960). *Aquatic Phycomycetes, 2nd revised edn.* Ann Arbor, University of Michigan Press.

Sparrow, F.K. (1973). Chytridiomycetes, Hyphochytridiomycetes. In *The Fungi* **IVB** (ed. G.C. Ainsworth, F.K. Sparrow & A.S. Sussman), pp. 85–110. Academic Press.

Stobbs, L.W., Cross, G.W. & Monocha, M.S. (1982). Specificity and methods of transmission of cucumber necrosis virus by *Olpidium radicale* zoospores. *Canadian Journal of Plant Pathology*, **4**, 134–42.

Suberkropp, K.F. & Cantino, E.C. (1972). Environmental control of motility and encystment in *Blastocladiella emersonii* zoospores at high population densities. *Transactions of the British Mycological Society*, **59**, 463–75.

Suberkropp, K.F. & Cantino, E.C. (1973). Utilization of endogenous reserves by swimming zoospores of *Blastocladiella emersonii*. *Archives of Microbiology*, **89**, 205–21.

Taylor, J.W. & Fuller, M.S. (1981). The Golgi apparatus, zoosporogenesis and development of the zoospore discharge apparatus of *Chytridium confervae*. *Experimental Mycology*, **5**, 35–59.

Teakle, D.S. (1962). Transmission of tobacco necrosis virus by a fungus, *Olpidium brassicae*. *Virology*, **18**, 224–31.

Teakle, D.S. & Gold, A.H. (1964). Prolonging the motility and virus-transmitting ability of *Olpidium* zoospores with chemicals. *Phytopathology*, **54**, 29–32.

Temmink, J.H.M., Campbell, R.N. & Smith, P.R. (1970). Specificity and site of in vitro acquisition of tobacco necrosis virus by zoospores of *Olpidium brassicae*. *Journal of General Virology*, **9**, 201–13.

Thomas, D.D. & Butler, D.L. (1989). Cationic interactions regulate the initiation and termination of zoospore activity in the water mould *Achlya heterosexualis*. *Journal of General Microbiology*, **135**, 1917–22.

Thomas, D.D. & Peterson, A.P. (1990). Chemotactic auto-aggregation in the water mould *Achlya*. *Journal of General Microbiology*, **136**, 847–53.

Tomlinson, J.A. & Garret, R.G. (1964). Studies on the lettuce big-vein virus and its vector *Olpidium brassicae* (Wor.) Dang. *Annals of Applied Biology*, **54**, 45–61.

Turian, G. & Kellenberger, E. (1956). Ultrastructure du corps paranucléaire, des mitochondries et de la membrane nucléaire des gamètes d'*Allomyces macrogynus*. *Experimental Cell Research*, **11**, 417–22.

Vlk, W. (1938). Über den Bau der Geissel. *Archives für Protistenkunde*, **90**, 157–60.

Waterhouse, G.M. (1962). Presidential address. The zoospore. *Transactions of the British Mycological Society*, **45**, 1–20.

Weston, W.H. (1919). Repeated zoospore emergence in *Dictyuchus*. *Botanical Gazette*, **63**, 287–96.

Weston, W.H. (1935). The bearing of recent investigations on the interrelationships of the aquatic Phycomycetes. *Proceedings of the Sixth International Botanical Congress, Amsterdam*, **1**, p. 266.

Whittaker, S.L., Shattock, R.C. & Shaw, D.S. (1992). The duplication cycle and DAPI–DNA contents in nuclei of germinating zoospore cysts of *Phytophthora infestans*. *Mycological Research*, **96**, 355–8.

Zhang, L., Mitra, A., French, R.C. & Langenberg, W.G. (1994). Fungal zoospore-mediated delivery of a foreign gene to wheat roots. *Phytopathology*, **84**, 684–7.

8

Interconnectedness and individualism in fungal mycelia

ALAN D.M. RAYNER

Introduction

Mycologists often like to emphasize the uniqueness of fungi. By so doing they reserve a special place for mycology in the study of living things. But is this place a pedestal, an ivory tower or a blind alley?

All too often, mycology has become viewed both from inside and outside as an esoteric field, with a terminology and way of thinking all of its own. Mycologists may therefore have come to feel that they have little to learn from or convey to other biologists and vice versa. As a result, there has been a general tendency to overlook the enormous importance of fungi, both in terms of their contribution to global ecosystem processes and in terms of the lessons they provide about the evolution and organization of living systems.

Ironically, it is in the perceived uniqueness of mycelial growth patterns that I think the fundamental environmental and evolutionary importance of fungi resides. Fungal mycelia are heterogeneous, networking systems that retain the potential for growth throughout their lives. They are capable of converting organic and mineral nutrients and water into biomass on scales measurable in units ranging from micrometres to kilometres. As such, they exhibit properties characteristic of all kinds of 'dynamic fields', both with respect to the configuration of their variably sealed hyphal boundaries and the relationships between populations of genomic organelles within these boundaries.

Mycelia therefore illustrate how organizational and behavioural principles familiar in human and animal societies operate equally well at cellular and subcellular scales. Fundamentally, these principles concern the varied counteraction of expansive (dissociative) and resistive (associative) processes which, *in concert with* genetic adaptation, is in large part responsible for the richly heterogeneous patterns of life.

How, then, has a century of mycology helped to illuminate or obscure the view of mycelial fungi as living systems finely balanced between associative and dissociative trends? How may this view contribute to a better understanding of fungi and their biological and practical relevance in the future? These are the questions which I aim to address in the following pages.

Before trying to provide some answers, I have to admit to my inability to itemize all the landmark events and personalities that have contributed to the picture of fungal life which I will be describing. I also have to confess that I doubt the wisdom of identifying landmarks as a means of understanding history, because no events or individuals ever really occur in isolation. As with any indeterminate (open-ended) process, the development of mycology has been highly serendipitous – subject to unforeseeable local circumstances and interactions. Whilst some ramifications have been reinforced, others have remained dormant or fallen into disuse. To try to identify any precise moment of insight or discovery, let alone to accredit it to a particular individual, is to misunderstand the process and the complex feedbacks that it involves. On the other hand, to come to view historical processes in a way that diminishes the significance of individual competition and responsibility, may be one of the deeper insights that comes from an appreciation of mycelial patterns.

Communicating behind the scenes

Think of a city. Impressive though its superstructure of buildings and the toings and froings of its human inhabitants might be, it could not function as a coherent system without its underlying infrastructure of communicating pipelines and cables. Though there may be outward signs of this infrastructure – lamp standards, fire hydrants, manhole covers, telephone kiosks, etc – it takes prior knowledge to recognize these signs for what they are.

Similar outward signs of infrastructure, in the form of fungal sporophores, occur in natural ecosystems. Underlying sporophores, whose roles in reproduction and drawing attention to themselves might be likened to those of glamorous film stars, are behind-the-scenes production teams. The production teams are those indeterminate, collectively organized systems of branching, apically extending, protoplasm-filled hyphal tubes that make up mycelia. These nutrient-absorbing *and* distributing systems (cf. Hartig, 1874) are responsible for locating energy supplies (resources) and converting them into a viable output that includes those determinate offshoots that disseminate genetic survival units in the form of spores. In this inconspicuous role, mycelia maintain cycles of growth, death and decay in ecosystems and interconnect the lives of other organisms in many, often surprising ways.

Missing the mycelium

Rather like a cinema audience that focuses on the film stars whilst forgetting the production team, mycologists have tended to be distracted by sporophores – giving them names and dressing them in fancy language – whilst taking the mycelial infrastructure for granted. Even Reginald Buller, regarded by many as the Patron Saint of mycelial biology (Gregory, 1984), devoted much of his time to spore production and perhaps viewed the mycelium primarily as a means to this end. This view possibly contributed to his and others' willingness to endorse the concept of genetic mosaicism within a species that, for good reasons or bad, may have impeded the development of fungal population biology (see later).

For mycologists more concerned with what fungi do than with what they are called, the mycelium should be hard to ignore! An example is provided by fungi that cause wood decay – where the mycelium is not only directly responsible for the degradation of lignocellulose, but is also often relatively easy to culture. It is not surprising therefore that these fungi, and above all the much feared, much vaunted dry rot fungus, *Serpula lacrimans*, should have been the subject of such classical studies of mycelial biology as those of Falck (1912). These studies clearly revealed the diversity of form and function that can occur within mycelia.

However, in spite of its importance to understanding the fundamental nature of fungi and how they colonize natural habitats, there seems to have been a general reluctance to study the fungal mycelium for its own sake. Even where, as in wood decay fungi, mycelial diversity was quickly recognized, it became used more as an aid to diagnosis (e.g. Nobles, 1965; Stalpers, 1978) than as an indicator of the wide range of activities involved in colonization. Comparatively recently, Burnett (1976) was forced to conclude that 'largely invisible, little studied, the vegetative mycelium of fungi provides an almost endless series of problems whose investigation is long overdue'.

The repercussions of this relative lack of concern for the mycelium have been profound, notably in fungal ecology. Here, traditional approaches have been unable to provide relevant information about where fungi (and their offspring) are, what they do there, how they arrived, whether they will persist and how they are likely to change. Instead these approaches have focused on compiling species abundance and diversity indices (without reference to intraspecific variation), counting fungi (without being clear about what constitutes an individual), weighing fungi (without acknowledging heterogeneity) and assessing metabolic activity (rather than ecological role).

The problem is epitomized by the fact that production of sporophores, which can be observed easily and identified using conventional taxonomic procedures, has traditionally provided a major basis for defining fungal distribution patterns in natural communities. As determinate offshoots, sporophores do not constitute the whole fungus and may at least partly be subject to selection processes that differ from those operating on mycelia. Their absence does not necessarily imply the absence of mycelium, and it is not generally possible without further work to establish whether separate collections of them arise from the same genetic source. They therefore do not in themselves provide a reliable or even a legitimate basis for locating fungi in space or in time – as illustrated by the artefactual nature of the coprophilous fungal succession exposed by Harper & Webster (1964). To map fungal distribution patterns, it is necessary to map fungal mycelia.

Misunderstanding the mycelium

Even when the mycelium demands attention, as when it causes disease or decay, or is cultured to provide food, drugs and other products, the tendency to regard it as the featureless forerunner of sporophores has been deeply ingrained. Correspondingly, mycelia have become widely regarded as if, ideally, they are reducible to purely additive assemblages of hyphal tips whose sole function is to grow in direct proportion to the amount that they assimilate, prior to reproduction. As such, mycelia may be assumed to exhibit fully predictable, readily calculable dynamics under any particular set of growth conditions. Much effort has gone into providing predictive mathematical models based on this assumption (e.g. Trinci, 1978; Prosser, 1991, 1993, 1994*a,b*), and the hyphal tip has long been focused on as the 'key' to the entire mycelial system.

The fact that mycelia do not behave as additive systems, and tend instead to change their organizational pattern and become heterogeneous, has therefore been seen as the consequence of 'imperfection' – some 'failure' on the part of the organism or its environment to maintain constant conditions. Moreover, this imperfection, and the inconsistencies it gives rise to, have been regarded as a major impediment to the practical

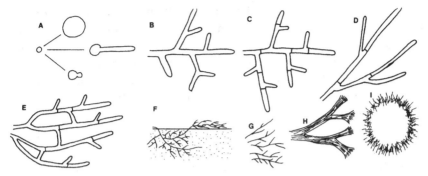

Fig. 8.1. Organizational shifts in mycelial systems. A. A spore germinating to yield a germ tube, a giant cell or a budding yeast. B. A coenocytic (unpartitioned) hyphal branching system. C. A septate hyphal system with Christmas tree-like branching. D. A septate system with delta-like branching. E. An anastomosed (partially networked) hyphal branching system. F. Assimilative and non-assimilative hyphae (represented as single lines) growing respectively within and above the nutrient source (stippled). G. Diffusely organized hyphae. H. Hyphae organized into cable-like linear aggregates (mycelial cords). I. An annular mycelium (fairy ring) resulting from degeneration of the central region.

exploitation and control of fungi. Consequently, many attempts have been made to eliminate heterogeneity at source, for example through the use of genetically defined strains and precise management of conditions.

The ultimate futility of such efforts becomes evident when it is realized that fungal mycelia are intrinsically heterogeneous, both in view of the environmental contexts in which they operate naturally and because of the way they are organized as dynamic systems.

Heterogeneity, uncertainty and indeterminacy

Constancy of conditions and resource supply is not a typical feature of most fungal habitats. Rather, there is variability, discontinuity and consequent unpredictability, both in space and in time. The mycelium negotiates uncertain, heterogeneous regimes as a versatile, indeterminate structure that uses local feedback to adjust the balance between explorative, assimilative, conservative and redistributive processes according to requirements.

The versatility of mycelia is evident in the marked shifts in organizational pattern that can occur spontaneously or in response to environmental change during the lifespan of an individual system (Fig. 8.1; see, for example, Rayner, 1991, 1994; Rayner, Griffith & Ainsworth, 1994*a*,

Rayner, Griffith & Wildman, 1994*b*; Rayner, Ramsdale & Watkins, 1995). When a spore germinates by taking up water and nutrients, it often expands isotropically at first and then 'breaks symmetry' to allow the emergence of one or more apically extending, protoplasm-filled hyphal tubes. Alternatively, a determinate developmental pattern may be maintained for greater or lesser periods, resulting in the formation of 'giant cells' or yeast-like phases.

Once polarity has been established, the hyphal tubes may be fully coenocytic or they may become internally partitioned by centripetal ingrowths or septa. Sooner or later, the tubes branch, either in a Christmas-tree-like or a delta-like pattern. The branches may diverge or they may converge and fuse (anastomose), so converting the initially radiate system at least partially into a network. Whilst some parts of the system may be intimately associated with the nutrient source, others may become sealed off or emerge beyond the immediate sites of assimilation. The branches may remain diffusely associated or they may aggregate to form protective, reproductive or migratory structures. The latter consist of cable-like arrays, mycelial cords and rhizomorphs, and can often extend an order of magnitude faster than individual hyphae. Whilst some parts of the boundary of an established mycelial system may continue to expand, others may stop growing and degenerative processes may set in, resulting, for example, in the annular patterns characteristic of fairy rings.

The biological utility of such a changeable dynamic structure becomes apparent when fungi are observed growing in the field or in laboratory systems that simulate at least some elements of the biotic and abiotic heterogeneity of natural habitats. For example, if the loose covering of leaf litter is removed from woodland soil, mycelial cord and rhizomorph networks often come into view that can readily be mapped *in situ* (Grainger, 1962; Thompson & Rayner, 1982, 1983) and seen to interconnect discrete resource units. The processes leading to the formation of such networks can be revealed by growing the relevant fungi between colonized 'inocula' and uncolonized 'baits' in trays of soil. Experiments of this kind with wood-decay fungi have revealed a variety of long-range and short-range 'foraging strategies' that produce patterns extraordinarily similar to, for example, the raid swarms of army ants and the stoloniferous systems of some plants (e.g. Dowson, Rayner & Boddy, 1986, 1988; Boddy, 1993).

Perhaps even more revealing are the patterns produced by mycelia grown in matrix systems of the kind illustrated in Fig. 8.2. These systems consist of sets of chambers that are isolated from one another with respect

Fig. 8.2. Ordered pattern produced by *Coprinus picaceus* when grown in a matrix of 25 × 4 cm² chambers filled alternately with high and low nutrient media. Holes have been cut in the partitions just above the level of the medium. The fungus has been inoculated into the central, high nutrient-containing chamber, whence it has produced alternating assimilative and explorative states. Hyphal systems linking between chambers have been reinforced into persistent mycelial cords whereas others unable to extend further have been prone to degenerate. (Photograph reproduced by courtesy of Louise Owen and Erica Bower.)

to diffusion through the growth medium, but interconnected by passageways that allow particular portions of the mycelium to grow between and across separate domains. The design therefore combines the discreteness that enables key stages of development to be analysed in a particular locale with the continuity which is fundamental to the operation of the mycelium as an integrated system.

Four processes underlie the ability of matrix systems both to enhance the diversity of phenotypic patterns and the degree to which these patterns are produced in an ordered sequence: (i) *microenvironmental selection* resulting from the influence of local conditions on the organizational properties of mycelia; (ii) *physical sieving* due to the chance selection and amplification of those phenotypic forms that pass through the gaps between one chamber and the next; (iii) *physical focusing* due to the ability of those parts of the mycelial system that emerge through gaps and so continue to extend to act as nodes through which growth resources are gathered and redistributed; (iv) *physical reinforcement* due to the reiterative use of mycelia connecting between chambers as distributive channels.

As a result of these processes, a large part, if not the whole of the developmental repertoire of a fungus may be revealed at once, yielding insights into the organism's niche and offering exciting prospects for more rational approaches to culturing fungi for practical purposes. For many such purposes it may be better to *increase* rather than to try to eliminate heterogeneity as a means of controlling fungal growth and avoiding inconsistencies.

In their indeterminacy and heterogeneity, mycelia display patterns that are generic to all kinds of distributive systems, from nerves and blood vessels to animal societies. Mycelial biology therefore promises to contribute usefully to and gain the development of self-organization and complexity theory which is currently taking place in biology as a whole (e.g. Goodwin, 1994). Whilst such ideas might seem radical against the backcloth of a recent mycological history committed to homogeneous idealism, they have had many precedents, throughout the century.

Buller (1931, 1933) famously got carried away with the idea of the mycelium as a social structure. He summarized his viewpoint as follows: 'we may conclude . . . that all living cells which make up an individual plant are connected together so as to form a single mass of protoplasm. A realization of this important fact helps us to understand not merely the phenomenon of protoplasmic streaming in the mycelium of certain Discomycetes, Pyrenomycetes and Hymenomycetes, but also how it is that a multicellular fungus can develop and react to external stimuli in a unitary manner' (Buller, 1933). In concord with this viewpoint, the occurrence, importance and mechanisms of translocation in mycelia have long attracted attention, leading to the design of such classical heterogeneous culture systems as the 'double dish' and the 'split plate' (Shütte, 1956; Thrower & Thrower, 1968; Lyon & Lucas, 1969; Jennings, 1984, 1987).

Somehow, however, what should have been the current providing the vital spark igniting an awareness of the fundamental nature of mycelial systems has always been undercurrent – a dark suspicion of unseemly activities not quite within grasp. Translocation has therefore been thought of as an 'optional extra' or 'complication' in the life of a fungus rather than an intrinsic consequence of the way mycelia are organized as coherent, dynamic systems. The hyphal tip has retained pride of place as the fundamental unit of fungal organization.

Mycelia as non-linear, 'self-plumbing' systems

The ability of mycelial systems to produce distinctive organizational states associated with predominantly assimilative, explorative, conservative and redistributive functions cannot be explained by linear growth models (Rayner *et al.*, 1994*a,b*, 1995). Non-linear systems theory, on the other hand, has the potential to explain these and all other heterogeneous properties of mycelia, as well as encompassing exponential growth at low initial input rates and linear extension at equilibrium capacity (Rayner, 1995).

Generally, non-linearity is due to counteraction between resistive and expansive processes at and within a system's operational boundary. When expansive processes are driven by energy inputs greater than can be accommodated by constant rates of boundary deformation, the systems become unstable. They are then prone to partition at successively finer scales into heterogeneous subdomains, for example by becoming turbulent and by branching.

The route to instability can be traced mathematically using equations in which the relationship between input to and expansion of a system changes from direct proportionality as the content increases and becomes progressively more constrained by boundary limits or resistances.

One way in which mycelia could develop non-linearly would be through their operation as hydrodynamic systems whose expansion is driven by uptake of water and resources from the environment and regulated by three basic parameters (Rayner, 1995). Two of the latter are the resistances of hyphal exteriors to deformation and passage of molecules; together, these comprise what may be defined as the degree of 'insulation' of mycelial systems (Rayner *et al.*, 1994*a*). The third parameter is the internal resistance to displacement of hyphal contents. This is a function of hyphal diameter and protoplasmic viscosity and continuity, the latter being determined by the occurrence of septa, cytoskeletal elements, anastomoses, septal sealing and death of compartments.

Ways in which variations in these parameters along the length of hyphal systems could affect four fundamental processes governing the primary

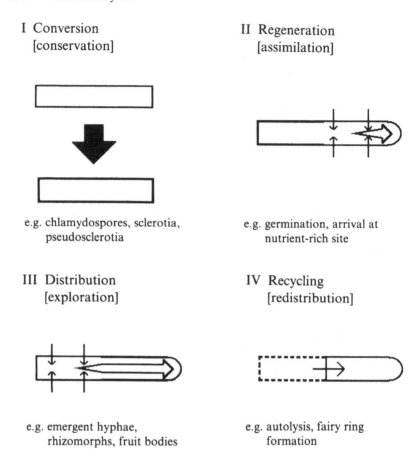

I Conversion
   [conservation]

II Regeneration
    [assimilation]

e.g. chlamydospores, sclerotia,
     pseudosclerotia

e.g. germination, arrival at
     nutrient-rich site

III Distribution
     [exploration]

IV Recycling
    [redistribution]

e.g. emergent hyphae,
     rhizomorphs, fruit bodies

e.g. autolysis, fairy ring
     formation

Fig. 8.3. Four fundamental processes in elongated hydrodynamic systems, as determined by boundary deformability, permeability and internal partitioning. Rigid boundaries are shown as straight lines, deformable boundaries as curves, impermeable boundaries by thicker lines, degenerating boundaries as broken lines and protoplasmic disjunction by an internal dividing line. Simple arrows indicate input across permeable boundaries into metabolically active protoplasm, tapering arrows represent throughput due to displacement. (Devised during discussions with Philip Drazin and David Griffel.)

functions of mycelia are illustrated in Fig. 8.3. In relating these processes and functions to the organizational patterns produced by mycelia in natural habitats, energetic considerations suggest that insulation should be minimized when external nutrient availability is maximal and vice versa. This way resource input can be maximized when hyphae emerge into high nutrient domains, and leakage minimized in low nutrient domains.

In hydrodynamic systems, and indeed any system whose pattern is determined by association–dissociation interplay, branching occurs when input exceeds throughput capacity (Rayner *et al.*, 1994*a*,*b*, 1995; Rayner, 1995). The latter may be defined as the ability to displace contents smoothly to existing sites of deformation or discharge on the system boundary. It is equivalent to the equilibrium or carrying capacity of a population modelled by the logistic equation (Rayner, 1995).

The frequency and pattern of branching will therefore depend on the degree of insulation of the lateral boundaries of hyphae and hence on whether they are in a regenerative (assimilative) or distributive (explorative) mode. Assimilative hyphae may be expected to branch like Christmas trees, because the sites of input to and outgrowth from the system coincide. By contrast, distributive hyphae will branch like river deltas since the sites of input are remote from the sites of emergence (Rayner, 1995).

Since branching depends on throughput capacity and hence on the resistance of hyphae to axial displacement of contents, septa and anastomoses will have an important influence on pattern formation. Septation facilitates compartmental differentiation (Gull, 1978) but impedes throughput and so will promote branching. However, a severely impeded system has a very much reduced capacity to explore. The association between septation and anastomosis (e.g. Boddy & Rayner, 1983; Ainsworth & Rayner, 1991) is therefore understandable. By replacing an in-series resistance system with an in-parallel one, anastomoses greatly increase throughput capacity, hence tending to limit branching, whilst allowing enhanced delivery to local sites on the boundary. This enhanced delivery may allow the mycelium to increase its operational scale, via the emergence of wide diameter hyphae, fast growing sectors, cable-like mycelial cords and rhizomorphs, etc., as well as to supply fruit body primordia.

However, networks resulting from anastomosis can also become strongly self-reinforcing; due to reiterative processes they become powerful sinks drawing resources from peripheral components and preventing further exploration. In effect, networking tends to restore the symmetry broken at germination, and it may then be that the only way in which exploration can be resumed is via the onset of degeneration, associated with recycling and redistribution. Such processes may underlie fairy ring formation (Dowson, Rayner & Boddy, 1989; Rayner *et al.*, 1994*a*,*b*; Rayner, 1995).

The relative importance of the four processes identified in Fig. 8.3 with respect to adaptation to niche can be understood in terms of ecological strategies theory (e.g. Pugh, 1980; Cooke & Rayner, 1984; Andrews, 1992). Briefly, R-selected fungi or organizational states, which capitalize on ecological disturbance by being equipped for rapid arrival and exploitation of readily assimilable resources, emphasize conversional and regenerative processes. S-selected fungi or states, adapted to environmental stress, either emphasize conversional processes, if they tend to remain *in situ*, or distributional and recycling processes if they migrate. Similarly, C-selected fungi or states, effective in combat, either retain resources captured at early stages of colonization, hence emphasizing conversional processes, or replace former residents, hence emphasizing distributional and recycling processes.

Control mechanisms and local feedbacks

It is possible to rationalize the dynamic processes involved in fungal development at individual, population and community levels, by combining adaptive or strategic considerations with organizational ones (Rayner, 1994*a,b*, 1995). By harnessing the organizational properties of a fluid-displacement system, mycelia can produce a diversity of patterns subject to the influence of boundary-defining parameters that dictate paths of least resistance. The mechanisms by which these parameters are varied determine how fungi adapt to changeable environmental circumstances.

Here it is important to recognize that boundary parameters can be controlled at several organizational levels. It is also relevant to appreciate the contrast between the much studied, strictly ordered, developmental programmes characteristic of determinate systems and the versatility of indeterminate systems.

Based on aspects of determinate development, it has become conventional to attribute all phenotypic properties primarily to the presence of genes and gene products, with the environment playing a secondary, modifying role. From this perspective, phenotypic attributes are judged according to their adaptive (= selective) value. Variation is assumed to arise via direct effects (e.g. on enzyme kinetics) of the environment on structure or physiological functioning (environmental variation) or via differences either in content (genetic variation) or expression (epigenetic variation) of genetic information.

However, a potentially fundamental feature of indeterminate systems is that they are subject to what might be termed 'hyperepigenetic variation' (Rayner *et al.*, 1995). Such variation involves autocatalytic processes which, once initiated, fall outside immediate genetic control and are extremely sensitive to local environmental conditions. These processes result in various kinds of reiteration, based on accumulated local 'experience', and the directions in which they lead are highly serendipitous, with phenotype becoming effectively uncoupled from genotype. For example, hyperepigenetic variation in human societies is based on learning, memory, the development of language and the circulation of ideas, and gives rise to 'cultural' evolution.

Historically, phenotypic transitions in mycelia that do not occur continuously over an environmental gradient have commonly been attributed to genetic variation. Heterokaryosis, mutation, aneuploidy and recombination have therefore all been invoked as possible mechanisms. A classical example is what became known, rather mysteriously, as the 'dual phenomenon' (Jinks, 1959).

However, there is little modern evidence that more than two kinds of nuclear genome and one kind of mitochondrial genome can persist in continuous protoplasm (e.g. Rayner, 1991; Rayner & Ross, 1991; see also below). Moreover, it is difficult to envisage how genetic variation could be regulated so as to produce functional alternative organizational states appropriate to circumstances.

Epigenetic or indeed hyperepigenetic processes that alter the properties of hyphal boundaries may therefore provide a better explanation for the heterogeneity of individual mycelial systems. That being so, the manner in which boundary parameters may be affected by and influence differential gene expression becomes important.

Mechanisms controlling the deformability of hyphal boundaries have received much attention. The production, chemical cross-linking and enzymic lysis of microfibrillar wall components are of potentially fundamental importance here (Bartnicki-Garcia, 1973; Wessels, 1986, 1991), as is the role of cytoskeletal elements as temporary scaffolding (Gow, 1994).

Mechanisms controlling the relative permeability of hyphal boundaries have received much less attention. However, as shown in Fig. 8.3, these mechanisms may be critical to patterns of uptake, throughput, conservation and redistribution of resources. Of particular interest here are water-resistant materials that coat, impregnate or line hyphal walls. These materials include the cysteine-rich polypeptides known as 'hydrophobins' whose expression has clearly been associated with production of emergent (non-assimilative) hyphal and mycelial phases (Wessels, 1991, 1992, 1994). Also likely to be important are aromatic and terpenoid products from the acetate, polyketide and shikimate pathways.

The production of hydrophobic wall components may generally be activated by depletion of external resource supplies and associated reduction in internal energy charge (e.g. Bushell, 1989*a,b*). The biosynthetic pathways involved have often been encompassed by the term, 'secondary metabolism' (as distinct from 'primary metabolism'). However, 'transductive metabolism' (as distinct from energy-gathering or 'inductive metabolism') might be a better term in view of their possible role in insulation (Rayner *et al.*, 1994*a,b*).

It may also be highly relevant that many hydrophobic metabolites are capable of reacting in complex ways with ground state and reactive oxygen species, particularly in the presence of phenoloxidases and peroxidases, via the production of free radicals. Many of the reactions, once initiated by enzyme action, occur autonomously, their course being determined by and extremely sensitive to local environmental conditions. For example, during melanin biosynthesis via the action of catechol oxidase (tyrosinase), alkaline pH and high oxygen tensions favour quinone production and polymerization, whilst low pH and low oxygen tensions favour depolymerization into phenolic subunits (Bell & Wheeler, 1986). Hyphae growing out into air would therefore be liable to develop a very different, more hydrophobic, boundary chemistry from those growing within a nutrient medium.

It is therefore interesting that changes in phenoloxidase and peroxidase activity have been clearly linked with a variety of developmental landmarks, including the initiation of fruit bodies, sclerotia and rhizomorphs (e.g. Willetts, 1978; Leatham & Stahmann, 1981; Ross, 1985; Worrall, Chet & Hüttermann, 1986; Ainsworth & Rayner, 1991). Moreover, free radical production is implicated in the senescence of *Podospora pauciseta* (Frese & Stahl, 1992; Marbach & Stahl, 1994) and evidence is accumulating for a key role for phenol-oxidizing activity in interspecific and intraspecific mycelial interactions (see below). Densely and effusely branched colonies of *Hypholoma fasciculare* have been shown to differ in the quantities of laccase that they release into the medium (highest in dense colonies) or sequester at hyphal boundaries. Effuse colonies arising as sectors from the same dense source colony can vary markedly in pigmentation and aerial mycelium production (Fellows & Rayner, unpubl.). A reciprocal relation between phenoloxidase activity and oxidative phosphorylation was reported in mycelial fungi by Lyr (1958, 1963). Indeed, Frese & Stahl (1992) have reported that the induction of laccase activity in *P. pauciseta* is correlated with inhibition of cytochrome *c*-oxidase. The metabolic uncoupler, 2,4-dinitrophenol induces morphogenetic changes in basidiomycete mycelia remarkably similar to those observed during interspecific interactions (Griffith, Rayner & Wildman, 1994*a,b,c*; see below). The role of reactive oxygen species in production of aerial conidiophores by *Neurospora crassa* has been emphasized by Hansberg & Aguirre (1990).

Oxidation–reduction and free radical chain reactions could therefore play an important hyperepigenetic role in determining the degree of insulation of hyphal boundaries. These reactions may also have other important consequences. By producing free radical scavengers, they may help protect against the toxicity of superoxide and singlet oxygen species – generated from electron transport chains and photosensitization of haem- and flavin-containing proteins in mitochondria, for example (Halliwell & Gutteridge, 1989). In wood-inhabiting fungi, they can result in the depolymerization (and repolymerization) of lignocellulose – indeed this has long been assumed to be their primary function (e.g. Ander & Eriksson, 1976). On the other hand, were those reactions to occur within the protoplasm rather than at hyphal boundaries, their effect would be to initiate degeneration and the sealing off of internal communication pathways.

There may therefore be some justification in relating the unstable, indeterminate development of mycelia to unstable, indeterminate chemistry, precipitated by the evolutionary crisis and opportunity caused when photosynthesis poisoned the atmosphere with oxygen. The mitochondrion (and endoplasmic reticulum containing cytochrome P-450), as the sensitive mediator(s) of electron flows in eukaryotic cells, may thereby assume an executive role in switches between inductive, transductive or degenerative pathways (Rayner & Ross, 1991; Rayner *et al.*, 1994*a,b*).

More immediate opportunities and crises are a frequent feature in the natural lives of mycelial networks when they encounter one another as they grow within and between their nutrient sources. The abilities of the networks to respond appropriately to local events on their boundaries, and thereby the attunement of the underlying control mechanisms, are then tested to the full, with individual genetic survival at stake.

Embattled networks

Close encounters between mycelial networks commonly lead to conflicts in which the participants dispute rights of possession like territorial armies, probing one another's weak points with thrust, counterthrust and biochemical weaponry. These conflicts have been studied, sporadically, throughout the last hundred years, going back at least as far as the work of

Reinhardt (1892). However, the implications of these conflicts in understanding how mycelia organize themselves as dynamic systems do not generally seem to have been grasped. Indeed, the relative scarcity of studies of mycelial interactions epitomizes the limited appreciation of the importance of mycelial organization in the determination of fungal life patterns.

Some of the most striking evidence of the outcome of mycelial conflict in nature is provided by the beautiful mosaic-like patterns often found in decaying wood, which directly reflect the spatiotemporal distribution of embattled individuals, both of the same and of different species (Rayner & Todd, 1979). Similar mosaics are familiar in communities of crustose lichens.

The boundary regions between different mycelial individuals in wood frequently appear as dark lines in cross section, and so have often been referred to as 'zone lines'. However, care is needed to discriminate between lines due to mycelial interaction and others of similar appearance that have a different origin.

The aetiology of zone lines was generally considered to be unresolved until the classical studies of Campbell (1933, 1934), who identified three basic causes: interaction between adjacent mycelia of different species; 'wound gum' produced at the interface between healthy and diseased wood, and pseudosclerotial plates. The latter consist of crust-like layers of melanized hyphae which can be induced to form within or around a single mycelium in response to damage, desiccation and/or exposure to light or oxygen (cf. Lopez-Real & Swift, 1977). Perhaps as a result of Campbell's studies of pseudosclerotial plates – firstly in *Armillaria* and *Xylaria*, and later in other fungi (Campbell & Munson, 1936; Campbell, 1938*a,b*) – these features became widely treated as synonymous with zone lines (Lopez-Real, 1975). Meanwhile, zones formed at interfaces between healthy and diseased wood became regarded as a host-defence response (e.g. Shain, 1979) and the distinctive properties of zones due to mycelial interaction (of which some were noted by Hubert, 1924) were apparently neglected.

Compared with pseudosclerotial plates, which, being organized as crusts of compact mycelium, commonly appear as single, intensely pigmented, sharply defined zones, interaction zone lines are more diffuse and heterogeneous. Often, interaction zone lines contain relatively undegraded material, they may be abutted by pseudosclerotial plates and they are commonly inhabited by microfungi that are able to survive in the 'no man's land' between warring neighbours (Rayner, 1976; Rayner & Todd, 1979, 1982*a*).

In paired cultures on agar, interspecific mycelial interaction responses are commonly evident before contact as an inhibition of hyphal extension of one or both participants. In some cases such inhibition may be attributed to relatively trivial factors, e.g. nutrient depletion and pH changes due to organic acid production (Nesemann, 1953; Rasanayagam & Jeffries, 1992), but more generally it has been thought to be due to the production of diffusible antibiotics. Indeed in the 1950s and 1960s, antibiosis was widely considered to be the principal mechanism underlying fungal and microbial antagonism (Brian, 1957; Park, 1960). The historical precedent set by the discovery of penicillin, and the general focus of fungal ecologists on soil-inhabiting fungi may have contributed to this view.

Many interspecific interactions only become evident, however, once hyphal systems have come into contact with one another, or at least within very close range. Such post-contact responses occur both at the level of individual hyphae and whole mycelia, in much the same way that encounters between two territorial armies consist both of individual skirmishes and mass manoeuvres.

Interactions between individual hyphae can be classified into two categories: hyphal interference and mycoparasitism. The difference between these categories may well be related to the rapidity with which a cell death reaction follows upon hyphal contact.

Hyphal interference involves a rapid cell death reaction characterized by vacuolation and degeneration of cellular organelles (Ikediugwu, 1976). The reaction resembles the hypersensitive response of plant cells (Stakman, 1915; Heath, 1976, 1984) to microbial infection and also the apoptosis (programmed cell death) of animal cells (Aylmore & Todd, 1986). Hyphal interference came to prominence as a widespread mechanism of hyphal interaction following the studies of Ikediugwu and Webster (1970*a*,*b*) on coprophilous fungi and the demonstration of its role in the biological control of *Heterobasidion annosum* by *Phlebiopsis gigantea* (Ikediugwu, Dennis & Webster, 1970). The underlying mechanism is not known, but seems likely, following on from earlier discussion, to involve the generation of free radicals and reactive oxygen species (cf. Buttke & Sandstrom, 1994).

Hyphal parasitism has long been known in fungi, and can involve both necrotrophic and biotrophic modes of nutrition (e.g. Barnett & Binder, 1973). Hyphal encoiling, penetration and haustorium-formation may be indicative respectively of increasingly intimate or specialized mycoparasitic associations (e.g. Ainsworth & Rayner, 1989; Laing & Deacon, 1991).

The induction of large-scale shifts in mycelial organization as a consequence of interspecific iteration has been widely observed, particularly amongst basidiomycetes (e.g. Harder, 1911; Zeller & Schmitz, 1919; Rayner, Griffith & Wildman, 1994c). Commonly, these shifts involve the emergence of massed ranks of hyphae, organized into ridges, sheets, fans, cords or rhizomorphs, beyond the immediate sites of nutrient assimilation at mycelial fronts. These shifts are fundamental to the outcome of interactions in that they determine the invasiveness and resistance to invasion of the participants. However, they have generally attracted little attention, other than causing some confusion as to whether they should be regarded as the result of 'inhibition' or 'stimulation' (cf. Porter, 1924; Porter & Carter, 1938), perhaps reflecting a mind-set that views the mycelium as a purely assimilative structure.

Clearly, the capacity of mycelia to undergo organizational shifts in response to interaction is yet further evidence of their versatility as complex, heterogeneous systems that are able to respond to unpredictable changes in their circumstances. A neighbouring mycelium is liable to pose a resistance to expansion for a variety of reasons, including physical obstruction, deprivation of nutrients and release of inhibitory chemicals. The emergence of relatively insulated hyphal systems in response to such a resistance would both be an expected consequence of the way mycelia are organized as hydrodynamic systems and an adaptive means of retaining and/or extending genetic territory.

It is therefore interesting that emergent mycelial phases produced during mycelial interactions exhibit instabilities characteristic of a driven non-linear system with impeded throughput capacity. Moreover, production of these phases is associated with characteristic changes in the activity of phenol-oxidizing enzymes and the release of hydrophobic metabolites that can be mimicked by exposure of mycelia to the uncoupling agent 2,4-dinitrophenol (Griffith *et al.*, 1994a,b,c).

To make love or to make war

When mycelia of the same or closely related species of ascomycetous and basidiomycetous fungi encounter one another, individual hyphal branches can fuse or anastomose with one another, so that their protoplasm becomes contiguous. This fact has important consequences for the survival prospects of the populations of genomic organelles, nuclei and mitochondria, that proliferate within the unstable boundaries of mycelial networks (Rayner, 1991).

Even in 'self'-fusions between hyphae having the same genetic origin, interference is possible between organelles in different epigenetic states. Some means of harmonizing or synchronizing gene expression may therefore be necessary if destabilization is to be avoided. This may account for the close physical association and conjugate division or nuclei in basidiomycete dikaryons (see below) and the 'nuclear replacement reaction' described by Noble (1937), Aylmore and Todd (1984) and Todd and Aylmore (1985). The latter reaction follows self-fusions in basidiomycete monokaryons and dikaryons, and involves the disintegration of nuclei within 'recipient' hyphal compartments and their replacement by daughter nuclei derived from 'donor' compartments.

Where anastomosing hyphae have a different genetic origin and so can be regarded as 'nonself', genetically disparate nuclei and/or mitochondria become associated with one another in the same protoplasm. The ability or inability of the associated organelles to co-exist stably then becomes of critical importance to the genetic structure of natural populations.

In many basidiomycetes, nonself fusion between sexually undifferentiated hyphae has been known since early in the twentieth century to be an essential feature of sexual outcrossing (Ainsworth, 1976). Homokaryotic 'primary mycelia' derived from haploid meiospores fuse and exchange nuclei and so give rise to a 'secondary mycelium' containing two kinds of nuclear genome. This secondary mycelium generally has a greater tendency than its progenitors to produce non-assimilative structures appropriate to the explorative, protective and reproductive stages of colonization that follow initial establishment. Often, it is a true dikaryon, having two conjugately dividing nucei per hyphal compartment, and in many cases it may also produce clamp connections that distinguish it from monokaryotic primary mycelia containing uninucleate compartments. However, the production of clampless monokaryons and clamp-bearing dikaryons is by no means a universal feature of outcrossing basidiomycetes. Despite the impression given by many elementary texts and even by research articles concerning the most-studied organisms, *Coprinus cinereus* and *Schizophyllum commune*, patterns of nuclear behaviour can vary markedly between and even within genera (Boidin, 1971; Kemp, 1975; Kühner, 1977). Both primary and secondary mycelia can contain multinucleate hyphal compartments. Clamp-connections may be absent from secondary mycelia, and, in genera such as *Stereum*, *Coniophora* and *Phanerochaete*, clamps can be present on both primary and secondary mycelia (Coates, Rayner & Todd, 1981; Ainsworth, 1987). Whereas karyogamy is delayed in most basidiomycetes until formation of the basidium initial, so that the secondary mycelium is heterokaryotic, *Armillaria* species have been shown to produce secondary mycelia containing allodiploid nuclei (Korhonen & Hintikka, 1974).

The pattern of emergence of the secondary mycelium from mated homokaryons depends on the rates and routes of nuclear migration following non-self fusion, and provides useful insights into the organizational properties of mycelial systems. However, this fact has received relatively little attention because of the widespread treatment of the mycelium as a homogeneous structure in which nuclear migration can simply be switched on or off (cf. Snider, 1965).

As was explicitly recognized by Buller (1931), and in keeping with the organization of mycelial systems as open-ended networks, two main routes are available for nuclear migration – radial and tangential. There are also possibilities for migration either through hyphae already existing at the time of fusion between systems and those formed subsequently. Radial and tangential migration through pre-existing hyphae allows homokaryons to be converted into heterokaryons but necessitates passage through dolipore septa which have to be eroded enzymatically. By contrast, nuclei migrating via the proliferation of anastomoses can travel tangentially and outwardly without negotiating septa, but secondary mycelium emergence will be confined to peripheral regions. An example occurs in *Coniophora puteana*, where it appears that the emerging secondary mycelium acts as an explorative 'sink', drawing resources from the homokaryotic progenitors (Ainsworth & Rayner, 1990).

Secondary mycelia are the predominant forms that can be found growing in natural populations of outcrossing basidiomycetes. When they encounter one another, they typically retain their genetic and physiological integrity as discrete individuals by means of non-self rejection or somatic (= vegetative) incompatibility reactions. These reactions follow hyphal fusion and lead slowly or rapidly (the rapid reactions are reminiscent of hyphal interference) to protoplasmic degeneration and melanization along interaction interfaces, most likely associated with oxidative stress (cf. Rayner, 1991). The resultant 'demarcation zones' were observed in paired cultures of *Fomitopsis pinicola* by Schmitz (1925) and Mounce (1929) and reported sporadically thereafter in a variety of other fungi (see Rayner & Todd, 1979, 1982*a,b*). Somatic incompatibility was shown rigorously to be a cause of zone lines in wood inhabited by natural populations of the same species of basidiomycetes by Rayner & Todd (1977) and Williams, Todd & Rayner (1981*a,b*). It is now recognized also to occur in litter-decomposing (Murphy & Miller, 1993) and mycorrhizal (Dahlberg & Stenlid, 1994,

1995) species and has been used widely in the study of the genetic structure of natural populations.

In spite of the widespread occurrence of somatic incompatibility, a persistent view became established between the mid-1930s and mid-1970s to the effect that naturally occurring basidiomycete mycelia can develop as physiologically and ecologically unified genetic mosaics. This view had its origin in the work of Buller (1931) which demonstrated the need for co-operation between mycelia originating from separate basidiospores of *Coprinus sterquilinus* in order to produce a sizeable fruit body. It was then championed by Raper (1966) and Burnett (1976), based largely on studies of the distribution of mating alleles in natural populations of *Coriolus versicolor* and *Piptoporus betulinus* made by Burnett & Partington (1957). However, outcrossing populations of *C. versicolor* and *P. betulinus* were eventually shown to consist of somatically incompatible individuals by Rayner and Todd (1977) and Adams, Todd and Rayner (1981).

A contributory factor in the confusion which developed (and still persists) between co-operative and individualistic views of basidiomycete interrelationships lies in the differing behaviour of primary and secondary mycelia. Whereas genetically different secondary mycelia reject one another and thereby maintain their individual integrity, homokaryons must override any potential for rejection if they are to mate and produce a secondary mycelium (Rayner & Todd, 1979; Rayner *et al.*, 1984).

This situation is emphasized by patterns of secondary mycelium emergence following matings between heterokaryons and homokaryons ('he-ho' matings). The ability of heterokaryons to mate with homokaryons was demonstrated by Buller (1931) and so is widely referred to as the 'Buller phenomenon'. Where the participants respectively possess strictly binucleate and uninucleate compartments, their interactions may also be referred to as dikaryon-monokaryon or 'di-mon' matings.

A characteristic feature of he-ho matings is that nuclear migration is unidirectional, from the heterokaryon, which resists nuclear access, into the homokaryon. Where both kinds of nuclei from the heterokaryon invade the homokaryon, the latter becomes subdivided into alternating sectors. These sectors contain heterokaryons that are somatically incompatible, even though they all contain one nuclear type in common and the other nuclear types can mate in homokaryotic combinations (Todd & Rayner, 1978; Coates *et al.*, 1981; Boddy & Rayner, 1982; Coates & Rayner, 1985a; Angwin & Hansen, 1993). The adage that two's company, three's a crowd holds true.

Unidirectional nuclear migration can also occur under other circumstances, as when one of two paired homokaryons is a 'blocker' (a trait sometimes associated with a 'senescent' morphology), or when one of these homokaryons originates from a non-outcrossing population. The occurrence of non-outcrossing and outcrossing populations in the same species has been detected, for example, in several species of *Stereum* (Ainsworth, 1987). When homokaryons from these populations are paired, nuclei from the non-outcrosser can invade and sometimes take over the outcrosser (Ainsworth *et al.*, 1990*a*).

Besides unidirectional migration, patterns of nuclear exchange between basidiomycete homokaryons can sometimes be strongly asymmetric, with one participant being more readily invaded than the other. An example is provided by the 'bow-tie' reaction in *Stereum hirsutum* (Coates & Rayner, 1985*b*; see below). Studies of such interactions have indicated that they result from different degrees of invasiveness of non-resident nuclei. Other studies indicate that once invasive nuclei have become established within an acceptor homokaryon, they migrate at equal rates, irrespective of their genotype. Such findings have been taken to imply that two distinctive kinds of process may be involved in nuclear migration. One of these processes, 'acceptor migration' is purely a function of the acceptor genotype, whilst the other, 'access migration', is a function of the invasive genotype (Rayner *et al.*, 1984).

Somatic incompatibility is also a widespread phenomenon in outcrossing populations of ascomycetes. However, in these fungi the persistent expression of a rejection response following fusion of sexually undifferentiated hyphae between homokaryons need not preclude mating, as it does in most basidiomycetes. This is because many ascomycetes produce specialized sex organs through which plasmogamy is effected, and so can keep their sex lives and somatic lives separate. This fact is fundamental to understanding differences between the regulation, timing and consequences of nonself rejection in ascomycetes and basidiomycetes but seems commonly to be disregarded, leading to misinterpretation (e.g. Esser & Blaich, 1973, 1994).

In ascomycetes, somatic incompatibility between homokaryons is usually inferred from one of two, not necessarily coincident, lines of evidence: demarcation zone formation and inability to produce a heterokaryon (heterokaryon incompatibility) (Glass & Kuldau, 1992).

Demarcation zone formation involves a similar kind of degenerative reaction to that observed between basidiomycete secondary mycelia, or between basidiomycete homokaryons that are not competent to mate. It was described in *Diaporthe* by Cayley (1923), who also made a considerable effort to identify the underlying genetic controls (Cayley, 1931).

Heterokaryon incompatibility was classically studied in *Aspergillus nidulans*, where its demonstration provided the basis for questioning the long-held view that heterokaryosis is widespread in natural populations of ascomycetous fungi (Caten & Jinks, 1966; Croft & Jinks, 1977).

Heterokaryon incompatibility has been much favoured as a 'precise' criterion for characterizing the genetic structure of ascomycete populations (cf. Leslie, 1993), even though it may have poor resolution (Rayner, 1992). Since basidiomycete homokaryons can mate to form heterokaryons, there has been much confusion in relating between apparently equivalent phenomena involving ascomycete homokaryons, sexually incompatible basidiomycete homokaryons and basidiomycete heterokaryons.

In outcrossing populations of both ascomycetes and basidiomycetes, the complexity of the genetic systems regulating somatic incompatibility is generally sufficient to ensure that virtually any individuals that are not closely related to one another will not integrate. The main difference lies in the fact that these individuals are basically homokaryotic in ascomycetes, whilst they are secondary mycelia in basidiomycetes. Expression of a rejection response following somatic fusions between ascomycete homokaryons does not imply that the homokaryons are reproductively isolated, only a failure to produce a fertile zygote from the ascogonium implies that. By contrast, rejection between basidiomycete homokaryons does imply infertility, and may result in speciation or the development of non-outcrossing sub-populations (Rayner *et al.*, 1984). The difference between the multiallelic mating systems of most basidiomycetes, which regulate heterokaryosis, and the biallelic mating systems of ascomycetes, which regulate ascus formation, may be related to this fact.

The dynamic interplay between war and peace

Interpretation of non-self interactions in ascomycete and basidiomycete species has been dominated by the tendency to view incompatibility as an absolute, fully genetically determined, 'plus and minus' phenomenon. The typical line of thinking can be caricatured as follows (cf. Esser & Blaich, 1973, 1994). Either two strains can mate or they cannot. Either they can form a heterokaryon or they cannot. Incompatibility must be regulated by the products of specific, precisely quantifiable genes or sets of genes that interact in a predictable and fully consistent manner. These genes can be identified by inbreeding programmes that eliminate all genetic variation other than that at the specific loci of interest. When these loci are identified, they can be singled out (tautologically) as *the* genes that *cause* incompatibility, irrespective of the wide variety of factors that may impinge on nonself interactions in natural populations.

However, numerous observations suggest that mycelia interact as complex systems in which the outcome of genomic associations is liable to be both varied and finely balanced between rejection and acceptance. For example, it is not uncommon, in homokaryotic interactions of both ascomycetes and basidiomycetes, to observe a reciprocal relationship between rejection and acceptance along an interaction interface, or to observe expression of rejection prior to heterokaryotic emergence (or vice versa) (Rayner, 1991). Genetic differences therefore potentiate rather than instigate rejection and acceptance, so that interaction outcomes are dependent both on genes and context.

The role of context in determining interaction outcomes can be appreciated by regarding mycelial systems as indefinitely expandable, heterogeneous genetic territories in which genomic relationships both influence and are influenced by resistances at and within hyphal boundaries. Encounters between these systems can then be likened to those between any kinds of expandable organizations capable of both competing and co-operating with one another (Fig. 8.4). Depending on the degree of integration, such encounters may result in stable co-existence, takeovers, degeneracy or persistent internal conflicts.

Stable co-existence can be achieved in one of two radically different ways. On the one hand, each system can retain its individual integrity either by not opening or by closing its boundaries in response to nonself, so sustaining a persistent 'stalemate' across territorial boundaries, as in the formation of somatic rejection zones. The systems therefore 'co-exist' in the sense that they continue to inhabit the same arena by dint of their mutual exclusivity. On the other hand, a synergistic partnership based on complementation and non-interference may be developed, as in the formation of basidiomycete secondary mycelia. Here co-existence implies integration, such that formerly separate systems coalesce.

Where, as often occurs between homokaryons derived from sympatric, outcrossing, basidiomycete populations, both rejection and acceptance are possible following a nonself encounter, the question arises as to the relative ease with which any resistance to acceptance can be obviated.

This question can be addressed by placing disparate 'donor' homokaryons at different positions within or at the boundary of an 'acceptor' homokaryon, and mapping the pattern of emergence of the resulting secondary mycelia. Studies of this kind in *Stereum hirsutum* have shown that moderately unrelated (non-sib) donors achieve more effective access, and hence occupy larger genetic territories, than donors closely related to the acceptor (Coates & Rayner, 1985*c*).

A different kind of study, with *Heterobasidion annosum*, revealed that homokaryons derived mitotically *via* conidia from a common hetero-karyon exhibited less initial somatic rejection when paired together than the original basidiospore-derived strains (Ramsdale & Rayner, 1995). This may suggest that nuclei within a heterokaryon become epigenetically mod-ified in such a way as to enable them to co-exist without interference, and that this modification is retained following production of homokaryotic or uninucleate conidia, but not following meiosis.

Other studies of *H. annosum* have indicated that production of uni-nucleate conidia may provide a means of escape from heterokaryons subject to conflict between disparate nuclear genomes. Homokaryons arising from such conidia may also retain certain heterokaryotic pheno-typic attributes, such as ability to form pseudosclerotial plates. Whereas laboratory-synthesized heterokaryons may commonly exhibit conflict, heterokaryons isolated from the field rarely do so, suggesting that there may be strong natural selection favouring non-conflicting genomic associations (Ramsdale & Rayner, 1994).

Whereas interactions between sympatric basidiomycete homokaryons carrying different mating alleles tend to result in outright rejection or out-right co-existence, sometimes even in different locations between the same combination of strains, more intermediate outcomes can occur in other kinds of interactions.

For example, it has been found that homokaryons of *Stereum hirsutum* containing the same mating-type alleles but differing at another locus (or loci) that regulates nuclear access can produce 'bow-tie' reactions in which degenerative genomic associations occur prior to the re-emergence of vigorously growing hyphae. The extent of these interactions is recipro-cally related to the degree of expression of a somatic rejection response and the relative effectiveness of invasive genomes in gaining access. The reactions can result in territorial ingress, genomic takeover, by one homo-karyon or the other (and sometimes both) *via* an intermediary, unstable heterokaryotic phase (Coates & Rayner, 1985*b*; Ainsworth & Rayner, unpubl.).

Analogous reactions to the bow-tie reaction in *S. hirsutum* have been detected in populations of xylariaceous fungi, including species of

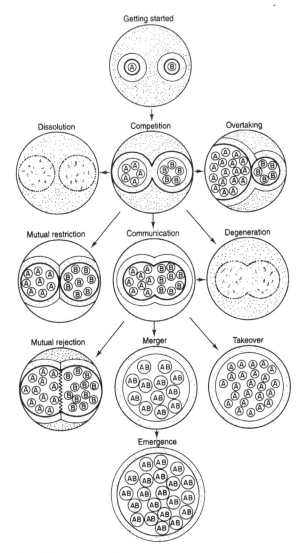

Fig. 8.4. Corporate interplays. Two organizations, containing inhabitants (A and B) with similar but not identical expertise and requirements for resources begin to establish themselves in the same arena, represented here as a circle with densely stippled contents. Each organization is surrounded by a region of 'influence' (lightly stippled) which it draws on to support its activities. If the net income from this region balances net expenditure, the organization will maintain but fail to expand its boundaries. If there is a 'profit margin' (income exceeds expenditure), the organization will grow. However, if there is a deficit, the organization will lose viability. Both organizations grow until their regions of influence overlap and begin also to become limited by the boundaries of the arena. The

Daldinia, Hypoxylon and *Biscogniauxia* (Sharland & Rayner, 1986, 1989*a,b*). Here again, territorial ingress may be effected *via* production of an unstable heterokaryotic phase, to an extent that depends on the timing and placement of a somatic rejection response. In some cases, somatic recombination can occur within the heterokaryon.

Genomic takeovers have also been demonstrated in diploid–haploid matings of *Armillaria* species, where somatic recombination has been shown to accompany invasion of the haploid mycelium by diploid nuclei (Rizzo & May, 1994; Carvalho, Smith & Anderson, 1995), and, as described earlier, in outcrossing – non-outcrossing interactions. All such takeovers highlight the primarily 'selfish' nature of associating genomes and the resultant need for some kind of 'harmonization' process if stable partnerships are ever to arise. The occurrence of such a process would explain why nuclear genomes of *H. annosum* re-associate more readily than they associate within heterokaryons (see above).

The contrast between the 'all-or-none' mechanisms promoting or restricting gene flow in sympatric basidiomycete populations and the less well-defined mechanisms operating between allopatric populations became evident in several early studies. These studies indicated that whereas strongly reproductively isolated populations or 'sibling species' developed sympatrically, these populations could be 'partially interfertile' with allopatric populations (e.g. Macrae, 1967). This situation may be explained by the operation of natural selection on variants prone to take-over or degeneracy following nonself encounters in sympatric populations (Rayner & Franks, 1987; Rayner, 1988; Ainsworth *et al.*, 1990*a*). Allopatric populations would not normally be subject to such selection,

resulting competition can reduce profit margins to the point where no further expansion is sustainable (leading to mutual restriction), or beyond this point and into deficit (leading to dissolution). However, if one organization proves to be more competitive, it will begin to monopolize the arena as it erodes the other's region of influence (overtaking). On the other hand, if the organizations open their boundaries to one another and start to communicate, they will have the opportunity to operate as one and so expand their mutual influence. Much now depends on the way the inhabitants interact. If they compete or interfere, the resulting incompatibility may culminate in extensive degeneration, mutual rejection or takeover. If they complement, they may not only allow the corporation to fill the arena to its fullest potential, but also enable it to expand the boundary of the arena itself, by means of innovative interactions. The same organizational principles apply in all kinds of living systems, from cells to societies, and are beautifully demonstrated by anastomosable mycelial systems and their inhabitant nuclear and mitochondrial genomes.

hence allowing a potentially wide range of genomic interaction outcomes to be expressed.

Some of the varied consequences of genomic associations resulting from fusions between allopatrically derived basidiomycete homokaryons have been studied in *Stereum* spp. and *Heterobasidion annosum*. Although such homokaryons would be expected to have different mating alleles, they interact unstably, sometimes in patterns reminiscent of those just described in bow-tie interactions and similar phenomena.

In *Stereum*, examples of takeover, degeneracy and subdivision into conflicting local domains have all been observed (Ainsworth & Rayner, 1989). In one particular interaction, between strains of *Stereum hirsutum* derived from England and Australia, reciprocal nuclear exchange was shown to give rise to extensive degeneration on the Australian side, whilst the English side retained a vigorous morphology, similar to that of the original homokaryon. The degenerative reaction was accompanied by the proliferation of crystalline aggregates of the sesquiterpene, (+)-torreyol (Fig. 8.5A, B) and subcultures from the Australian side gave rise to colonies with varied and unstable phenotypes (Ainsworth *et al.*, 1990*b*).

An example of subdivision into conflicting local domains was provided by an interaction between a strain of *Stereum hirsutum*, originating from what at the time was the Soviet Union, and a strain of *Stereum complicatum* from the United States (Ainsworth *et al.*, 1992). Reciprocal nuclear exchange was followed by the emergence of two distinctive kinds of heterokaryons, 'H-types' and 'C-types', which respectively resembled either the *S. hirsutum* or the *S. complicatum* homokaryotic progenitors. The *S. complicatum* side gave rise only to a C-type heterokaryon, whereas the *S. hirsutum* side gave rise to both C-types and H-types, resulting in a complex pattern of subdivision (Fig. 8.5C, D). Molecular fingerprinting analysis revealed that whereas a full complement of *S. complicatum* nuclear DNA sequences could be detected in both kinds of heterokaryons, certain *S. hirsutum* sequences were absent from C-types. Subsequent observations of patterns of protoplast regeneration from C-types and H-types have revealed higher rates of post-germination mortality from H-types, and a low incidence of homokaryotic, *S. hirsutum* regenerants, from C-types (Ramsdale & Rayner, unpubl.). These observations suggest that *S. hirsutum* nDNA is modified in C-types so as to become a dependent passenger, thereby obviating the genomic conflict that results when *S. complicatum* and unmodified *S. hirsutum* nuclear genomes are associated.

Allopatric, sexually compatible homokaryons of *H. annosum* have been found to exchange nuclei in such a way that the non-resident genomes come

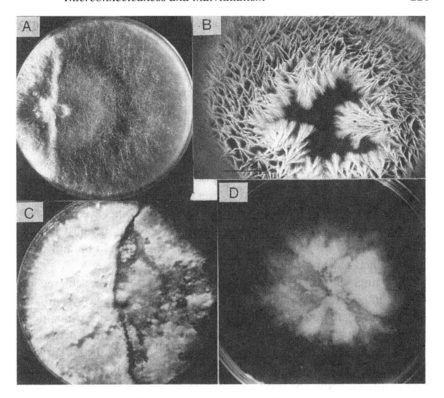

Fig. 8.5. Unstable outcomes of exchange of nuclear DNA between allopatrically derived homokaryons of *Stereum*. A. Degenerative development of an Australian strain (right) when paired with an English strain, associated with the outgrowth in alternating zones of densely and sparsely branched crystalline aggregates of the sesquiterpene (+)-torreyol. B. Crystallization of sesquiterpene (+)-torreyol, illustrating the fact that complex, mycelium-like patterns can be based on simple organizational principles. C. Complex phenotypic pattern formed in a *S. hirsutum* strain isolated from what was originally the Soviet Union when paired with a *S. complicatum* strain from the USA (right). D. Subculture from the Soviet strain shown in D, showing outgrowth of somatically incompatible resident and invasive phenotypes. (A, B from Ainsworth *et al.*, 1990*b*; C, D from Ainsworth *et al.*, 1992.)

to outnumber (by as much as 9:1) resident genomes on either side of the interaction interface. Remarkably, the resultant heterokaryons retain the phenotypic attributes of their progenitor homokaryons when subcultured, suggesting either that the invasive genomes were not expressed or the operation of some form of 'imprinting'. The possibility of a role for cytoplasmic genes was eliminated by showing that homokaryotic conidia derived from the heteropkaryons and containing invasive nuclei produced mycelial phenotypes typical of the non-resident. These observations contrasted with behaviour in sympatrically derived heterokaryons. Here nuclear genomes were maintained in larger numbers within their resident mycelium than in the mycelium they invaded, and the proportions of nuclei were maintained in a strict 'dominance hierarchy' reflecting their rate of proliferation in resident cytoplasm (Ramsdale & Rayner, 1994, 1995).

All these observations point to the delicate balance between association and dissociation of nonself nuclear genomes in common cytoplasm, and the sensitivity of this balance to local circumstances within the context of a heterogeneous mycelial system. From this standpoint, it is perhaps more surprising that stable associations can ever be achieved between genetically disparate genomes than that unstable interaction outcomes occur.

For nonself genomes to associate stably, two conditions must be fulfilled. Each kind of genome must proliferate at an equal rate and their gene products must not interfere.

Equal rates of proliferation of nuclear genomes can be achieved by physical integration or coupling, as in diploid and dikaryotic systems, or through the regulation of division within a cytoplasmic context that does not differentiate between nuclei containing disparate genetic information. The latter situation may occur in sympatric *H. annosum* heterokaryons that maintain a constant proportion of resident and non-resident nuclei despite differences in the proliferation rate of these nuclei within homokaryons and immediately following mating (Ramsdale & Rayner, 1994).

Prevention of interference between gene products may become increasingly impossible, perhaps in a highly non-linear fashion, with the degree of disparity between genomes. Interference between products affecting electron flows may be particularly critical here in view of the capacity for dysfunction in these products to lead to the generation of free radicals and reactive oxygen species. For example, conflict between nuclear genes encoding mitochondrial components could have this effect. Here, it is of interest that with the exception of certain strains showing unusual mating behaviour, e.g. in *Neurospora tetrasperma* (Lee & Taylor, 1993) and in *Stereum hirsutum* (M. Ramsdale, M.L. Rayner & A.D.M. Rayner, unpubl.), extensive mitochondrial transfer between paired homokaryons has not been demonstrated (e.g. May & Taylor, 1988). It might therefore be expected that variation in the outcome of allopatric matings on either side of interaction interfaces, such as in *Stereum* spp., could be due to differences in resident mtDNA. However, this has not been found to be the case (Ainsworth *et al.*, 1992), and it may be instead that differential epigenetic effects on the expression of invading nuclear genomes plays a role.

One way of avoiding interference between gene products would be to limit or prevent their expression; this could enable nuclear genomes to invade a recipient mycelium without precipitating a degenerative reaction, much in the same way that virulent biotrophic parasites can establish in susceptible (= compatible) plants by not inducing hypersensitivity (e.g. Heath, 1984). The extent to which such silencing would be necessary would depend on the degree of disparity, and in extreme cases, e.g. in allopatric combinations, virtually complete genomic shutdown may be appropriate. This could explain the patterns observed in allopatric matings of *Stereum* and *Heterobasidion*. Once having infiltrated an opposing mycelium in this way, re-expression of the invasive genomes could then result in the elimination of the resident.

Another way of avoiding the consequences of genomic conflict may be achieved by restricting the supply of oxygen that could otherwise amplify the conflict into a degenerative reaction. Some observations suggest that limited aeration may actually allow stable heterokaryons to form in some natural environments. For example, it has been found that isolates of the ascomycete, *Eutypa spinosa*, from wet parts of beech trees grow out initially as heterokaryons whereas isolates from drier, aerated regions develop as homokaryons (Hendry, 1993).

The fact that heterokaryotic mycelia tend to exhibit morphogenetic patterns expected of better-insulated systems tallies with the need to restrict oxygen supply to potentially conflicting genomes. Indeed, the ability to produce insulating compounds with oxygen-scavenging or antioxidant properties as a means of protection from oxidative stress may be fundamental to the ability of nonself genomes to associate stably in terrestrial eukaryotes. This would add a new dimension to the maxim that unprotected sex is unsafe sex, liable to culminate in the dissolution of the individual! It would also reinforce the view that the interactions of genomic organelles can only be understood within the context of cellular boundaries, just as much as the interactions of individual organisms can only be grasped within the context of population boundaries.

Conclusions: balancing association and dissociation

So, are fungi to be regarded, after all, as unique and separate forms of life? Certainly, their mycelial habit results from a mode of cell proliferation that is unusual in other life forms, but which does also occur in the prokaryotic Actinomycetes, in pollen tubes and in certain kinds of algae. It also provides unique opportunities for examining the varied ways in which genomic populations can interact within a complex cellular system. However, when it comes to the organizational principles that enable it to negotiate heterogeneous and therefore locally unpredictable environments, the mycelium is far from unique. Rather, it epitomizes the indeterminate processes that enable open-ended systems of all kinds to vary the balance between associative and dissociative mechanisms and so to respond in an energy-efficient manner to local variations in circumstances. Such systems are able both to change and to be changed by their local environment, and so are continually creative as they gain, distribute and discharge energy. They cannot be understood as simple, clockwork-like assemblies of fully discrete, particle-like units that can be individually exposed to natural selection and discarded if they don't fit in. Emphasis on systems where discrete units of selection, in the form of genes or individuals, can readily be defined and allocated to separate generations have given rise to a view of evolution that emphasizes adaptational processes. From this perspective competition appears to be of paramount importance in the refinement of super-efficient, centrally administered organizations. By contrast, the perspective gained from concentrating on indeterminate systems reveals competition to be a hindrance to innovation. The latter depends very much on 'communication' and 'give and take', and in mycelia may be founded most fundamentally on the give and take of electrons! An understanding of indeterminacy also highlights the importance of serendipity, being in the appropriate place at the appropriate time, and of a diffuse power structure regulated by local feedback, if systems are to thrive in an uncertain world. Human and other societies, like mycelia, are indeterminate systems . . . .

References

Adams, T.J.H., Todd, N.K. & Rayner, A.D.M. (1981). Antagonism between dikaryons of *Piptoporus betulinus*. *Transactions of the British Mycological Society*, **76**, 510–13.

Ainsworth, A.M. (1987). Occurrence and interactions of outcrossing and non-outcrossing populations of *Stereum, Phanerochaete* and *Coniophora*. In *Evolutionary Biology of the Fungi* (ed. A.D.M. Rayner, C.M. Brasier & D. Moore), pp. 285–299. Cambridge: Cambridge University Press.

Ainsworth, A.M., Beeching, J.R., Broxholme, S.J., Hunt, B.A., Rayner, A.D.M. & Scard, P.T. (1992). Complex outcome of reciprocal exchange of nuclear DNA between two members of the basidiomycete genus *Stereum*. *Journal of General Microbiology*, **138**, 1147–57.

Ainsworth, A.M. & Rayner, A.D.M. (1989). Hyphal and mycelial responses associated with genetic exchange within and between species of the basidiomycete genus *Stereum*. *Journal of General Microbiology*, **135**, 1643–59.

Ainsworth, A.M. & Rayner, A.D.M. (1990). Mycelial interactions and outcrossing in the *Coniophora puteana* complex. *Mycological Research*, **94**, 627–34.

Ainsworth, A.M. & Rayner, A.D.M. (1991). Ontogenetic stages from coenocyte to basidiome and their relation to phenoloxidase activity and colonization processes in *Phanerochaete magnoliae*. *Mycological Research*, **95**, 1414–22.

Ainsworth, A.M., Rayner, A.D.M., Broxholme, S.J. & Beeching, J.R. (1990*a*). Occurrence of unilateral genetic and genomic replacement between strains of *Stereum hirsutum* from non-outcrossing and outcrossing populations. *New Phytologist*, **115**, 119–28.

Ainsworth, A.M., Rayner, A.D.M., Broxholme, S.J., Beeching, J.R., Pryke, J.A., Scard, P.T., Berriman, J., Powell, K.A., Floyd, A.J. & Branch, S.K. (1990*b*). Production and properties of the sesquiterpene, (+)-torreyol, in degenerative mycelial interactions between strains of *Stereum*. *Mycological Research*, **94**, 799–809.

Ainsworth, G.C. (1976). *Introduction to the History of Mycology*. Cambridge: Cambridge University Press.

Ander, P. & Eriksson, K.E. (1976). The importance of phenol-oxidase activity in lignin degradation by the white-rot fungus *Sporotrichum pulverulentum*. *Archives of Microbiology*, **109**, 1–8.

Andrews, J.H. (1992). Fungal life-history strategies. In *The Fungal Community*, 2nd edn (ed. G.C. Carroll & D.T. Wicklow), pp. 119–45. New York: Marcel Dekker.

Angwin, P.A. & Hansen, E.M. (1993). Pairing tests to determine mating compatibility in *Phellinus weirii*. *Mycological Research*, **97**, 1469–75.

Aylmore, R.C. & Todd, N.K. (1984). Hyphal fusion in *Coriolus versicolor*. In *The Ecology and Physiology of the Fungal Mycelium* (ed. D.H. Jennings & A.D.M. Rayner), pp. 103–25. Cambridge University Press.

Aylmore, R.C. & Todd, N.K. (1986). Cytology of non-self hyphal fusions and somatic incompatibility in *Phanerochaete velutina*. *Journal of General Microbiology*, **132**, 581–91.

Barnett, H.L. & Binder, F.L. (1973). The fungal host–parasite relationship. *Annual Review of Phytopathology*, **11**, 273–92.

Bartnicki-Garcia, S. (1973). Fundamental aspects of hyphal morphogenesis. *Symposia of the Society for General Microbiology*, **23**, 245–67.

Bell, A.A. & Wheeler, M.H. (1986). Biosynthesis and functions of fungal melanins. *Annual Review of Phytopathology*, **24**, 411–51.

Boddy, L. (1993). Saprotrophic cord-forming fungi: warfare strategies and other ecological aspects. *Mycological Research*, **94**, 641–55.

Boddy, L. & Rayner, A.D.M. (1982). Population structure, intermycelial interactions and infection biology of *Stereum gausapatum*. *Transactions of the British Mycological Society*, **78**, 337–51.

Boddy, L. & Rayner, A.D.M. (1983). Mycelial interactions, morphogenesis and ecology of *Phlebia radiata* and *Phlebia rufa* from oak. *Transactions of the British Mycological Society*, **80**, 437–48.

Boidin, J. (1971). Nuclear behaviour in the mycelium and the evolution of Basidiomycetes. In *Evolution in the Higher Basidiomycetes* (ed. R.H. Petersen), pp. 129–48. University of Knoxville: Tennessee Press.

Brian, P.W. (1957). The ecological significance of antibiotic production. *Symposia of the Society for General Microbiology*, **7**, 168–88.

Buller, A.H.R. (1931). *Researches on Fungi* 4. London: Longman.

Buller, A.H.R. (1933). *Researches on Fungi* 5. London: Longman.

Burnett, J.H. (1976). *Fundamentals of Mycology*, 2nd edn. London: Arnold.

Burnett, J.H. & Partington, M. (1957). Spatial distribution of mating-type factors. *Proceedings of the Royal Physical Society of Edinburgh*, **26**, 61–8.

Bushell, M.E. (1989*a*). Biowars in the bioreactor. *New Scientist*, **124**, 42–5.

Bushell, M.E. (1989*b*). The process physiology of secondary metabolite production. *Symposia of the Society for General Microbiology*, **44**, 95–120.

Buttke, T.M. & Sandstrom, P.A. (1994). Oxidative stress as a mediator of apoptosis. *Immunology Today*, **15**, 7–10.

Campbell, A.H. (1933). Zone lines in plant tissues. I. The black lines formed by *Xylaria polymorpha* (Pers.) Grev. in hardwoods. *Annals of Applied Biology*, **20**, 123–45.

Campbell, A.H. (1934). Zone lines in plant tissues. II. The black lines formed by *Armillaria mellea* (Vahl) Quél. *Annals of Applied Biology*, **21**, 1–22.

Campbell, A.H. (1938*a*). Contribution to the biology of *Collybia radicata* (Relh.) Berk. *Transactions of the British Mycological Society*, **22**, 151–9.

Campbell, A.H. (1938*b*). On the 'sclerotium' of *Collybia fusipes* (Bull.) Berk. *Transactions of the British Mycological Society*, **22**, 244–51.

Campbell, A.H. & Munson (1936). Zone lines in plant tissues. III. The black lines formed by *Polyporus squamosus* (Huds.) Fr. *Annals of Applied Biology*, **23**, 453–64.

Carvalho, D.B., Smith, M.L. & Anderson, J.B. (1995). Genetic exchange between diploid and haploid mycelia of *Armillaria gallica*. *Mycological Research*, **99**, (in press).

Caten, C.E. & Jinks, J.L. (1966). Heterokaryosis: its significance in wild homothallic ascomycetes and fungi imperfecti. *Transactions of the British Mycological Society*, **49**, 81–93.

Cayley, D.M. (1923). The phenomenon of mutual aversion between monospore mycelia of the same fungus (*Diaporthe perniciosa*, Marchal) with a discussion of sex-heterothallism in fungi. *Journal of Genetics*, **13**, 353–70.

Cayley, D.M. (1931). The inheritance of the capacity for showing mutual aversion between mono-spore mycelia of *Diaporthe perniciosa* (Marchal). *Journal of Genetics*, **24**, 1–63.

Coates, D. & Rayner, A.D.M. (1985*a*). Heterokaryon – homokaryon interactions in *Stereum hirsutum*. *Transactions of the British Mycological Society*, **84**, 637–45.

Coates, D. & Rayner, A.D.M. (1985*b*). Genetic control and variation in expression of the 'bow-tie' reaction between homokaryons of *Stereum hirsutum*. *Transactions of the British Mycological Society*, **84**, 191–205.

Coates, D. & Rayner, A.D.M. (1985*c*). Interactions between mating and somatic incompatibility in the basidiomycete, *Stereum hirsutum*. *New Phytologist*, **99**, 473–83.

Coates, D., Rayner, A.D.M. & Todd, N.K. (1981). Mating behaviour, mycelial antagonism and the establishment of individuals in *Stereum hirsutum*. *Transactions of the British Mycological Society*, **76**, 41–51.

Cooke, R.C. & Rayner, A.D.M. (1984). *Ecology of Saprotrophic Fungi*. London: Longman.

Croft, J.H. & Jinks, J.L. (1977). Aspects of the population genetics of *Aspergillus nidulans*. In *Genetics and Physiology of Aspergillus* (ed. J.E. Smith & J.A. Pateman), pp. 339–360. London: Academic Press.

Dahlberg, A. & Stenlid, J. (1994). Size, distribution and biomass of genets in populations of *Suillus bovinus* (L.: Fr.) Roussel revealed by somatic incompatibility. *New Phytologist*, **128**, 225–34.

Dahlberg, A. & Stenlid, J. (1995). Spatiotemporal patterns in ectomycorrhizal populations. *Canadian Journal of Botany* (in press).

Dowson, C.G., Rayner, A.D.M. & Boddy, L. (1986). Outgrowth patterns of mycelial cord-forming basidiomycetes from and between woody resource units in soil. *Journal of General Microbiology*, **121**, 203–11.

Dowson, C.G., Rayner, A.D.M. & Boddy, L. (1988). Foraging patterns of *Phallus impudicus*, *Phanerochaete laevis* and *Steccherinum fimbriatum* between discontinuous resource units in soil. *FEMS Microbiology Ecology*, **53**, 291–8.

Dowson, C.G., Rayner, A.D.M. & Boddy, L. (1989). Spatial dynamics and interactions of the woodland fairy ring fungus, *Clitocybe nebularis*. *New Phytologist*, **111**, 501–9.

Esser, K. & Blaich, R. (1973). Heterogenic incompatibility in plants and animals. *Advances in Genetics*, **17**, 107–52.

Esser, K. & Blaich, R. (1994). Heterogenic incompatibility in fungi. In *The Mycota. I. Growth, Differentiation and Sexuality* (ed. K. Esser & P.A. Lemke), pp. 211–32. Berlin, Heidelberg: Springer-Verlag.

Falck, R. (1912). Die *Merulius* – Faüle des Bauholzes. *Hausschwammforsch*, **6**, 1–405.

Frese, D. & Stahl, U. (1992). Oxidative stress and ageing in the fungus *Podospora anserina*. *Mechanisms in Ageing and Development*, **65**, 277–88.

Glass, N.L. & Kuldau, G.A. (1992). Mating type and vegetative incompatibility in filamentous ascomycetes. *Annual Review of Phytopathology*, **30**, 201–24.

Goodwin, B. (1994). *How the Leopard Changed its Spots*. London: Weidenfield & Nicolson.

Gow, N.A.R. (1994). Tip growth and polarity. In *The Growing Fungus* (ed. N.A.R. Gow & G.M. Gadd), pp. 277–99. London: Chapman & Hall.

Grainger, J. (1962). Vegetative and fructifying growth in *Phallus impudicus*. *Transactions of the British Mycological Society*, **45**, 145–55.

Gregory, P.H. (1984). The fungal mycelium: an historical perspective. *Transactions of the British Mycological Society*, **82**, 1–11.

Griffith, G.S., Rayner, A.D.M. & Wildman, H.G. (1994*a*). Interspecific interactions and mycelial morphogenesis of *Hypholoma fasciculare* (Agaricaceae). *Nova Hedwigia*, **59**, 47–75.

Griffith, G.S., Rayner, A.D.M. & Wildman, H.G. (1994*b*). Interspecific interactions, mycelial morphogenesis and extracellular metabolite production in *Phlebia radiata* (Aphyllophorales). *Nova Hedwigia*, **59**, 331–44.

Griffith, G.S., Rayner, A.D.M. & Wildman, H.G. (1994*c*). Extracellular metabolites and mycelial morphogenesis of *Hypholoma fasciculare* and *Phlebia radiata* (Hymenomycetes). *Nova Hedwigia*, **59**, 311–29.

Gull, K. (1978). Form and function of septa in filamentous fungi. In *The Filamentous Fungi* **III** (ed. J.E. Smith & D.R. Berry), pp. 78–93. London: Arnold.

Halliwell, B. & Gutteridge, J.M.C. (1989). *Free Radicals in Biology and Medicine*, 2nd edn. Oxford: Clarendon Press.

Hansberg, W. & Aguirre, J. (1990). Hyperoxidant states cause microbial cell differentiation by cell isolation from dioxygen. *Journal of Theoretical Biology*, **142**, 201–21.

Harder, R. (1911). Uber des Verhalten von Basidiomyceten und Ascomyceten in Mischkulturen. *Naturwissenschaftliche Zeitschrift für Forst und Landwirtschaft*, **9**, 129–60.

Harper, J.E. & Webster, J. (1964). An experimental analysis of the coprophilous fungal succession. *Transactions of the British Mycological Society*, **47**, 511–30.

Hartig, R. (1874). *Wichtige Krankheiten der Waldbaüme*. Berlin, Springer-Verlag.

Heath, M.C. (1976). Hypersensitivity, the cause or the consequence of rust resistance? *Phytopathology*, **66**, 935–6.

Heath, M.C. (1984). Relationship between heat-induced fungal death and plant necrosis in compatible and incompatible interactions involving the bean and cowpea rust fungi. *Phytopathology*, **74**, 1370–6.

Hendry, S.J. (1993). Strip-cankering in relation to the ecology of Xylariaceae and Diatrypaceae in beech (*Fagus sylvatica* L.). PhD Thesis, University of Wales.

Hubert, E.E. (1924). The diagnosis of decay in wood. *Journal of Agricultural Research*, **29**, 523–67.

Ikediugwu, F.E.O. (1976). The interface in hyphal interference by *Peniophora gigantea* against *Heterobasidion annosum*. *Transactions of the British Mycological Society*, **66**, 291–6.

Ikediugwu, F.E.O., Dennis, C. & Webster, J. (1970). Hyphal interference by *Peniophora gigantea* against *Heterobasidion annosum*. *Transactions of the British Mycological Society*, **54**, 307–309.

Ikediugwu, F.E.O. & Webster, J. (1970*a*). Antagonism between *Coprinus heptemerus* and other coprophilous fungi. *Transactions of the British Mycological Society*, **54**, 180–204.

Ikediugwu, F.E.O. & Webster, J. (1970*b*). Hyphal interference in a range of coprophilous fungi. *Transactions of the British Mycological Society*, **54**, 205–10.

Jennings, D.H. (1984). Water flow through mycelia. In *The Ecology and Physiology of the Fungal Mycelium* (ed. D.H. Jennings & A.D.M. Rayner), pp. 143–64. Cambridge: Cambridge University Press.

Jennings, D.H. (1987). Translocation of solutes in fungi. *Biological Reviews*, **62**, 215–43.

Jinks, J.L. (1959). The genetic basis of 'duality' in imperfect fungi. *Heredity*, **13**, 525–8.

Kemp, R.F.O. (1975). Breeding biology of *Coprinus* species in the section *Lanatuli*. *Transactions of the British Mycological Society*, **65**, 375–88.

Korhonen, K. & Hintikka, V. (1974). Cytological evidence for somatic diploidization in dikaryotic cells of *Armillariella mellea*. *Archiv für Microbiologie*, **95**, 187–92.

Kühner, R. (1977). Variation of nuclear behaviour in the Homobasidiomycetes. *Transactions of the British Mycological Society*, **68**, 1–16.

Laing, S.A.K. & Deacon, J.W. (1991). Video microscopial comparison of mycoparasitism by *Pythium oligandrum*, *P. nunn* and an unnamed *Pythium* species. *Mycological Research*, **95**, 469–79.

Leatham, G.F. & Stahmann, M.A. (1981). Studies on the laccase of *Lentinus edodes*: specificity, localization and association with developing fruit bodies. *Journal of General Microbiology*, **125**, 147–57.

Lee, S.B. & Taylor, J.W. (1993). Uniparental inheritance and replacement of mitochondrial DNA in *Neurospora tetrasperma*. *Genetics*, **134**, 1063–75.

Leslie, J.F. (1993). Fungal vegetative compatibility. *Annual Review of Phytopathology*, **31**, 127–50.

Lopez-Real, J.M. (1975). Formation of pseudosclerotia ('zone lines') in wood decayed by *Armillaria mellea* and *Stereum hirsutum*. I. Morphological aspects. *Transactions of the British Mycological Society*, **64**, 465–71.

Lopez-Real, J.M. & Swift, M.J. (1977). The formation of pseudosclerotia ('zone lines') in wood decayed by *Armillaria mellea* and *Stereum hirsutum*. III. Formation in relation to composition of gaseous atmosphere in wood. *Transactions of the British Mycological Society*, **68**, 321–5.

Lyon, A.J.E. & Lucas, R.L. (1969). The effect of temperature on the translocation of phosphorus by *Rhizopus stolonifer*. *New Phytologist*, **68**, 963–9.

Lyr, H. (1958). Die Induktion der Laccase-Bildung bei *Collybia velutipes* Curt. *Archiv für Mikrobiologie*, **28**, 310–24.

Lyr, H. (1963). Enzymatisches Detoxifikation chlorierter Phenole. *Phytopathologische Zeitschrift*, **38**, 342–54.

Macrae, R. (1967). Pairing incompatibility and other distinctions among *Hirschioporus (Polyporus) abietinus*, *H. fusco-violaceus* and *H. laricinus*. *Canadian Journal of Botany*, **45**, 1371–98.

Marbach, K. & Stahl, U. (1994). Senescence of mycelia. In *The Mycota. I. Growth, Differentiation and Sexuality* (ed. K. Esser & P.A. Lemke), pp. 195–210. Berlin, Heidelberg: Springer-Verlag.

May, G. & Taylor, J.W. (1988). Patterns of mating and mitochondrial DNA inheritance in the agaric basidiomycete *Coprinus cinereus*. *Genetics*, **118**, 213–20.

Mounce, I. (1929). Studies in forest pathology. II. The biology of *Fomes pinicola* (Sw.) Cooke. *Bulletin of the Canadian Department of Agriculture*, **111**, 1–77.

Murphy, J.F. & Miller, O.K. (1993). The population biology of two litter-decomposing agarics on a southern Appalachian mountain. *Mycologia*, **85**, 769–76.

Nesemann, G. (1953). Uber die antagonistische Beeinflussung von Wachstum und Atmung bei einigen höheren Pilzen. *Archiv für Mikrobiologie*, **19**, 319–52.

Noble, M. (1937). The morphology and cytology of *Typhula trifolii* (Rostr.). *Annals of Botany* (N.S.), **1**, 67–98.

Nobles, M.K. (1965). Identification of cultures of wood-inhabiting hymenomycetes. *Canadian Journal of Botany*, **43**, 1097–139.

Park, D. (1960). Antagonism – the background to soil fungi. In *The Ecology of Soil Fungi* (ed. D. Parkinson & J.S. Waid), pp. 148–59. Liverpool: Liverpool University Press.

Porter, C.L. (1924). Concerning the character of certain fungi as exhibited by their growth in the presence of other fungi. *American Journal of Botany*, **11**, 168–88.

Porter, C.L. & Carter, J.C. (1938). Competition among fungi. *Botanical Review*, **4**, 165–82.

Prosser, J.I. (1991). Mathematical modelling of vegetative growth of filamentous fungi. In *Handbook of Applied Biology 1* (ed. D.H. Arora, B. Rai, K.G. Mukerji & G.R. Knudsen), pp. 591–623. New York: Marcel Dekker.

Prosser, J.I. (1993). Growth kinetics of mycelial colonies and aggregates of ascomycetes. *Mycological Research*, **97**, 513–28.

Prosser, J.I. (1994a). Kinetics of filamentous growth and branching. In *The Growing Fungus* (ed. N.A.R. Gow & G.M. Gadd), pp. 301–18. London: Chapman & Hall.

Prosser, J.I. (1994*b*). Mathematical modelling of fungal growth. In *The Growing Fungus* (ed N.A.R. Gow & G.M. Gadd), pp. 319–35. London: Chapman & Hall.

Pugh, G.J.F. (1980). Strategies in fungal ecology. *Transactions of the British Mycological Society*, **75**, 1–14.

Ramsdale, M. & Rayner, A.D.M. (1994). Distribution patterns of number of nuclei in conidia from heterokaryons of *Heterobasidion annosum* (Fr.) Bref. and their possible interpretation in terms of genomic conflict. *New Phytologist*, **128**, 123–34.

Ramsdale, M. & Rayner, A.D.M. (1995). Phenotype-genotype relationships in allopatric heterokaryons of *Heterobasidion annosum*. *New Phytologist* (in press).

Raper, J.R. (1966). *Genetics of Sexuality in Higher Fungi*. New York: Ronald Press.

Rasanayagam, S. & Jeffries, P. (1992). Production of acid is responsible for antibiosis by some ectomycorrhizal fungi. *Mycological Research*, **96**, 971–6.

Rayner, A.D.M. (1976). Dematiaceous hyphomycetes and narrow dark zones in decaying wood. *Transactions of the British Mycological Society*, **67**, 546–9.

Rayner, A.D.M. (1988). Life in a collective: lessons from the fungi. *New Scientist*, **120**, 49–53.

Rayner, A.D.M. (1991). The challenge of the individualistic mycelium. *Mycologia*, **83**, 48–71.

Rayner, A.D.M. (1992). Monitoring genetic interactions between fungi in terrestrial habitats. In *Genetic Interactions Among Micro-organisms in the Natural Environment* (ed. E.M.H. Wellington & J.D. Van Elsas), pp. 267–85. Oxford: Pergamon.

Rayner, A.D.M. (1994). Pattern-generating processes in fungal communities. In *Beyond the Biomass* (ed. K. Ritz, J. Dighton & K.E. Giller), pp. 247–58. Chichester: Wiley-Sayce.

Rayner, A.D.M. (1995). Has chaos theory a place in environmental mycology? In *Fungi and Environmental Change* (ed. J.C. Frankland, N. Magan & G.M. Gadd). Cambridge: Cambridge University Press.

Rayner, A.D.M., Coates, D., Ainsworth, A.M., Adams, T.J.H., Williams, E.N.D. & Todd, N.K. (1984). The biological consequences of the individualistic mycelium. In *The Ecology and Physiology of the Fungal Mycelium* (ed. D.H. Jennings & A.D.M. Rayner), pp. 509–40. Cambridge: Cambridge University Press.

Rayner, A.D.M. & Franks, N.R. (1987). Evolutionary and ecological parallels between ants and fungi. *Trends in Ecology and Evolution*, **2**, 127–33.

Rayner, A.D.M., Griffith, G.S. & Ainsworth, A.M. (1994*a*). Mycelial interconnectedness. In *The Growing Fungus* (ed. N.A.R. Gow & G.M. Gadd), pp. 21–40. London: Chapman & Hall.

Rayner, A.D.M., Griffith, G.S. & Wildman, H.G. (1994*b*). Differential insulation and the generation of mycelial patterns. In *Shape and Form in Plants and Fungi* (ed. D.S. Ingram), pp. 293–312. London: Academic Press.

Rayner, A.D.M., Griffith, G.S. & Wildman, H.G. (1994*c*). Induction of metabolic and morphogenetic changes during mycelial interactions among species of higher fungi. *Biochemical Society Transactions*, **22**, 391–6.

Rayner, A.D.M., Ramsdale, M. & Watkins, Z.R. (1995). Origins and significance of genetic and epigenetic instability in mycelial systems. *Canadian Journal of Botany* (in press).

Rayner, A.D.M. & Ross, I.K. (1991). Sexual politics in the cell. *New Scientist,* **129,** 30–3.

Rayner, A.D.M. & Todd, N.K. (1977). Intraspecific antagonism in natural populations of wood-decaying basidiomycetes. *Journal of General Microbiology,* **103,** 85–90.

Rayner, A.D.M. & Todd, N.K. (1979). Population and community structure and dynamics of fungi in decaying wood. *Advances in Botanical Research,* **7,** 333–420.

Rayner, A.D.M. & Todd, N.K. (1982*a*). Population structure in wood-decomposing basidiomycetes. In *Decomposer Basidiomycetes: Their Biology and Ecology* (ed. J.C. Frankland, J.N. Hedger & M.J. Swift), pp. 109–28. Cambridge: Cambridge University Press.

Rayner, A.D.M. & Todd, N.K. (1982*b*). Ecological genetics of wood-decomposing basidiomycetes. In *Decomposer Basidiomycetes: Their Biology and Ecology* (ed. J.C. Frankland, J.N. Hedger & M.J. Swift), pp. 129–142. Cambridge: Cambridge University Press.

Reinhardt, O.M. (1892). Das Wachstum der Pilzenhyphen. *Jahrbucher für Wissenschaftliche Botanik,* **23,** 479–566.

Rizzo, D.M. & May, G. (1994). Nuclear replacement during mating in *Armillaria ostoyae* (Basidiomycotina). *Microbiology,* **140,** 2115–24.

Ross, I.K. (1985). Determination of the initial steps in differentiation in *Coprinus congregatus.* In *Developmental Biology of Higher Fungi* (ed. D. Moore, L.A. Casselton, D.A. Wood & J.C. Frankland), pp. 353–73. Cambridge: Cambridge University Press.

Schmitz, H. (1925). Studies in wood decay. V. Physiological specialization in *Fomes pinicola* Fr. *American Journal of Botany,* **12,** 163–77.

Shain, L. (1979). Dynamic responses of differentiated sapwood to injury and infection. *Phytopathology,* **69,** 1143–7.

Sharland, P.R. & Rayner, A.D.M. (1986). Mycelial interactions in *Daldinia concentrica. Transactions of the British Mycological Society,* **86,** 643–50.

Sharland, P.R. & Rayner, A.D.M. (1989*a*). Mycelial interactions in outcrossing populations of *Hypoxylon. Mycological Research,* **93,** 187–98.

Sharland, P.R. & Rayner, A.D.M. (1989*b*). Mycelial ontogeny and interactions in non-outcrossing populations of *Hypoxylon. Mycological Research,* **93,** 273–81.

Shütte, K.H. (1956). Translocation in fungi. *New Phytologist,* **55,** 164–82.

Snider, P.J. (1965). Incompatibility and nuclear migration. In *Incompatibility in Fungi* (ed. K. Esser & J.R. Raper), pp. 52–70. Berlin: Springer.

Stakman, E.C. (1915). Relation between *Puccinia graminis* and plants highly resistant to its attack. *Journal of Agricultural Research,* **4,** 193–200.

Stalpers, J.A. (1978). Identification of wood-inhabiting Aphyllophorales in pure culture. *Studies in Mycology, Baarn,* **16,** 1–248.

Thompson, W. & Rayner, A.D.M. (1982). Structure and development of mycelial cord systems of *Phanerochaete laevis* in soil. *Transactions of the British Mycological Society,* **78,** 193–200.

Thompson, W. & Rayner, A.D.M. (1983). Extent, development and functioning of mycelial cord systems in soil. *Transactions of the British Mycological Society,* **81,** 333–45.

Thrower, L.B. & Thrower, S.L. (1968). Movement of nutrients in fungi. I. The mycelium. *Australian Journal of Botany,* **16,** 71–80.

Todd, N.K. & Aylmore, R.C. (1985). Cytology of hyphal interactions and reactions in *Schizophyllum commune*. In *Developmental Biology of Higher Fungi* (ed. D. Moore, L.A. Casselton, D.A. Wood & J.C. Frankland), pp. 231–48. Cambridge: Cambridge University Press.

Todd, N.K. & Rayner, A.D.M. (1978). Genetic structure of a natural population of *Coriolus versicolor* (L. ex Fr.) Quél. *Genetical Research*, **32**, 55–65.

Trinci, A.P.J. (1978). The duplication cycle and vegetative development in moulds. In *The Filamentous Fungi* 3 (ed. J.E. Smith & D.R. Berry), pp. 132–63. London: Arnold.

Wessels, J.G.H. (1986). Cell wall synthesis in apical hyphal growth. *International Review of Cytology*, **104**, 37–79.

Wessels, J.G.H. (1991). Fungal growth and development: a molecular perspective. In *Frontiers in Mycology* (ed. D.L. Hawksworth), pp. 27–48. Wallingford: CAB International.

Wessels, J.G.H. (1992). Gene expression during fruiting of *Schizophyllum commune*. *Mycological Research*, **96**, 609–20.

Wessels, J.G.H. (1994). Developmental regulation of fungal cell wall formation. *Annual Review of Phytopathology*, **32**, 413–37.

Willetts, H.J. (1978). Sclerotium formation. In *The Filamentous Fungi* 3 (ed. J.E. Smith & D.R. Berry), pp. 197–213. London: Arnold.

Williams, E.N.D., Todd, N.K. & Rayner, A.D.M. (1981*a*). Propagation and development of fruit bodies of *Coriolus versicolor*. *Transactions of the British Mycological Society*, **77**, 409–14.

Williams, E.N.D., Todd, N.K. & Rayner, A.D.M. (1981*b*). Spatial development of populations of *Coriolus versicolor*. *New Phytologist*, **89**, 307–19.

Worrall, J.J., Chet, I. & Hüttermann, A. (1986). Association of rhizomorph formation with laccase activity in *Armillaria* spp. *Journal of General Microbiology*, **132**, 2527–33.

Zeller, S.M. & Schmitz, H. (1919). Studies in the physiology of the fungi. VIII. Mixed cultures. *Annals of the Missouri Botanical Garden*, **6**, 183–92.

9
Fungal secondary metabolism: regulation and functions

ARNOLD L. DEMAIN

Introduction

Secondary metabolites, also known as idiolites, are special compounds, often possessing chemical structures quite different from primary metabolites from which they are produced such as sugars, amino acids, and organic acids. Idiolites from micro-organisms are not essential for the growth of the producing culture but serve diverse survival functions in nature. These special metabolites, in contrast to the general nature of primary metabolites, are produced only by some species of a genus, and by some strains of a species. Most of the organisms making secondary metabolites undergo complex schemes of morphological differentiation. Moulds make 17% of all described antibiotics (Miyadoh, 1993). Approximately 10 000 microbial secondary metabolites have been discovered (Omura, 1992). Their unusual chemical structures include β-lactam rings, cyclic peptides containing 'unnatural' and non-protein amino acids, unusual sugars and nucleosides, unsaturated bonds of polyacetylenes and polyenes, and large macrolide rings. Idiolites are typically produced as slightly differing components of a particular chemical family as a result of low specificity of some enzymes of secondary metabolism. The ratio of components is often shifted by the addition of a precursor of one of the components (this is called 'directed biosynthesis'). The main types of biosynthetic pathways involved are those forming peptides, polyketides, isoprenes, oligosaccharides, aromatic compounds and β-lactam rings. Knowledge of the pathways varies from cases in which the amino acid sequences of the enzymes and nucleotide sequences of the genes are known, e.g. for penicillins and cephalosporins, to those in which even the enzymatic steps are still unknown.

Regulation of fungal secondary metabolism

The intensity of secondary metabolism can often be increased by addition of limiting precursors. An example is the stimulation of penicillin G production by phenylacetic acid. Secondary metabolism occurs best at submaximal growth rates. The distinction between the growth phase ('trophophase') and production phase ('idiophase') is sometimes very clear but in many cases, idiophase overlaps trophophase. The timing between the two phases can be manipulated, i.e. the two phases are often distinctly separated in a complex medium favouring rapid growth, but overlap partially or even completely in a chemically-defined medium supporting slower growth. A secondary metabolite is not 'secondary' because it is produced after growth, but because it is not involved in the growth of the producing culture. Thus, elimination of production of a secondary metabolite by mutation will not stop or slow down growth; indeed, it may increase the growth rate.

The factors controlling the onset of secondary metabolism are complex and not well understood (Demain, 1992). Growth rate is important, but we do not know the mechanism(s) involved. Deficiencies in certain nutritional factors are important but again we are ignorant of the basic mechanisms involved. The temporal nature of secondary metabolism is certainly genetic in nature but expression can be influenced greatly by environmental manipulations.

The delay often seen before onset of secondary metabolism was probably established by evolutionary pressures. Many secondary metabolites have antibiotic activity and could kill the producing culture if made too early. The resistance of antibiotic producers to their own metabolites is well-known (Demain, 1974; Cundliffe, 1989). Antibiotic-producing species possess suicide-avoiding mechanisms such as (I) enzymatic detoxification of the antibiotic, (ii) alteration of the antibiotic's normal target in the cell, and (iii) modification in permeability to allow the antibiotic to be pumped out of the cell and restrict its re-entry. Such mechanisms are often inducible, but in some cases are constitutive. In the case of inducible resistance, death could result if the antibiotic is produced too early and induction is slow. Delay in secondary metabolite production until the starvation phase makes sense to the producing micro-organism if the product is being used as a competitive weapon in nature or endogenously as an effector of differentiation. In nutritionally rich habitats such as the intestines of mammals, where enteric bacteria thrive, secondary metabolite production is not as important as in soil and water, where nutrients limit microbial

Table 9.1. *Carbon sources interfering with secondary metabolism*

Fungal Idiolite	Interfering Carbon Source	Non-Interfering Carbon Source
Benzodiazapene alkaloids	glucose	sorbitol, mannitol
Cephalosporins	glucose, glycerol, maltose	sucrose, galactose
Enniatin	glucose	lactose
Ergot alkaloids	glucose	polyols, organic acids
Penicillin	glucose, fructose, galactose, sucrose	lactose

growth. Thus secondary metabolites tend not to be produced very often by enteric bacteria such as *Escherichia coli* but are formed extensively by soil and water inhabitants such as bacilli, actinomycetes and fungi. Nutrient deficiency in nature often induces morphological and chemical differentiation, i.e. sporulation and secondary metabolism, respectively; both are beneficial for survival in the wild. Thus the regulation of the two types of differentiation is often related.

Most secondary metabolites are formed via enzymatic pathways rather than by a ribosomal mechanism. The enzymes occur as individual proteins, free or complexed, or as parts of large multifunctional polypeptides carrying out a multitude of enzymatic steps, e.g. polyketide synthases and peptide synthetases. Expression of the genes of secondary metabolism is under strong control by nutrients, inducers, products, metals and growth rate. In most cases, regulation is at the level of transcription as revealed by the absence of mRNA encoding idiolite synthases until growth rate has decreased.

Regulation by the carbon source

Glucose, usually an excellent carbon source for growth, interferes with the formation of many secondary metabolites. Polysaccharides, e.g. starch, oligosaccharides, e.g. lactose and oils, e.g. soybean oil, methyloleate, are often preferable for fermentations yielding secondary metabolites. In media containing a mixture of a rapidly used carbon source and slowly used carbon sources, the former is used first to produce cells but little to no secondary metabolite is formed. After the rapidly-assimilated compound is depleted, the 'second-best' carbon source is used for the idiophase. Examples of interfering carbon sources are given in Table 9.1.

Cyclic AMP (3,5-cyclic adenosine monophosphate; cAMP) is not usually involved in the repression of secondary metabolism by carbon sources. For example, cAMP did not reverse glucose repression of benzodiazapene alkaloid biosynthesis in *Penicillium cyclopium* (Luckner, Nover & Böhm, 1977).

In many secondary metabolite pathways, the enzymes subject to repression by the carbon source are known. Direct inhibition of the action of secondary metabolic enzymes may also be involved in carbon source control. Glucose and a number of its phosphorylated metabolites inhibit δ-(L-α-aminoadipyl)-L-cysteinyl-L-valine (ACV) synthetase, the first enzyme of cephalosporin biosynthesis, in crude extracts of *Cephalosporium acremonium* (Zhang & Demain, 1992). Because no inhibition was observed with the purified enzyme, it appeared that this phenomenon might be caused by competition for ATP between primary metabolism (Embden–Meyerhof pathway) and secondary metabolism. Accordingly, it was found that inhibition of ACV synthetase by sugars could be prevented by adding more ATP. On the other hand, both the crude and the purified forms of ACV synthetase were inhibited by glyceraldehyde-3-phosphate. This is due to the ability of glyceraldehyde-3-phosphate to chemically complex and remove cysteine, a substrate of ACV synthetase. The *in vivo* significance of such inhibitory phenomena remains to be determined.

Regulation by the nitrogen source

Many secondary metabolic pathways are negatively affected by nitrogen sources favourable for growth, e.g. ammonium salts. As a result, complex fermentation media often include an insoluble protein source, and chemically defined media a slowly assimilated amino acid as the nitrogen source, to encourage high production of secondary metabolites. Processes subject to regulation by the nitrogen source are shown in Table 9.2. Only a bit of information is available concerning the mechanisms underlying the negative effects of NH_4^+ and certain amino acids. In *C. acremonium*, at least two enzymes of the cephalosporin biosynthetic pathway, ACV synthetase and deacetoxycephalosporin C synthase/hydroxylase ('expandase'), are repressed by ammonium salts (Zhang, Wolfe & Demain, 1987a).

Nitrogen regulation of gibberellic acid production is exercised via glutamine in *Gibberella fujikuroi* (Munoz & Agosin, 1993). Whereas both NH_4^+ and glutamine addition stop gibberellic acid formation, only glutamine could stop biosynthesis when the conversion of NH_4^+ to glutamine was blocked with L-methionine-DL-sulfoxamine.

Table 9.2. *Nitrogen sources interfering with secondary metabolism*

Fungal Idiolite	Interfering Carbon Source	Non-Interfering nitrogen source
Aflatoxin	nitrate	NH_4^+
Alternariol	nitrate, L-glutamate, urea	
Bikaverin	glycine	
Cephalosporin	NH_4^+, L-lysine	L-asparagine, L-arginine
Gibberellic acid	NH_4^+, glutamine	
Patulin	NH_4^+	
Penicillin	NH_4^+, L-lysine	L-glutamate
Trihydroxy-toluene	NH_4^+	

Ammonium regulation of patulin biosynthesis in *Penicillium urticae* appears to resemble classical regulation of nitrogen-repressible enzymes in *Neurospora crassa* and *Aspergillus nidulans*. In those organisms, the homologous genes *nit-2* and *areA*, respectively, encode proteins which bind to TATCNN sequences, separated by 5 to 40 bp, in the upstream region of genes of nitrogen assimilation (Fu & Marzluf, 1990). Similar sequences have been found in the upstream 5'-flanking regions of two patulin biosynthetic genes (Ellis & Gaucher, 1993). Furthermore, gene *nrfA* of *P. urticae* is similar to *nit-2* and *areA*, encoding a deduced zinc finger DNA-binding protein virtually identical to those of *N. crassa* and *A. nidulans*.

The above paragraphs have dealt with a general nitrogen source effect. There is also a more specific type of control in which a particular amino acid (or biosynthetic group of amino acids) represses and/or inhibits production of a secondary metabolite because the primary metabolite(s) and the idiolite are derived from the same branched pathway and the amino acid(s) exerts negative feedback regulation on the biosynthetic pathway before the branch point. An example is the negative effect of lysine on penicillin and cephalosporin synthesis which is caused by lysine inhibiting homocitrate synthase (Demain & Masurekar, 1974).

Regulation by the phosphorus source

Phosphate interferes in secondary metabolic pathways such as biosynthesis of cephalosporin C, bikaverin and ergot alkaloids. Such fermentations have to be conducted at levels of free phosphate which are suboptimal for growth (usually below 10 μM). A number of enzymes are known to be repressed by phosphate, including dimethylallyltryptophan synthetase (Krupinski, Robbers & Floss, 1976), chanoclavine-1-cyclase (Erge, Maier & Groger, 1973) of ergot alkaloid biosynthesis, and ACV synthetase, isopenicillin N synthase ('cyclase') and expandase in cephalosporin processes (Zhang, Wolfe & Demain, 1989).

Although only little is known about the mechanism of phosphate control of fungal secondary metabolism, the possibility that phosphate regulation works by affecting enzyme activity via phosphorylation by protein kinases and dephosphorylation by phosphoprotein phosphatases (Krebs & Beavo, 1979), certainly exists. Although this possibility has not yet been examined in secondary metabolism, it should be noted that morphological differentiation in the fungus *Dictyostelium discoideum* is mediated by a protein kinase (Leichtling *et al.*, 1984).

Control by metals

Metals often exert control over fungal secondary metabolism. Zinc stimulates aflatoxin production, and manganese increases patulin production via an effect on transcription (Scott *et al.*, 1986). Optimal production of pneumocandins by *Zalerion arboricola* requires magnesium-limiting conditions (Tkacz, Giacobbe & Monaghan, 1993) which slow down assimilation of the carbon source mannitol and allow it to remain for a longer period, thus extending the duration of antibiotic production. Fungi protect themselves against iron starvation by producing siderophores when the iron concentration drops below about 1 μM.

Induction of secondary metabolite synthases

In a number of fungal pathways, primary metabolites induce synthases and thus increase production of the final product. These include tryptophan for dimethylallyltryptophan synthetase in ergot alkaloid biosynthesis (Krupinski *et al.*, 1976), phenylalanine in benzodiazapene alkaloid formation (Luckner *et al.*, 1977), and methionine for ACV synthetase, cyclase and expandase in the cephalosporin pathway of *C. acremonium* (Zhang *et al.*, 1987*b*).

Control by growth rate

Growth rate control appears to be important in secondary metabolism and may be the overriding factor in the cases where nutrient limitation is needed for production of secondary metabolites. In the production of patulin by *P. urticae*, the shift from trophophase to idiophase occurs at the point of nitrogen source depletion when growth slows down and the first enzyme, 6-methylsalicyclic acid synthase, is derepressed. The resultant 6-methylsalicyclic acid induces transcription of genes encoding later enzymes in the pathway. Since derepression also occurs when the growth rate is reduced by other means, it appears that the low growth rate is more important than nitrogen limitation (Gaucher *et al.*, 1981). As another example, *G. fujikuroi* produces two secondary metabolites, each at a particular growth rate: bikaverin at 0.05 h^{-1} and gibberellin at 0.01 h^{-1} (Bu'Lock *et al.*, 1974).

In general, growth rate-dependent enzyme synthesis is known to be brought about by at least three mechanisms. In one case, the stability of mRNA is increased or decreased by the growth-rate dependent formation of a nuclease. A second mechanism involves a 16-nucleotide sequence in the coding region of a structural gene which is complementary to its ribosome-binding sequence; the sequence produces a *cis*-acting antisense RNA. At low growth rates, the antisense RNA competes with the ribosome-binding site for ribosome binding, whereas at high growth rates it does not. Thus, the enzyme is produced at high growth rates and is not produced at low growth rates, opposite from the situation in secondary metabolism. A third mechanism involves sequential production of regulatory proteins which modify promoter recognition of RNA polymerase and allow transcription of genes of morphological differentiation and chemical differentiation.

Feedback regulation

Some fungal secondary metabolites feedback inhibit their own biosynthetic enzymes. In alkaloid biosynthesis, the first enzyme (dimethylallyltryptophan synthetase) is inhibited, whereas in the pathway to mycophenolic acid, the final enzyme, which is an *O*-methyltransferase, is inhibited.

Enzyme decay

Production of fungal secondary metabolites eventually stops due to feedback inhibition (see above) and decay of the synthase(s). For example, cessation of patulin production by *P. urticae* is caused by decay of the initial enzyme, 6-methylsalicylic acid synthase, which has an *in vivo* half-maximal lifetime of only seven hours (Neway & Gaucher, 1981).

Natural functions of fungal secondary metabolites

Over the years, a number of erroneous views have appeared in the literature such as secondary metabolites being laboratory artefacts, not being produced in nature or lacking natural functions. It has always amazed me that the importance of secondary metabolites in ecological interactions between plant vs herbivore, insect vs insect and plant vs plant has been universally accepted (Mann, 1978), but their importance in microbial interactions has been almost universally denied. By this time, however, enough direct and indirect evidence has accumulated so that we can put aside such old-fashioned views and use the knowledge gained to further the exploitation of these remarkable natural products to advance medicine, agriculture and the environment. The following paragraphs will summarize the functions of fungal secondary metabolites as they are known today.

There is no doubt that fungal secondary metabolites are natural products. Nutrient limitation is the usual situation in nature resulting in very low growth rates which favour secondary metabolism. Over 40% of filamentous fungi produce antibiotics when the organisms are freshly isolated from nature. If one examines soil, straw and agricultural products, they often contain antibacterial and antifungal substances produced by fungi. We may call these 'mycotoxins', but they are also antibiotics. One of our major public health problems is the production of such toxic metabolites in the field and during storage of crops. The natural production of ergot alkaloids by the sclerotial (dormant overwintering) form of *Claviceps* on the seed heads of grasses and cereals has led to widespread and fatal poisoning ever since the Middle Ages (Vining & Taber, 1979). Natural soil and wheat-straw contain patulin (Norstadt & McCalla, 1969) and aflatoxin is known to be produced on corn in the field (Hesseltine, Rogers & Shotwell, 1981). Corn grown in the tropics or semitropics always contains aflatoxin (Hesseltine, 1986). At least five mycotoxins of *Fusarium* have been found to occur naturally in corn: moniliformin, zearalenone, deoxynivalenol, fusarin C and fumonisin (Sydenham *et al.*, 1990). Trichothecin is found in anise fruits, apples, pears and wheat (Ishii *et al.*, 1986). Microbially produced siderophores have been found in soil (Castignetti & Smarrelli, 1986).

Antibiotics have been produced in unsterilized, unsupplemented soil, in unsterilized soil supplemented with clover and wheat straws, in mustard, pea and maize seeds and in unsterilized fruits (for review see Demain, 1980). A further indication of natural antibiotic production is the possession of antibiotic-resistance plasmids by most soil bacteria.

The widespread nature of secondary metabolite production and the preservation of the multigenic biosynthetic pathways in nature indicate that secondary metabolites serve survival functions in organisms that produce them. There are a multiplicity of such functions, some dependent on antibiotic activity and others independent of such activity. Indeed in the latter case, the molecule may possess antibiotic activity but may be used by the producing micro-organism for an entirely different purpose.

The view that secondary metabolites act by improving the survival of the producer in competition with other living species has been expressed by Williams and co-workers (Williams *et al.*, 1989; Stone & Williams, 1992). They contend that this is done by acting at specific receptors (DNA, cell wall synthesizing enzymes, etc.) in competing organisms. Their arguments are as follows: (i) Only organisms lacking an immune system are prolific producers of secondary metabolites which act as alternative defence mechanisms; (ii) the compounds have sophisticated structures, mechanisms of action, and complex and energetically expensive pathways; (iii) soil isolates produce natural products, most of which have physiological properties; (iv) the compounds are produced in nature and act in competition between micro-organisms, plants and animals; (v) clustering of biosynthetic genes is often found, which would be selected for only if the product conferred a selective advantage; also there is an absence of 'junk genes' in these clusters; (vi) resistance and regulatory genes occur in these clusters; (vii) resistance genes are clustered in non-producers strongly suggesting that such resistance is important in chemical warfare between micro-organisms in nature; (viii) there is a temporal relationship between antibiotic formation and sporulation (Hopwood, 1988) due to sensitivity of cells during sporulation to competition and the need for protection when a nutrient runs out. The authors call this 'pleiotropic switching', i.e. a way to concurrently express both components of a two-pronged defence strategy when survival is threatened.

Competitive weapons

According to Cavalier-Smith (1992), secondary metabolites are most useful to the organisms producing them as agents of chemical warfare and the selective forces for their production have existed even before the first cell. The antibiotics are more important than macromolecular toxins, e.g. colicins, animal venoms, because of their diffusibility into cells and broader modes of action.

Microbial antagonism

One of the first pieces of evidence indicating that one micro-organism produces an antibiotic against other micro-organisms and that this provides for survival in nature was published by Bruehl, Millar & Cunfer (1969). These workers found that *Cephalosporium gramineum*, the fungal cause of stripe disease in winter wheat, produces a broad spectrum antifungal antibiotic of unknown structure. Over a three-year period, more than 800 isolates were obtained from diseased plants, each of which was capable of producing the antibiotic in culture. On the other hand, ability to produce the antibiotic was lost during storage on slants at 6°C. Thus antibiotic production was selected for in nature but was lost in the test tube. The selection was found to be exerted during the saprophytic stage in soil. These workers further showed that antibiotic production in the straw-soil environment aids in the survival of the producing fungus and markedly reduces competition by other fungi.

Another example involves the parasitism of one fungus on another. The parasitism of *Monocillium nordinii* on the pine stem rust fungi *Cronartium coleosporioides* and *Endocronartium harknessii* is due to its production of the antifungal antibiotics monorden and monocillins (Ayer *et al.*, 1980).

Gliocladium virens inhibits the growth of *Pythium ultimum*, a phytopathogen, in the soil by production of the antibiotic, gliovirin (Howell & Stipanovic, 1983). A non-producing mutant of *G. virens* was overgrown in culture by *P. ultimum* and did not protect cotton seedlings from damping off disease in soil infested with *P. ultimum*. A superior-producing mutant was more inhibitory than the parent culture and showed parental efficiency in disease suppression even though the growth rate was lower than that of the parent.

Fungi vs higher plants

Fungi produce a large number of phytotoxins of varied structure such as sesquiterpenoids, sesterterpenoids, diketopiperazines, peptides, spirocyclic lactams, isocoumarins and polyketides (Strobel *et al.*, 1991). The AM-toxins are peptidolactones, e.g. alternariolide, produced by *Alternaria mali* which form brown necrotic spots in infected apples (Lee *et al.*, 1976). The phytotoxins produced by plant pathogens *Alternaria helianthi* and *Alternaria chrysanthemi* (the pyranopyrones deoxyradicinin and radicinin, respectively) are not only pathogenic to the Japanese chrysanthemum but also to fungi (Robeson & Strobel, 1982). The phytotoxin of *Rhizopus chinensis*, the causative agent of rice seedling blight, is a 16-membered macrolide antifungal agent, rhizoxin (Iwasaki *et al.*, 1984). The fungal pathogen responsible for onion pink root disease, *Pyrenochaeta terrestris*, produces three pyrenocines, A, B and C. Pyrenocine A is the most phytotoxic to the onion and is the only one of the three that has marked antibacterial and antifungal activity (Sparace, Reeleder & Khanizadeh, 1987).

The plant pathogenic basidiomycete, *Armillaria ostoyae*, which causes great forest damage, produces a series of toxic antibiotics when grown in the presence of plant cells (*Picea abies* callus) or with competitive fungi. The antibiotics have been identified as sesquiterpene aryl esters which have antifungal, antibacterial and phytotoxic activities (Peipp & Sonnenbichler, 1992).

Secondary metabolites play a crucial role in the evolution and ecology of plant pathogenic fungi (Scheffer, 1991). Some have evolved from opportunistic low-grade pathogens to high-grade virulent host-specialized pathogens by gaining the genetic potential to produce a toxin under single gene control. This ability to produce a secondary metabolite has allowed fungi to exploit the monocultures and genetic uniformity of modern agriculture resulting in disastrous epidemics and broad destruction of crops.

Of course, with all these weapons directed by microbes against plants, the latter do not take such insults 'lying down'. They produce antimicrobials after exposure to plant pathogenic micro-organisms in order to protect themselves; these are called phytoalexins (Darvill & Albersheim, 1984). They are low molecular weight, weakly active and indiscriminate, i.e. they inhibit prokaryotes and eukaryotes including higher plant cells and mammalian cells. There are approximately 100 known phytoalexins. They are not a uniform chemical class and include isoflavonoids, sesquiterpenes, diterpenes, furanoterpenoids, polyacetylenes, dihydrophenanthrenes, stilbenes and other compounds. Their formation is induced via invasion by fungi, bacteria, viruses and nematodes. Almost all of these which have been tested show some antibiotic activity (Mitscher, 1975). They are thought to function as chemical signals to protect plants against competitors, predators and pathogens, as pollination-insuring agents and as compounds attracting biological dispersal agents (Swain, 1977; Bennett, 1981). The compounds which are responsible for induction of phytoalexin biosynthesis are called 'elicitors'. The fungi respond by modifying and breaking down the phytoalexins.

Fungi vs insects

Certain fungi have entemopathogenic activity, i.e. they infect and kill insects via their production of secondary metabolites. One such compound is bassianolide, a cyclodepsipeptide produced by the fungus,

Beauveria bassiana, which elicits atonic symptoms in silkworm larvae (Kanaoka, Isogai & Suzuki, 1975). Another pathogen, *Metarhizium anisopliae*, produces the peptido-lactone toxins known as destruxins (Lee *et al.*, 1975).

The function of the aflatoxin group of mycotoxins by aspergilli could be that of spore dispersal via an insect vector (Bennett, 1981). Aflatoxins are potent insecticides and *Aspergillus flavus* and *A. parasiticus* are pathogens of numerous insects. The fungi are brought to many plants by the insects and if the insect is killed by an aflatoxin, a massive inoculum of spores is delivered to the plant. Already a strong correlation has been established between insect damage of crops (in storage and in the field) and aflatoxin contamination of the crops.

Fungi vs higher animals

Competition may also exist between fungi and large animals. Janzen (1977) argues that the reason fruits rot, seeds mould and meats spoil is that it is 'profitable' for microbes to make seeds, fresh fruit and carcasses as objectionable as possible to large organisms in the shortest amount of time. Among their means is the production of secondary metabolites such as antibiotics and toxins. In agreement with this concept are the observations that livestock generally refuse to eat mouldy feed. Kendrick (1986) states that animals which come upon a mycotoxin-infected food will do one of four things: (i) smell the food and reject it; (ii) taste the food and reject it; (iii) eat the food, get ill and avoid the same in the future; or (iv) eat the food and die. In each case, the fungus is more likely to live than if it produced no mycotoxin.

Metal transport agents

Certain secondary metabolites such as the siderophores (also known as sideramines) act as metal transport agents which function in uptake, transport and solubilization of iron. Iron-transport factors in many cases are antibiotics. They are on the borderline between primary and secondary metabolites since they are usually not required for growth but do stimulate growth under iron-deficient conditions. Micro-organisms have 'low' and 'high' affinity systems to solubilize and transport ferric iron. The high affinity systems involve siderophores. The low affinity systems allow growth in the case of a mutation abolishing siderophore production (Neilands, 1984). The low affinity system works unless the environment contains an iron chelator, e.g. citrate, which binds the metal and makes it unavailable to the cell; under such conditions, the siderophore stimulates growth. Over a hundred microbial siderophores have been described. Antibiotic activity is due to the ability of these compounds to starve other species of iron when the latter lack the ability to take up the Fe–sideramine complex. Such antibiotics include desferritriacetylfusigen produced by *Aspergillus deflectus* (Anke, 1977).

Fungal–plant symbiosis and plant growth stimulants

Almost all plants depend on soil micro-organisms for mineral nutrition, especially that of phosphate. The most beneficial micro-organisms are those which are symbiotic with plant roots, i.e. the mycorrhizas which are highly specialized associations between soil fungi and roots. The ectomycorrhizas, used by 3–5% of plant species, feature the fungus developing around the roots. These are symbiotic growths of fungi on plant roots in which the fungal symbionts penetrate intracellularly and replace partially the middle lamellae between the cortical cells of the feeder roots. The endomycorrhizas, which form on the roots of 90% of plant species, enter the root cells and form an external mycelium which extends into the soil (Gianinazzi & Gianinazzi-Pearson, 1988). Mycorrhizal roots can absorb much more phosphate than roots which have no symbiotic relationship with fungi. Mycorrhizal fungi lead to reduced damage by pathogens such as nematodes, *Fusarium*, *Pythium*, and *Phytophthora*. Such symbiotic relationships often involve antibiotics. In the case of ectomycorrhizas, the fungi produce antibiotics which protect the plant against pathogenic bacteria or other fungi. One such antibacterial agent was extracted from ectomycorrhizas composed of *Cenococcum graniforme* and white pine, red pine and Norway spruce (Krywolap, Grand & Casida, 1964). Two other antibiotics, diatretyne nitrile and diatretyne 3, were extracted from ectomycorrhizas formed by *Leucopaxillus cerealis* var. *piceina*; they make feeder roots resistant to the plant pathogen, *Phytophthora cinnamomi* (Marx, 1969).

A related type of plant–microbe interaction involves the production of plant growth stimulants by fungi. Two secondary metabolites, altechromones A and B, produced by *Alternaria* sp. isolated from oats are plant-growth stimulators (Kimura *et al.*, 1992).

Sexual hormones

Many fungal secondary metabolites function as sexual hormones (Gooday, 1987). The most well known are the trisporic acids, which are metabolites of *Mucorales*. When vegetative hyphae of the two mating types of these heterothallic organisms approach one another, they form zygophores (sexual hyphae). The trisporic acids are the factors that induce zygophore formation. Trisporic acids are formed from mevalonic acid in a secondary metabolic pathway of which the early steps are present in both (+) and (–) sexes. However, different later steps are missing in the individual sexes and thus both strains must be present and in contact to complete the pathway to the trisporic acids. In *Gibberella zeae* (*Fusarium roseum*), the secondary metabolite, zearalenone, is a regulator of sexual reproduction (Wolf & Mirocha, 1977). The secondary metabolite, sirenin, is involved in sexual reproduction in *Allomyces*, a phycomycete. It acts as a chemotaxic hormone which brings together uniflagellate motile male and female gametes. Sirenin is a sesquiterpene diol made by female gametangia and gametes; it is extremely active, the sexual process requiring less than 0.1 ng/ml for activity (Nutting, Rapoport & Machlis, 1968). In the phycomycete *Achlya*, antheridiol, a steroidal secondary metabolite, is produced by vegetative female mycelia and initiates the formation of male gametangia. The compound is active at concentrations as low as 10^{-11} M (Barksdale *et al.*, 1974). The male gametangia produce another secondary metabolite (hormone B) which leads to oogonium formation in female mycelia. *Tremella mesenterica*, a jelly fungus of the heterobasidomycete group, produces the peptide tremorgen A-10 which induces germ tubes in mating type **a** (Sakagami *et al.*, 1987). A compound inducing sexual development in *Aspergillus nidulans* has been isolated but not yet identified (Champe, Rao & Chang, 1987). Crude preparations containing the factor (called psi) are active at levels as low as 50 ng per test.

Male *Aphomia sociella* L., the bumble bee waxmoth, contains a sex pheromone in its wing gland, the major part of which is R(-)-mellein (= ochracin; = 3,4-dihydro-9-hydroxy-3-methylisocoumarin). The compound, which evokes searching behaviour in females, is produced by a mould, *Aspergillus ochraceus*, found in the intestine of the last-instar larvae and in the bumble bee nest (Kunesch *et al.*, 1987). Apparently such insect–fungus relationships are widespread. Substances I_A, I_B and I_C are peptidic sexual agglutination factors of *Saccharomyces cerevisiae* (Sakurai *et al.*, 1977). Rhodotorucine A is a peptide produced by type *A* cells of *Rhodosporidium toruloides* which induces mating tube formation in yeast of the *a* mating type (Kamiya *et al.*, 1978).

Microbial secondary metabolites can exert regulation of cellular activities in higher organisms (Nisbet & Porter, 1989). It has been hypothesized that cell-to-cell communication first evolved in micro-organisms, long before the appearance of specialized cells of vertebrates (glands, neurons, immune cells, blood cells) (Roth, Leroith & Collier, 1986). Thus hormones, neuropeptides, biological response modifiers, and their receptors, may have been first made by micro-organisms. Indeed, steroid fungal sex hormones and mammalian sex hormones are similar in structure.

Effectors of differentiation

Development includes two phenomena, growth and differentiation. The latter is the progressive diversification of structure and function of cells in an organism or the acquisition of differences during development (Bennett, 1983). Differentiation encompasses both morphological differentiation (morphogenesis) and chemical differentiation (secondary metabolism). Secondary metabolites are not only made by chemical differentiation processes but also function in morphological and chemical differentiation.

Sporulation

Of the various functions postulated for secondary metabolites, the one which has received the most attention is the view that these compounds, especially antibiotics, are obligate compounds in the transition from vegetative cells to spores. For example, there appears to be a direct relationship between formation of ergot alkaloids and conidiation in *Claviceps purpurea* (Pazoutova, Pokorny & Rehacek, 1977). Despite the apparent connections between formation of antibiotics and spores, it has become clear that antibiotic production is not obligatory for spore formation. The most damaging evidence to the obligatory antibiotic-spore hypothesis is the existence of mutants which form no antibiotic but still sporulate. Such mutants have been found in the case of patulin production by *Penicillium urticae*. Although secondary metabolites such as antibiotics are not obligatory for sporulation, some stimulate the sporulation process. Factors inducing sporulation have been isolated from fungi. A sesquiterpenoid with an eremophilane skeleton was found to be a sporogenic factor (sporogen-A01) produced by *Aspergillus oryzae* (Tanaka *et al.*, 1984). Five cerebrosides were isolated from *Schizophyllum commune* which induce fruiting body formation in the same organism. All five were indentified, the major component being (4*E*, 8*E*)-*N*-2'-hydroxyhexadecanoyl-1-0-β-D-glucopyranosyl-9-methyl-4,8-sphingadienine (Kawai & Ikeda, 1985). An antifungal agent, lunatoic acid, produced by *Cochliobolus lunatus* is an inducer of chlamydospore formation (Marumo *et al.*, 1982).

Germination of spores

The close relationship between sporulation and antibiotic formation suggests that certain secondary metabolites involved in germination might be produced during sporulation and that the formation of these compounds and spores could be regulated by a common mechanism or by similar mechanisms. A number of secondary metabolites are involved in maintaining spore dormancy in fungi. One is discadenine [3-(3-amino-3-carboxypropyl)-6-(3-methyl-2-butenylamino)purine] in *Dictyostelium discoideum*, *D. purpureum* and *D. mucoroides* (Taya, Yamada & Nishimura, 1980); this compound is made from 5'-AMP (Dahlberg & Van Etten, 1982). The function of germination inhibitors appears to be that of inhibiting germination under densely crowded conditions. The auto-inhibitor of urediniospore germination in *Puccinia coronata* var. *avenae* (oat crown rust fungus) is methyl-*cis*-3,4-dimethoxycinnamate (Tsurushima *et al.*, 1990). The auto-inhibitor of conidial germination in *Colletotrichum graminicola* has been identified as mycosporine-alanine (Leite & Nicholson, 1992).

Another relationship between secondary metabolites and differentiation involves the stimulation of germination. Germination stimulators in rust fungi include nonanal and 6-methylhept-5-en-2-one which act on urediniospores (Rines, French & Daasch, 1976). In addition to producing extracellular ferric ion-transport and solubilizing factors (siderophores), fungi produce cell-bound siderophores which are involved in conidial germination (Charlang *et al.*, 1982). For example, a siderophore is required for conidial germination in *Neurospora crassa* (Horowitz *et al.*, 1976). These siderophores are considered to be iron storage forms in fungal spores analogous to the ferritins of animals and the phytoferritins of plants (Matzanke *et al.*, 1987).

Other relationships between differentiation and secondary metabolites

Cyclic AMP is a secondary metabolite in the slime mould, *Dictyostelium discoideum*. It is the chemotactic agent which, after initiation of development by starvation, attracts the amoebae-type cells and aggregates them leading to the formation of the elongated, multi-cellular 'slug' structures. Each cell of the slug differentiates into either a stalk cell or a spore of the fruiting body. Differentiation depends on c-AMP plus a low molecular weight factor of unknown structure, known as differentiation inducing factor (DIF). A high ratio of DIF to c-AMP appears to produce a stalk cell whereas a low ratio produces a spore (Kay, 1982).

Campbell (1984) and associates (Campbell *et al.*, 1982) have made some interesting observations on secondary metabolite production by fungal colonies growing on solid media which further implicate these molecules in differentiation. In general, they find certain secondary metabolites to be produced only by certain differentiating structures. In *Penicillium patulum*, 6-methylsalicylic acid (6-MSA) is produced only after aerial mycelia are formed. The same is true in *Penicillium brevicompactum*, with respect to formation of mycophenolic acid, brevianamides A and B, asperphenamate and ergosterol. Asperphenamate and ergosterol are the first to be formed, followed by mycophenolic acid, all three being made before conidial heads appear. The brevianamides appear only after the conidial heads appear and during conidiation. (In *P. patulum*, 6-MSA synthesis begins prior to the formation of conidial heads.) When conidiophores of *P. brevicompactum* are producing brevianamides, they rotate when exposed to UV or visible light; no rotation occurs if brevianamides are not present. Upon rotation, water is pumped up the conidiophore. Thus it has been proposed that brevianamides are involved in water translocation from the substrate up the conidiophore into the penicillus.

Acknowledgements

Work on secondary metabolites is funded by NASA grant (NAG 9–602). I thank Duane L. Pierson, Saroj K. Mishra and David Koenig for encouragement. I am indebted to Aiqi Fang for assistance in preparation of this review.

References

Anke, H. (1977). Metabolic products of micro-organisms. 163. Desferritriacetylfusigen, an antibiotic from *Aspergillus deflectus*. *Journal of Antibiotics*, **30**, 125–8.

Ayer, W.A., Lee, S.P., Tsuneda, A. & Hiratsuka, Y. (1980). The isolation, identification, and bioassay of the antifungal metabolites produced by *Monocillium nordinii*. *Canadian Journal of Microbiology*, **26**, 766–73.

Barksdale, A.W., Morris, T.C., Seshadri, R., Aranachalam, T., Edwards, J.A., Sundeen, J. & Green, J.M. (1974). Response of *Achlya ambisexualis* E87 to the hormone antheridiol and certain other steroids. *Journal of General Microbiology*, **82**, 295–9.

Bennett, J.W. (1981). Genetic perspective on polyketides, productivity, parasexuality, protoplasts, and plasmids. In *Advances in Biotechnology* 3, *Fermentation Products* (ed. C. Vezina & K. Singh), pp. 409–415. Toronto: Pergamon Press.

Bennett, J.W. (1983). Differentiation and secondary metabolism in mycelial fungi. In *Secondary Metabolism and Differentiation in Fungi* (ed. J.W. Bennett & A. Ciegler), pp. 1–32. New York: Marcel Dekker.

Bruehl, G.W., Millar, R.L. & Cunfer, B. (1969). Significance of antibiotic production by *Cephalosporium gramineum* to its saprophytic survival. *Canadian Journal of Plant Science*, **49**, 235–46.

Bu'Lock, J.D., Detroy, R.W., Hostalek, Z. & Munim-al-Shakarchi, A. (1974). Regulation of secondary biosynthesis in *Gibberella fujikuroi*. *Transactions of the British Mycological Society*, **62**, 377–89.

Campbell, I.M. (1984). Secondary metabolism and microbial physiology. *Advances in Microbial Physiology*, **25**, 1–60.

Campbell, I.M., Doerfler, D.L., Bird, B.A., Remaley, A.T., Rosato, L.M. & Davis, B.N. (1982). Secondary metabolism and colony development in solid cultures of *Penicillium brevicompactum* and *Penicillium patulum*. In *Overproduction of Microbial Products* (ed. V. Krumphanzl, B. Sikyta & Z. Vanek), pp. 141–51. London: Academic Press.

Castignetti, D. & Smarrelli, J.Jr (1986). Siderophores, the iron nutrition of plants, and nitrate reductase. *FEBS Letters*, **209**, 147–51.

Cavalier-Smith, T. (1992). Origins of secondary metabolism. In *Secondary Metabolites: Their Function and Evolution* (ed. D.J. Chadwick & J. Whelan), pp. 64–87. New York: John Wiley & Sons.

Champe, S., Rao, P. & Chang, A. (1987). An endogenous inducer of sexual development in *Aspergillus nidulans*. *Journal of General Microbiology*, **133**, 1383–7.

Charlang, G., Horowitz, R.M., Lowy, P.H., Ng, B., Poling, S.M. & Horowitz, N.H. (1982). Extracellular siderophores of rapidly growing *Aspergillus nidulans* and *Penicillium chrysogenum*. *Journal of Bacteriology*, **150**, 785–7.

Cundliffe, E. (1989). How antibiotic-producing organisms avoid suicide. *Annual Review of Microbiology*, **43**, 207–33.

Dahlberg, K.R. & Van Etten, J.L. (1982). Physiology and biochemistry of fungal sporulation. *Annual Review of Phytopathology*, **20**, 281–301.

Darvill, A.G. & Albersheim, P. (1984). Phytoalexins and their elicitors – a defense against microbial infection in plants. *Annual Review of Plant Physiology*, **35**, 243–75.

Demain, A.L. (1974). How do antibiotic-producing microorganisms avoid suicide? *Annals of the New York Academy of Sciences*, **235**, 601–12.

Demain, A.L. (1980). Do antibiotics function in nature? *Search*, **11**, 148–51.

Demain, A.L. (1992). Microbial secondary metabolism: a new theoretical frontier for academia, a new opportunity for industry. In *Secondary Metabolites: Their Function and Evolution* (ed. D.J. Chadwick & J. Whelan), pp. 3–23. New York, John Wiley & Sons.

Demain, A.L. & Masurekar, P.S. (1974). Lysine inhibition of *in vivo* homocitrate synthesis in *Penicillium chrysogenum*. *Journal of General Microbiology*, **82**, 143–51.

Ellis, C. & Gaucher, G.M. (1993). A putative regulatory gene associated with secondary metabolism in *Penicillium urticae*. In *Abstracts, Annual Meeting of the Society for Industrial Microbiology*, p. 72.

Erge, D., Maier, W. & Groger, D. (1973). Untersuchungen über die enzymatische Unwandlung von Chanoclavin-I. *Biochemie und Physiologie der Pflanzen*, **164**, 234–47.

Fu, Y.-H. & Marzluf, G.A. (1990). *nit-2*, the major positive-acting regulatory gene of *Neurospora crassa*, encodes a sequence-specific DNA-binding protein. *Proceedings of the National Academy of Sciences, USA*, **87**, 5331–5.

Gaucher, G.M., Lam, K.S., Grootwassink, J.W.D., Neway, J. & Deo, Y.M. (1981). The initiation and longevity of patulin biosynthesis. *Developments in Industrial Microbiology*, **22**, 219–32.

Gianinazzi, S. & Gianinazzi-Pearson, V. (1988). Mycorrhizae: a plant's health insurance. *ChimicaOggi*, 56–8.

Gooday, G.M. (1987). Hormones and sexuality in fungi. In *Secondary Metabolism and Differentiation in Fungi* (ed. J.W. Bennett & A. Ciegler), pp. 239–66. New York: Marcel Dekker.

Hesseltine, C.W. (1986). Global significance of mycotoxins. In *Mycotoxins and Phycotoxins* (ed. P.S. Steyn & R. Vleggaar), pp. 1–18. Amsterdam: Elsevier Science Publishers BV.

Hesseltine, C.W., Rogers, R.F. & Shotwell, O.L. (1981). Aflatoxin and mold flora in North Carolina in 1977 corn crop. *Mycologia*, **73**, 216–28.

Hopwood, D.A. (1988). Towards an understanding of gene switching in *Streptomyces*; the basis of sporulation and antibiotic production. *Proceedings of the Royal Society, London*, **235**, 121–38.

Horowitz, N.H., Charlang, G., Horn, G. & Williams, N.P. (1976). Isolation and identification of the conidial germination factor of *Neurospora crassa*. *Journal of Bacteriology*, **127**, 135–40.

Howell, C.R. & Stipanovic, R.D. (1983). Gliovirin, a new antibiotic from *Gliocladium virens* and its role in the biological control of *Pythium ultimum*. *Canadian Journal of Microbiology*, **29**, 321–4.

Ishii, K., Kobayashi, J., Ueno, Y. & Ichinoe, M. (1986). Occurrence of trichothecin in wheat. *Applied and Environmental Microbiology*, **52**, 331–3.

Iwasaki, S., Kobayashi, H., Furukawa, J., Namikoshi, M., Okuda, S., Sato, Z., Matsuda, I. & Noda, T. (1984). Studies on macrocyclic lactone antibiotics. 7. Structure of a phytotoxin 'rhizoxin' produced by *Rhizopus chinensis*. *Journal of Antibiotics*, **37**, 354–62.

Janzen, D.H. (1977). Why fruits rot, seeds mold and meat spoils. *American Naturalist*, **111**, 691–713.

Kamiya, M., Sakurai, A., Tamura, S., Takahashi, N., Abe, K., Tsuchiya, E. & Fukui, S. (1978). Isolation of rhodotorucine A, a peptidyl factor inducing the mating tube formation in *Rhodosporidum toruloides*. *Agricultural and Biological Chemistry*, **42**, 1239–43.

Kanaoka, M., Isogai, A. & Suzuki, A. (1975). Syntheses of bassianolide and its two homologs enniatin C and decabassianolide. *Agricultural and Biological Chemistry*, **43**, 1079–83.

Kawai, G. & Ikeda, Y. (1985). Structure of biologically active and inactive cerebrosides prepared from *Schizophyllum commune*. *Journal of Lipid Research*, **26**, 338–43.

Kay, R.R. (1982). cAMP and spore differentiation in *Dictyostelium discoideum*. *Proceedings of the National Academy of Sciences, USA*, **79**, 3228–31.

Kendrick, B. (1986). Biology of toxigenic anamorphs. *Pure and Applied Chemistry*, **58**, 211–18.

Kimura, Y., Mizumo, T., Nakajima, H. & Hamasaki, T. (1992). Altechromones A and B, new plant growth regulators produced by the fungus *Alternaria* sp. *Bioscience, Biotechnology and Biochemistry*, **56**, 1664–5.

Krebs, E.G. & Beavo, J.A. (1979). Phosphorylation-dephosphorylation of enzymes. *Annual Review of Biochemistry*, **48**, 923–59.

Krupinski, V.M., Robbers, J.E. & Floss, H.G. (1976). Physiological study of ergot: induction of alkaloid synthesis by tryptophan at the enzymatic level. *Journal of Bacteriology*, **125**, 158–65.

Krywolap, G.N., Grand, L.F. & Casida, Jr, L.E. (1964). The natural occurrence of an antibiotic in the mycorrhizal fungus *Cenococcum graniforme. Canadian Journal of Microbiology*, **10**, 323–8.

Kunesch, G., Zagatti, P., Pouvreau, A. & Cassini, R. (1987). A fungal metabolite as the male wing gland pheromone of the bumble bee wax moth, *Aphomia sociella* L. *Zeitschrift für Naturforschung*, **42c**, 657–9.

Lee, S., Aoyagi, H., Shimohigashi, Y., Izumiya, N., Ueno, T. & Fukami, H. (1976). A synthesis of cyclotetradepsipeptides, AM-toxin and its analogs. *Tetrahedron Letters*, 843–6.

Lee, S., Izumiya, N., Suzuki, A. & Tamura, S. (1975). Synthesis of a cyclohexadepsipeptide, protodestruxin. *Tetrahedron Letters*, 883–6.

Leichtling, B.H., Majerfeld, I.H., Spitz, E., Schaller, K.L., Woffendin, C., Kakinuma, S. & Rickenberg, H.V. (1984). A cytosolic cyclic AMP-dependent protein kinase in *Dictyostelium discoideum. Journal of Biological Chemistry*, **259**, 662–8.

Leite, B. & Nicholson, R.L. (1992). Mycosporine-alanine: a self-inhibitor of germination from the conidial mucilage of *Colletotrichum graminicola. Experimental Mycology*, **16**, 76–86.

Luckner, M., Nover, I. & Böhm, H. (1977). *Secondary Metabolism and Cell Differentiation*. New York: Springer-Verlag.

Mann, J. (1978). Secondary metabolism and ecology. In *Secondary Metabolism* (ed. J. Mann), pp. 279–304. Oxford: Clarendon Press.

Marumo, S., Nukina, M., Kondo, S. & Tomiyama, K. (1982). Lunatoic acid A, a morphogenic substance inducing chlamydospore-like cells in some fungi. *Agricultural and Biological Chemistry*, **46**, 2399–401.

Marx, D.H. (1969). The influence of ectotrophic mycorrhizal fungi on the resistance of pine roots to pathogenic infections. II. Production, identification and biological activity of antibiotics produced by *Leucopaxillus cerealis* var. *piceina. Phytopathology*, **59**, 411–17.

Matzanke, B.F., Bill, E., Trautwein, A.X. & Winkelmann, G. (1987). Role of siderophores in iron storage in spores of *Neurospora crassa* and *Aspergillus ochraceus. Journal of Bacteriology*, **169**, 5873–6.

Mitscher, L.A. (1975). Antimicrobial agents from higher plants. *Recent Advances in Phytochemistry*, **9**, 243–82.

Miyadoh, S. (1993). Research on antibiotic screening in Japan over the last decade: a producing microorganisms approach. *Actinomycetologia*, **7**, 100–6.

Munoz, G.A. & Agosin, E. (1993). Glutamine involvement in nitrogen control of *Gibberella fujikuroi. Applied and Environmental Microbiology*, **59**, 4317–22.

Neilands, J.B. (1984). Siderophores of bacteria and fungi. *Microbiological Sciences*, **1**, 9–14.

Neway, J. & Gaucher, G.M. (1981). Intrinsic limitations on the continued production of the antibiotic patulin by *Penicillium urticae. Canadian Journal of Microbiology*, **27**, 206–15.

Nisbet, L. & Porter, N. (1989). The impact of pharmacology and molecular biology on the exploitation of microbial products. *Symposium of the Society of General Microbiology*, **7**, 309–42.

Norstadt, F.A. & McCalla, T.M. (1969). Microbial populations in stubble-mulched soil. *Soil Science*, **107**, 188–93.

Nutting, W.H., Rapoport, H. & Machlis, L. (1968). The structure of sirenin. *Journal of the American Chemical Society*, **90**, 6434–8.

Omura, S. (1992). Trends in the search for bioactive microbial metabolites. *Journal of Industrial Microbiology*, **10**, 135–56.

Pazoutova, S., Pokorny, V. & Rehacek, Z. (1977). The relationship between conidiation and alkaloid production in saprophytic strains of *Claviceps purpurea*. *Canadian Journal of Microbiology*, **23**, 1182–7.

Peipp, H. & Sonnenbichler, J. (1992). Occurrence of antibiotic compounds in cultures of *Armillaria ostoyae* growing in the presence of an antagonistic fungus or host plant cells. *Biological Chemistry Hoppe-Seyler*, **373**, 675–83.

Rines, H.W., French, R.C. & Daasch, L.W. (1976). Nonanal and 6-methyl-5-hepten-2-one; endogenous germination stimulators of uredospores of *Puccinia graminis* var. *tritici* and other rusts. *Journal of Agricultural and Food Chemistry*, **22**, 96–100.

Robeson, D.J. & Strobel, G.A. (1982). Deoxyradicinin, a novel phytotoxin from *Alternaria helianthi*. *Phytochemistry*, **21**, 1821–3.

Roth, J., Leroith, D. & Collier, E.S. (1986). The evolutionary origins of intercellular communication and the Maginot Lines of the mind. *Annals of the New York Academy of Sciences*, **463**, 1–11.

Sakagami, Y., Isogai, A., Suzuki, A., Tamura, S., Tsuchiya, E. & Fukui, S. (1987). Amino acid sequence of tremorgen A-10, a peptidyl hormone inducing conjugation tube formation in *Tremella mesenterica* Fr. *Agricultural and Biological Chemistry*, **42**, 1301–2.

Sakurai, S., Tamura, S., Yanagishima, N. & Shimoda, C. (1977). Structure of a peptidyl factor, α substance-I_A, inducing sexual agglutinability in *Saccharomyces cerevisiae*. *Agricultural and Biological Chemistry*, **41**, 395–8.

Scheffer, R.P. (1991). Role of toxins in evolution and ecology of plant pathogenic fungi. *Experientia*, **47**, 804–11.

Scott, R.E., Jones, A., Lam, K.S. & Gaucher, G.M. (1986). Manganese and antibiotic biosynthesis. I. A specific manganese requirement for patulin production in *Penicillium urticae*. *Canadian Journal of Microbiology*, **32**, 259–67.

Sparace, S.A., Reeleder, R.D. & Khanizadeh, S. (1987). Antibiotic activity of the pyrenocines. *Canadian Journal of Microbiology*, **33**, 327–30.

Stone, M.J. & Williams, D.H. (1992). On the evolution of functional secondary metabolites (natural products). *Molecular Microbiology*, **6**, 29–34.

Strobel, G., Kenfield, D., Bunkers, G., Sugawara, F. & Clardy, J. (1991). Phytotoxins as potential herbicides. *Experientia*, **47**, 819–26.

Swain, T. (1977). Secondary compounds as protective agents. *Annual Review of Plant Pathology*, **28**, 479–501.

Sydenham, E.W., Gelderblom, W.C.A., Thiel, P.G. & Marasas, W.F.O. (1990). Evidence for the natural occurrence of fumonisin B_1, a mycotoxin produced by *Fusarium moniliforme*, in corn. *Journal of Agricultural and Food Chemistry*, **38**, 285–90.

Tanaka, S., Wada, K., Katayama, M. & Marumo, S. (1984). Isolation of sporogen-A01, a sporogenic substance, from *Aspergillus oryzae*. *Agricultural and Biological Chemistry*, **49**, 3189–91.

Taya, Y., Yamada, T. & Nishimura, S. (1980). Correlations between acrasins and spore germination inhibitors in cellular slime molds. *Journal of Bacteriology*, **143**, 715–19.

Tkacz, J.S., Giacobbe, R.A. & Monaghan, R.L. (1993). Improvement in the titer of echinocandin-type antibiotics: a magnesium-limited medium supporting the biphasic production of pneumocandins A_0B_0. *Journal of Industrial Microbiology*, **11**, 95–103.

Tsurushima, T., Ueno, T., Fukami, H., Tani, T. & Mayama, S. (1990). Germination self-inhibitor from uredospores of the oat crown rust fungus. *Abstract PS86-11, International Union of Microbiological Societies Congress, Osaka*.

Vining, L.C. & Taber, W.A. (1979). Ergot alkaloids. In *Economic Microbiology 3, Secondary Products of Metabolism* (ed. A.H. Rose), pp. 389–420. London: Academic Press.

Williams, D.H., Stone, M.J., Hauck, P.R. & Rahman, S.K. (1989). Why are secondary metabolites (natural products) biosynthesized? *Journal of Natural Products*, **52**, 1189–208.

Wolf, J.C. & Mirocha, C.J. (1977). Control of sexual reproduction in *Gibberella zeae* (*Fusarium roseum* 'Graminearum'). *Applied and Environmental Microbiology*, **33**, 546–50.

Zhang, J., Wolfe, S. & Demain, A.L. (1987a). Effect of ammonium as nitrogen source on production of δ-(L-α-aminoadiply)-L-cysteinyl-D-valine synthetase by *Cephalosporium acremonium* C-10. *Journal of Antibiotics*, **40**, 1746–50.

Zhang, J., Banko, G., Wolfe, S. & Demain, A.L. (1987b). Methionine induction of ACV synthetase in *Cephalosporium acremonium*. *Journal of Industrial Microbiology*, **2**, 251–5.

Zhang, J., Wolfe, S. & Demain, A.L. (1989). Phosphate regulation of ACV synthetase and cephalosporin biosynthesis in *Streptomyces clavuligerus*. *FEMS Microbiology Letters*, **57**, 145–50.

Zhang, J. & Demain, A.L. (1992). Regulation of ACV synthetase activity in the beta-lactam biosynthetic pathway by carbon sources and their metabolites. *Archives of Microbiology*, **158**, 364–9.

10

The nature and extent of mutualism in the mycorrhizal symbiosis

D.J. READ

The century which encompasses the history of the British Mycological Society also, to within a decade, is the period over which the mycorrhizal symbiosis has been recognized. Although there was a 'pre-history' of research on the subject (Trappe & Berch, 1985) the paper of Frank (1885) provided the first general account of the mycorrhizal phenomenon and of its distribution in nature. The purpose of the present paper is not to review the history of this subject. Historical analyses of individual contributions (Harley, 1985) and of scientific developments over the century (Harley & Smith, 1983) have already been written. Rather, it is intended to take stock and to look forward with a view to determining how best to build upon the foundation of knowledge already laid.

Molecular (Simon et al., 1993) and fossil (Stubblefield, Taylor & Trappe, 1987; Remy et al., 1994; Taylor et al., 1995) evidence taken in combination (Figs 10.1A,B) indicate that symbiotic interactions between fungi and plant roots developed very early in the process of colonization of the terrestrial environment. Evolutionary considerations would therefore suggest that the relationship between host and fungus has been inherently of the doubly beneficial or 'mutualistic' kind, selection having favoured its persistence in the overwhelming majority of species to the present day when over 90% of the world's vascular plants belong to families that are commonly mycorrhizal (Trappe, 1987). Most of these plants are colonized by fungi of the order Glòmales (Morton & Benny, 1990) which are characterized today, as they were in Triassic times (Fig. 10.1B) by the intra-cellular production of the much branched hyphal structures or 'arbuscules' that lead to the association being referred to as an 'arbuscular' (AM) or 'vesicular-arbuscular' (VA) mycorrhiza.

Subsequent events led to the selection of further distinct types of mycorrhizal symbiosis amongst which the ectomycorrhiza, formed

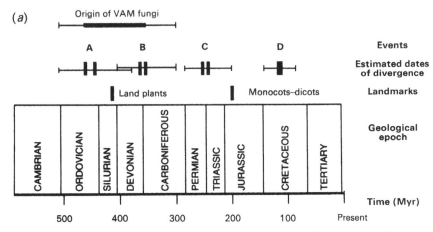

Fig. 10.1(*a*) Comparative analysis of origins and divergence of VA fungi and land plants based upon a calibrated molecular clock in which rates of substitution of gene sequences in the small subunit rRNA are used to provide dates. The VA fungi are seen to originate between 462 and 353 million years ago (My) when *Glomales* diverged from *Endogonales* (A). This period coincides with approximate origin of the first Rhynie-type land plants. Periods B, C and D represent estimated dates of divergence of *Gigasporaceae*, *Acaulosporaceae* and of *Entrophospora*, respectively. (From Simon *et al.*, 1993).

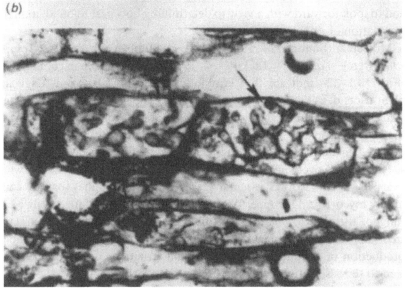

Fig. 10.1(*b*) Much-branched intra-cellular arbuscule-like structure in fossilized root of a fern ally recovered from Triassic rocks of Antarctica. (From Stubblefield *et al.*, 1987).

predominantly between trees and basidiomycetes or ascomycetes is important. Beautifully preserved fossil ectomycorrhizas formed by a *Rhizopogon*-like fungus have been found in chert of the Eocene period *ca* 50 million years old (Le Page *et al.*, 1996). The ericoid endomycorrhizal association between ericaceous shrubs and ascomycetes and the orchid mycorrhiza formed by basidiomycetes are probably of more recent origin.

This chapter explores the nature of the mycorrhizal relationship as we currently understand it and considers the extent to which our knowledge of its functions has kept pace with the ever lengthening lists of records of 'occurrence' of the various types of symbiosis. It emphasizes the need for critical appraisal, by experiments carried out under conditions that are as ecologically realistic as possible, of the status of recorded plant–fungus associations. The view, so readily derived from the evolutionary perspective that 'occurrence' of an association is necessarily indicative of 'mutualism' is brought into question.

Many pot and laboratory experiments have demonstrated that mycorrhizal fungi can enhance the ability of a plant to capture phosphate from the soil (Harley & Smith, 1983) but it is increasingly recognized that the environmental conditions employed in them bear little relation to the field situation where individuals co-exist in communities of mixed species, almost all of which are colonized by these fungi. It is far more difficult to demonstrate positive impacts of mycorrhizal associations under these circumstances and attempts to do so have, not surprisingly, produced ambiguous results (Fitter, 1985). There are, however, increasing numbers of experiments, carried out either under natural conditions, or using manipulations designed to simulate nature, which point to a diverse range of functions of the symbiosis. Attention will be concentrated upon these studies because they are of the type which provide the best opportunity for increased understanding of the true status of the mycorrhizal symbiosis.

The nature and extent of mutualism in the VA mycorrhizal symbiosis

Recent studies by Fitter and his group of two very different plants, bluebell, *Hyacinthoides non-scripta* and the grass, *Vulpia ciliata*, illuminate the contrasting nature of the possible beneficial impacts of VA colonization under field conditions. While *H. non-scripta* is a perennial vernal geophyte of woodland soil which has a coarse and poorly branched root system likely to provide only inefficient exploration of soil, *V. ciliata* is a winter annual of open sandy soil with a fine fibrous root system, that should not

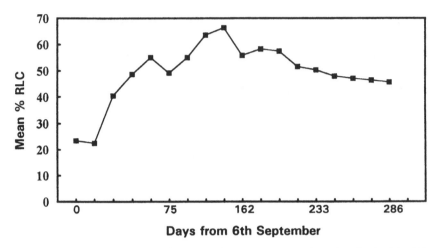

Fig. 10.2(a) Seasonal pattern of percentage root length colonized by VA fungi (% RLC) in a field population of bluebell, *Hyacinthoides non-scripta*.

suffer such deprivation. These studies provide examples, respectively, of a nutritional and a non-nutritional basis for mutualism in the VA symbiosis.

Using an approach based upon regular sampling of undisturbed plants of *H. non-scripta* throughout their annual life cycle in a deciduous woodland, Merryweather & Fitter (1995) demonstrated that the inflows of P necessary to satisfy the requirements of the plant can only be achieved by colonization. In this plant, which produces a new bulb and root system every year, there was a rapid increase in the proportion of root length colonized by VA fungi over the period from root emergence in September, to reach a maximum value in excess of 70% in January and February (Fig. 10.2(*b*)). Thereafter, as evidenced by declining numbers of entry points, (Fig. 10.2(*b*)) new colonization slowed. From the time of root emergence, P inflow increased rapidly at a similar rate to that of colonization, although until December, values were negative indicating that net loss of P was occurring. Maximum inflows were reached during the photosynthetic phase but these subsequently declined at the same rate as that of colonization. When curves were fitted to data for P inflow (Fig. 10.3(*a*)) and % root length colonized (Fig. 10.3(*b*)), they demonstrated a very similar inherent pattern, there being a significant correlation between the two variables.

The individual plants lose significant amounts of P, particularly at the end of the growing season, in seeds, old leaves and roots as they are shed. Glasshouse grown plants lacking mycorrhizal colonization are unable to capture sufficient P from the soil to enable their P budget to be balanced.

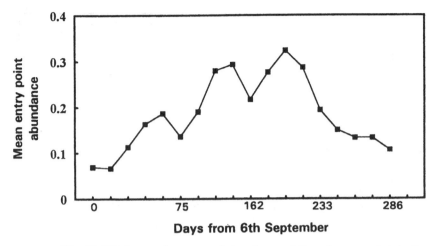

Fig. 10.2(*b*) Seasonal pattern of abundance of fungal entry points in the field calculated as mean number of penetrations per intersection. (From Merryweather & Fitter, 1995).

They therefore end the season with a large P deficit which could not be sustained in the field.

These results suggest that in the field the major effect of VA colonization is to provide the poorly branched root-system of bluebell with enhanced ability to explore soil for phosphate and that survival of the plant in nature is dependent upon these P-scavenging activities. They confirm the prediction of Baylis (1975) that species with coarse root systems are likely to be particularly responsive to VA colonization and may be dependent upon mycorrhiza for survival in the field.

This type of root architecture contrasts markedly with that of the much branched fibrous systems typical, for example, of grasses. The fact that grasses are usually extensively colonized by VA fungi despite the fibrosity of their root system raises questions as to functional status of mycorrhizal fungi in this type of plant. As in the case of forbs, there are many glasshouse experiments with grasses demonstrating that enhancement of phosphorus capture can lead to increases of yield, but such effects have been difficult to observe in natural communities.

An alternative explanation for the apparently beneficial impact of VA fungus colonization upon the annual grass, *Vulpia ciliata*, has recently been proposed (Newsham, Fitter & Watkinson, 1994, 1995). Using the fungicide benomyl to control VA colonization of the roots in the field, it was shown (West, Fitter & Watkinson, 1993) that there was no relationship between extent of occurrence of VA fungi and the rates of P uptake,

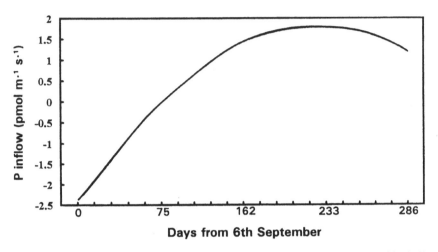

Fig. 10.3(a) Curve of fitted values of inflow of phosphorus (P) to bluebell roots growing in the field over a season.

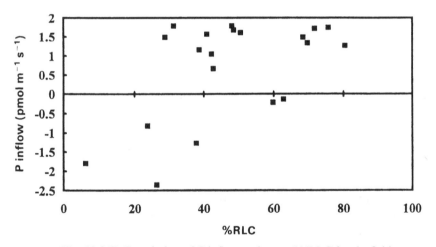

Fig. 10.3(b) Correlation of P inflow and mean % RLC for the field population (Linear regression: $r^2 = 28.8$ P = (0.015). (From Merryweather & Fitter, 1995).

but that the sterilant also controlled weakly pathogenic fungi such as the cosmopolitan *Fusarium oxysporum* which were known to reduce fecundity of the plant. It was therefore suggested by Newsham *et al.* (1994) that the benefits of VA fungi arose through their ability to protect the plant from pathogens. Similar effects have been observed previously in horticultural and agricultural crops (Linderman, 1994) and have been recognized as a feature of ectomycorrhizal relationships (see below) but have not previously been implicated as a factor determining plant response in natural communities.

This effect was examined further in field-grown populations of *V. ciliata* exposed to different concentrations of benomyl so as to control the extent of colonization by VA and pathogenic fungi (Newsham *et al.*, 1995). Fecundity was shown to be largely unresponsive to fungicide application, despite the fact that benomyl significantly reduced the abundance of both types of fungi in roots. However, the abundance of root pathogenic fungi, especially *F. oxysporum*, was negatively correlated with fecundity, even though plants displayed no disease symptoms. The poor relationship between fecundity and benomyl application contrasted markedly with the effects of benomyl on VA and pathogenic fungi, and with the negative effects of root pathogens on fecundity. These effects could be explained if the two groups of fungi interacted, so that when both were reduced in abundance by fungicides, the net effect on fecundity was slight. Such a hypothesis was supported by statistical analysis of the data showing that there was a positive effect of VA fungi, but only in relation to the negative effect of the pathogens.

This interaction was resolved using a transplant approach. Seedlings of *V. ciliata* were grown in a growth chamber with a factorial combination of inoculum of *F. oxysporum* or the *Glomus* sp., both isolated from *V. ciliata* at the field site, and then planted into a natural population of the grass in the field. Here, after 62 days of growth, there was clear evidence that colonization by *Glomus* gave a protective effect. Plants inoculated with that fungus performed as well as control plants, even when simultaneously inoculated with *F. oxysporum*, whereas those inoculated with *F. oxysporum* alone grew significantly less well (Table 10.1). The *Glomus* sp. had a negligible effect on the performance of the plants in the absence of the pathogen. Analysis of P status of the tissues showed that there was no correlation between shoot P concentration and the abundance in roots of either pathogenic or mycorrhizal fungal hyphae. The differences between treatments instead seem to have been due to a reduction in the frequency of pathogenic hyphae within roots brought about by VA fungal colonization.

Experiments such as these demonstrate that mycorrhizal colonization may have subtle impacts upon plant performance and that these may only be revealed by a combination of sensitive experimental design and careful data analysis.

The possibility is emerging that in addition to having impacts upon the

Table 10.1. *Effects of a factorial combination of* Fusarium oxysporum *(F) and a* Glomus sp. *(G) on shoot biomass and root length of* Vulpia ciliata *plants grown with both fungi in the laboratory, transplanted into the field and sampled from the field after 62 days growth there*

Variable	Treatment				Main effects		Interaction
	-G -F	+G -F	-G +F	+G +F	G	F	G × F
Log (ln) shoot biomass (mg)	2.4a	2.2a	1.4b	2.2a	$F = 3.5$	$F = 17.3***$	$F = 4.8*$
Root length (cm)	217a	203a	111b	228a	$F = 9.4**$	$F = 9.0**$	$F = 9.0**$

Note: Means are of 16 replicates; where followed by different letters they differ at $P<0.05$. Significant main and interaction effects in Anova are indicated by: * $P<0.05$, ** $P<0.01$, *** $P<0.001$.

current generation of plants, colonization by VA fungi may be of direct benefit to offspring of a mycorrhizal parent. It has been shown (Koide *et al.*, 1988; Bryla & Koide, 1990) that seed of mycorrhizal wild oat (*Avena fatua*) plants grown in P deficient soil, contained significantly more P, mostly as phytate, than did that produced by their non-mycorrhizal counterparts. In the course of germination these additional P reserves facilitated significantly enhanced rates of leaf expansion as well as greater root surface phosphatase and ATP-ase activities. When the offspring plants, grown in the non-mycorrhizal condition, were mature, effects of mycorrhizal colonization were still evident in terms of their having greater leaf area, shoot and root P content, root N content and root-shoot ratio. The spikelets of these second generation plants derived from mycorrhizal parents contained more seed, each of which were of higher N and P content, than those derived from non-mycorrhizal parents. These results suggest that the effects of mycorrhizal colonization of the first generation of plants might still be evident in mature third generation individuals. They emphasize the need for consideration of mycorrhizal impacts over the long as well as the short term.

All of the experiments described so far deal with the impacts of mycorrhizal fungi upon individual plant species. There is increasing recognition, however, that plants colonized by VA fungi in natural ecosystems normally co-exist in assemblages of mixed species. This has prompted alternative approaches to the question of the role of VA fungi in the dynamics of plant communities. Grime *et al.* (1987) reassembled a calcareous grassland sward in microcosms of sterilized dune sand in which were planted, either in the mycorrhizal or non-mycorrhizal condition, seedlings of the main turf-forming species, *Festuca ovina*. After establishment of the grass an ecologically realistic assemblage of species was sown, as seed, into the sward and the survivorship of each species was monitored over subsequent months. The presence of mycorrhizal inoculum provided for enhancement of survivorship of several of the plants which are normally colonized by these fungi in nature (Table 10.2). In contrast, two ruderal species, *Rumex acetosa* and *Arabis hirsuta*, which would normally invade such turf only after disturbance and are of families *Polygonaceae* and *Cruciferae*, respectively, not considered to be mycorrhizal, showed reduced survivorship in the microcosms containing VA fungi. This experiment thus demonstrated that the fungi were exerting a selective effect upon community composition, the recruitment of some species being enhanced while that of others was reduced.

In nature, seedlings of compatible host species become infected by VA

Table 10.2. *Survivorship (%) of forbs after 6 months in mycorrhizal (M) and non-mycorrhizal (NM) microcosms. Significant increases of survivorship were obtained in most forbs grown in the mycorrhizal condition, the exceptions being* Arabis hirsuta *and* Rumex acetosa *which show the reverse trend*

Species	M	NM	
Centaurium erythraea		64	2
Galium verum	58	11	
Hieraceum pilosella		49	6
Leontodon hispidus	42	13	
Plantago lanceolata		71	10
Sanguisorba minor	53	6	
Scabiosa columbaria		84	16
Arabis hirsuta	8	42	
Rumex acetosa	11	60	

Table 10.3. *Weighted averages (with 95% confidence limits) of the time required for the appearance of the first colonization (FCA) and for total colonization (TCA) of the original root axes and of lateral roots of four of the major component species of a phosphorus-deficient calcareous grassland community. Seedlings were placed into small gaps in the undisturbed turf at the time of radicle emergence and harvested sequentially over 20 days*

	FCA (days)		TCA (days)	
Host species	Root axis	Laterals	Root axis	Laterals
Arrhenatherum elatius	3.8±0.4	5.9±1.4	10.4±1.8	17.2±2.8
Festuca ovina	3.8±1.2	6.9±0.6	13.5[a]	14.9±1.8
Plantago lanceolata	2.1±0.6	4.3±0.5	16.4[a]	16.9±4.4
Medicago lupulina	3.0±0.5	5.1±0.8	8.9±2.0	12.2±2.2

fungi very soon after germination (Read, Koucheki & Hogdom, 1976). By sowing seeds of four of the dominant species of a calcareous grassland in Derbyshire, UK, into undisturbed turf at the time of radicle emergence, and sequentially harvesting them, it was shown (Table 10.3) that intensive infection was present within a few days (Read & Birch, 1988). The developing radicle and later the emerging lateral roots are repeatedly penetrated by hyphae of the external network, there being up to 100 entry points per centimetre of root length. In these experiments, infection of a given species proceeded at the same rate irrespective of whether the community into which the seedling was planted contained the same species or was an assemblage of different species, suggesting that inter- as well as intra-specific linkages between plants must occur.

Direct evidence for the presence of physiologically functional VA hyphal interconnections in intra- or interspecific combinations of plants has been achieved by means of an autoradiographic analysis of intact systems, in which plants which had also been the source of infection for associated seedlings were fed with $^{14}CO_2$ (Francis & Read, 1984; Read, Francis & Finlay, 1985). The subsequent quantification of this radioactivity has revealed that more transfer occurs into shaded than into fully illuminated plants, suggesting that the transfer processes are determined by source–sink relationships.

Ecologists examining the structure of calcareous grassland communities have recognized the distinction between groups of species that were or were not able to colonize closed turf (Fenner, 1978; Grubb, 1976, 1977). As a result of their inability to establish in small gaps, members of the latter group are inevitably consigned to the ruderal habit. However, no clear mechanisms for their exclusion has been proposed. The study of Grime *et al.* (1987) indicated that in two such species, *R. acetosa* and *A. hirsuta*, a sensitivity to the presence of an established VA mycorrhizal mycelium might be a factor, and further experiments were designed to examine such a possibility (Francis & Read, 1994, 1995). Chambers were designed in which a nylon mesh cylinder with a 37 μm pore size enabled separation of root from mycelial effects (Fig. 10.4). Plants known to be hosts to VA fungi, usually *F. ovina* and *Plantago lanceolata*, were grown in the M or NM condition in the outer compartment of such chambers, for sufficient time to enable the mycelium of the VA fungus to grow from M plants into the central compartment. This compartment was designed to represent a 'gap' or regeneration niche *sensu* Grubb (1977) in established vegetation in which the only major variable was presence or absence of a VA mycelial network. Seeds of a range of test species were sown into the central compartment, at the time of radicle emergence, and their subsequent development was followed in sequential harvests, so that the impact of the VA fungi upon them could be evaluated. Species selected for use in these groups were representative of a range of families considered to be largely non-mycorrhizal, so-called 'non-hosts', or of uncertain status. They included *Arabis hirsuta* (*Cruciferae*), *Arenaria serpyllifolia* (*Caryophyllaceae*), *Echium vulgare* (*Boraginaceae*), *Reseda luteola* (*Resedaceae*), and *Rumex acetosella* (*Polygonaceae*). For comparative purposes some species that are both mycorrhizal and turf-compatible were examined. These included *Plantago lanceolata* (*Plantaginaceae*) and *Centaurium erythraea* (*Gentianaceae*).

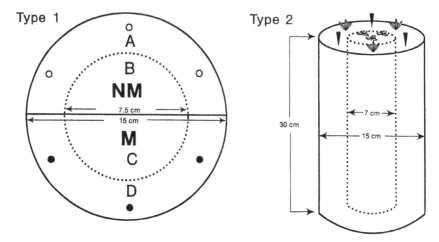

Fig. 10.4. Diagrammatic views of chambers used to investigate the impacts of actively growing VA mycelium upon plants. In Type 1 chambers mycorrhizal (M) 'donor' plants (closed circles) were planted in compartment D, and equivalent non-mycorrhizal (NM) plants (open circles) in compartment A. VA mycelia grew from the donors in compartment D through a 37 μm nylon mesh cylinder (dashed line) to occupy 'receiver' compartment C. A given test species was grown from germination to harvest in compartments C and B. Type 2 chambers had an undivided central compartment. In this case M and NM plants were grown in separate chambers.

The plants show very different responses to the presence in their root environment of mycelium of VA fungi. Those such as *P. lanceolata* and *C. erythraea* respond positively to the fungus and grow scarcely at all in its absence (Fig. 10.5(*a*)). In contrast, 'non-hosts' such as *Arabis hirsuta* show the converse growth response (Figs 10.5(*b*), 7(*c*)). Similarly, while reduction of survivorship is seen in non-hosts, e.g. *Arenaria serpyllifolia*, grown in the presence of VA mycelium (Fig. 10.6(*a*)) the reverse effect is seen in *C. erythraea* (Fig. 10.6(*b*)). *Echium vulgare* (Fig. 10.7(*a*)) and *Reseda luteola* (Fig. 10.7(*b*)) despite being colonized by the fungus which produced vesicles but no arbuscules in the roots of these plants, showed similar negative responses. These species are indeed reported in the literature (see, e.g. Harley & Harley, 1987) as having VA mycorrhiza. They provide striking examples of the need for functional analysis of the relationship between fungus and plant, and for caution when extrapolating from 'occurrence' of colonization to pronouncements about the status of the infection. The relationship observed here and in several other species examined by Francis & Read (1995) under conditions which are considered to resemble closely the natural regeneration niche were definitely of the 'antagonistic' rather than 'mutualistic' type.

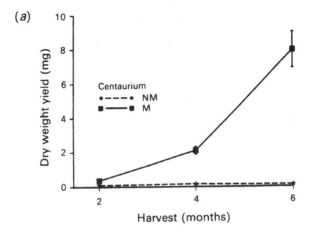

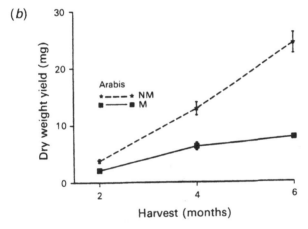

Fig. 10.5(a) Yields (mg dry weight) of a typical true host plant *Centaurium erythraeum* grown from the time of radicle emergence with (M) and without (NM) the presence of VA fungal mycelium in compartments C and B, respectively, of a type 1 chamber system (see Fig. 10.4). Vertical bars represent 95% confidence limits.
(b) As 5(a) but for the 'non-host' *Arabis hirsuta*. (From Francis & Read, 1994.)

The basis of the antagonistic effect VA fungi have upon ruderal species remains to be elucidated. In some cases adverse effects upon root development of non-hosts has been observed in the absence of colonization by the fungus suggesting that there is a chemical interaction (Allen, Allen & Friese, 1989; Francis & Read, 1984), while in others inhibition is

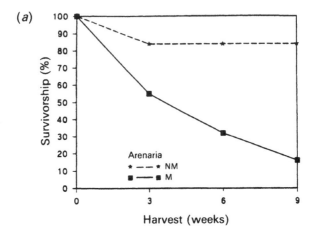

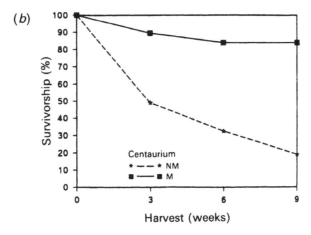

Fig. 10.6(a) Survivorship (%) of the 'non-host' plant *Arenaria serpyllifolia* grown as described in Fig. 10.5.
(*b*) As 6(*a*) but using the 'host' plant *Centaurium erythraea*. Some losses of both species occurred prior to soil penetration probably due to drought. The differences observed from 3 to 9 weeks are attributable to presence or absence of VA mycelium.

associated with penetration of the root and prolific production of vesicles (Francis & Read, 1995). In the latter case conventional microscopic analysis would leave many observers to record 'occurrence' of mycorrhiza despite the fact that arbuscules are not seen. This emphasizes the importance of functional analysis of the symbiosis and highlights the dangers associated with any assumption that occurrence of colonization

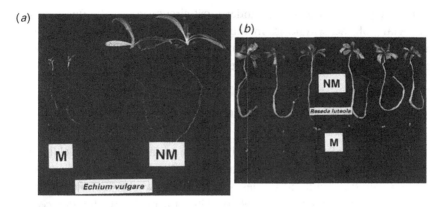

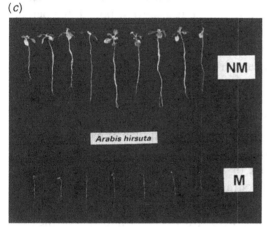

Fig. 10.7. Representative plants of *Echium vulgare* (*a*) *Reseda luteola* (*b*) *Arabis hirsuta* (*c*) grown for 63 days in the presence (M) or absence (NM) of VA mycelium, showing the severe antagonistic effects of the fungus upon these species. (From Francis & Read, 1995.)

is necessarily equivalent to 'mutualism'. There are clearly circumstances in which VA fungi are antagonistic to plants which they colonize and it is not unreasonable to hypothesize that this antagonism is sufficiently severe in some species to explain their exclusion from communities of mycorrhizal plants – their 'turf-incompatibility'. If this proves to be the case VA mycorrhizal fungi will be seen as major determinants both of the structure and biodiversity of plant communities.

The various experiments described above examined the symbiosis from the viewpoint of the plant response. The question of possible benefit to the fungus remains. Since VA fungi have only the most restricted ability to produce a vegetative mycelium or sporulate in the absence of a host plant, it can safely be concluded that under most circumstances the fungus is a beneficiary of the symbiosis. In those cases where 'non-hosts' are colonized, the status of the fungus remains unclear. Its resolution must await analysis of the carbon balance of these.

The nature and extent of mutualism in the ectomycorrhizal symbiosis

The significance of the ectomycorrhizal symbiosis at the global level lies in the fact that while it is found in a relatively small number of plant families, these contain the dominant species of some of the world's most important terrestrial ecosystems (Read, 1991a). Thus members of the *Pinaceae*, *Fagaceae* and *Myrtaceae* which dominate respectively, boreal, temperate and many sub-tropical forests are largely made up of species that are ecto-mycorrhizal. In the tropics some important families such as *Dipterocarpaceae*, as well as certain tribes of the *Leguminosae*, are also characteristically ectomycorrhizal (Alexander, 1989). The fungi involved in this symbiosis are almost exclusively basidiomycetes and ascomycetes. The ectomycorrhiza has three diagnostic structural features, a mantle of hyphae which ensheaths the root, a labyrinthine system of hyphal elements (the Hartig net) that penetrates between the outer cortical cells, and a more or less extensive network of hyphal elements often organized into linear organs variously termed 'strands', 'chords' or 'rhizomorphs', which explores the soil domain.

The likelihood that these fungi were involved in the nutrition of the tree has been recognized over the century since Frank's (1894) pioneering attempt to relate structural and functional aspects of the ectomycorrhiza of *Pinus*. In the ensuing period the work of other groups, notably those led by Melin working with intact mycorrhizal systems, and of Harley with excised roots (see Harley & Smith, 1983) have clearly established that ecto-mycorrhizal fungi have the ability to capture nutrient elements, in particular phosphate and ammonium ions from their environment, transport them to the hyphal mantle and facilitate their transfer to the plant. It has also been demonstrated in afforestation schemes that normally ectomycor-rhizal plants, such as pine, fail or grow only poorly when introduced into environments which lack ectomycorrhizal symbionts. At the same time, few if any ectomycorrhizal fungi are capable of producing an active free living mycelium or of fruiting in the absence of an autotrophic host plant.

A considerable amount of evidence has thus been obtained from both experimental and field studies to support the contention that the ectomycorrhizal symbiosis is mutualistic. However, our view of the functional basis of the ectomycorrhizal relationship has developed in a way which parallels to some extent that described for the VA symbiosis. An ever-broadening range of attributes has been demonstrated and increasingly these are assessed in terms of their possible contribution to the success of the host or fungus in real ecological situations, rather than in the narrower physiological context. In addition to providing realization that ectomycorrhizal fungi have a broader and more direct nutritional role than was hitherto appreciated, these studies have revealed that the hosts may gain enhanced resistance to pathogens and toxins, and that their performance in terms of yield may be reduced as well as increased by colonization. These aspects are considered below.

Nutritional role of ectomycorrhizal fungi

Forests dominated by ectomycorrhizal plants typically occur on impoverished acidic soils in which the major fund of nutrients is contained in a well-defined and largely superficial layer of organic material. Ectomycorrhizal roots proliferate preferentially and often almost exclusively in one zone of this organic layer the so-called fermentation horizon (FH) (Read, 1991a,b). As the name suggests, the convention has been to regard this as the domain of saprotrophic organisms, in which activities of ectomycorrhizal fungi were restricted to absorption of phosphate and ammonium ions released by the populations of saprotrophs. Despite the early demonstration by Melin (1925) that ectomycorrhizal fungi grown in pure culture themselves have considerable saprotrophic capability the possibility that they might express these attributes to the advantage of their hosts, and the disadvantage of the 'decomposer' population in the FH horizon has, with a few exceptions, e.g. Gadgil & Gadgil (1975), been overlooked.

Most ecosystems that contain a significant component of ectomycorrhizal species suffer from nitrogen rather than phosphorus limitation, often over 90% of their fund of this element being sequestered in the organic residues through which ectomycorrhizal roots and their fungal mycelia proliferate. Frank (1894) first recognized the possibility that these fungi were able directly to exploit the nitrogen contained in such residues but the possibility has not been fully explored in an experimental manner until recently.

Studies using peptides and proteins having different molecular weights and being in a number of chemical conformations ranging from pure soluble forms to protein-polyphenol complexes, have now revealed that ectomycorrhizal fungi have a broad spectrum of abilities to mobilize nitrogen contained in complex organic substrates. Some, termed 'protein' fungi readily break down soluble proteins into their constituent amino acids and assimilate these, without decarboxylation as nitrogen sources (Abuzinadah & Read, 1986a,b; Read, Leake & Langdale, 1989). Many of those fungi which exhibit proteolytic potential also have some ability, albeit much reduced, to liberate nitrogen from protein complexes with polyphenols such as tannic acid or with phenolic extracts of leaf tissue (Bending & Read, 1995a). The rapid loss of accessibility of nitrogen associated with precipitation of protein is likely to place a premium upon effective scavenging for nitrogen contained in macromolecular resources, and it is evident from studies in which the fungi are enabled to grow from their host plants across natural non-sterile substrates such as peat, that a considerable investment of carbon is made to facilitate such exploitation. Estimates of the proportion of host photosynthate allocated to support this growth range from 10 to 30% (Finlay & Söderström, 1992).

When grown under these circumstances, those fungi which are capable of producing hyphal aggregates appear to adopt two strategies in their nutrient foraging activities. The first involves the production of a thinly dispersed fan of mycelium which extends as a broad front at rates of 2–3 mm day over those homogeneously nutrient-poor parts of the substrate, while the second, which is expressed when the fungus makes contact with resource rich material, leads to the intensive proliferation of hyphae over the substrate which produces dense visually evident 'patches' of mycelium (Finlay & Read, 1986; Read, 1991b). Comparable dimorphic patterns of growth which can be regarded as representing respectively 'exploratory' and 'exploitative' foraging strategies, have been observed in rhizomorph-forming wood decomposing basidiomycetes (Rayner & Boddy, 1988). It has now been shown that the exploitative mode can be induced by placing discrete units of resource-rich material into otherwise nutrient-depleted peat in advance of the growing front of 'exploratory' hyphae. Proliferation begins as soon as the front makes contact with the new substrate and a 'patch' is formed. It was known from earlier studies that when these patches formed spontaneously in observation chambers they acted as sinks to which carbon was selectively allocated (Finlay & Read, 1986). By autoradiographic analysis of induced 'patches' after carbon was fed as $^{14}CO_2$ to the host plant, Bending & Read (1995a) have now shown that this allocation occurs over a finite period. Its duration will obviously depend upon the size of the resource. In their case, carbon allocation ceased after

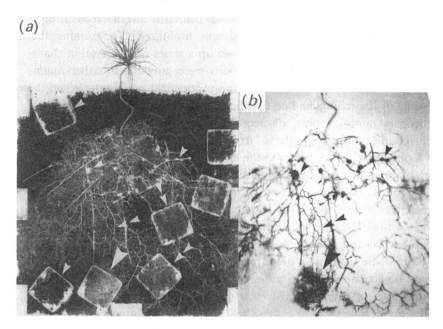

Fig. 10.8(a) Observation chamber showing extension of vegetative mycelium of *Suillus bovinus* from mycorrhizal roots of *Pinus sylvestris* across homogeneous peat and various stages of colonization of trays containing pre-weighed samples of organic matter collected from the fermentation horizon (FHOM) of a pine forest. The stages are uncolonized FHOM (single arrowhead), recently colonized FHOM (*ca* 20 days after initial contact with tray – large arrow head) and senescent patches (*ca* 40 days after initial contact – double arrowheads).
(*b*) Autoradiography of chamber shown in 8(*a*) revealing selective allocation of carbon to ectomycorrhizal roots (single arrowheads), to rhizomorphs of vegetative mycelial system crossing the peat (double arrowheads) and to the one tray containing FHOM colonized for *ca* 20 days (large arrow head). Failure to allocate carbon to trays colonized for more than 40 days is evident. (From Bending & Read, 1995*a*.)

ca 40 days (Fig. 10.8) by which time there was evidence of change of colour of mycorrhizal mycelium and appearance of sporulating saprotrophs on the substrates.

Questions as to whether or to what extent the 'costs' of this apparently large investment of carbon are balanced by 'benefit' obviously arise. Earlier unreplicated observations of mycorrhizal systems in which mycelial proliferation was induced by addition of comminuted litter to peat indicated that the formation of patches was coincident with re-greening of previously chlorotic host plants of *Pinus* and *Larix* and enhancement of their shoot nitrogen contents (Read, 1991*a*). This led to the hypothesis that the exploitation phase was primarily one in which nitrogen contained in the resource material was mobilized. To examine this possibility Bending & Read (1995*a*) set up a series of observation chambers in which plants of *Pinus sylvestris* were grown with either *Suillus bovinus* or *Thelephora terrestris* as ectomycorrhizal symbionts. When the mycelia of these fungi had occupied a significant proportion of the homogeneously humified peat, discrete weighed samples of organic matter which had been freshly collected from the fermentation horizon (FHOM) of a pine forest, were added to the systems on plastic trays. The chronology of intensive mycelial exploitation of the FHOM was followed (Fig. 10.9) and its nutrient status monitored before the initiation and after the senescence of the patch. These analyses revealed that nitrogen concentrations of the FHOM were reduced after patch formation by 23% in the case of *S. bovinus* and 13% in that of *T. terrestris*.

Since these quantities of nitrogen export were larger than those that could be accounted for by mineralization it was concluded that a major proportion of the total N must be mobilized from organic forms. Subsequent investigation of 'patches' formed by the fungus *Paxillus involutus* in FHOM have indicated that these are indeed zones of enhanced protease activity (Bending & Read, 1995*b*).

Clearly other elements, including phosphorus and potassium, can also be captured by the fungus from such substrates and there will be circumstances where enhancement of their availability will be advantageous to the plant. However, because nitrogen is not only the element which is in shortest overall supply in boreal systems but is also required by the plant in quantities approximately ten times greater than those of phosphorus, the beneficial effects of increased access to this element are likely to outweigh those of any other nutrient.

Results such as those do not provide proof that ectomycorrhizal fungi are directly responsible for mobilization of nutrients from organic matter but are strongly suggestive of this type of activity. They thus challenge us to rethink the conventional view that the process of mineralization is exclusively facilitated by a distinct population of 'decomposers' and highlight the need for integrated studies of interaction between 'decomposer' and mycorrhizal fungi in the organic horizons of forest soil.

Enhancement of resistance of ectomycorrhizal plants to toxicity

Until recently, emphasis in research on ectomycorrhiza was almost exclusively placed upon enhancement of capture of nutritionally important

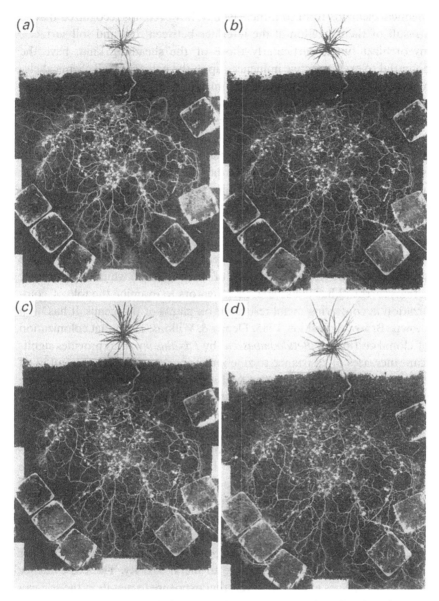

Fig. 10.9. Sequential views of a single observation chamber showing stages of colonizations of trays containing pre-weighed samples of FHOM by the mycelium of *S. bovinus* growing from *P. sylvestris*. Analysis of nutrient composition of FHOM before and after colonization facilitates quantification of mobilization and export of nutrients by the fungus (see text for details). (*a*) Day 0, (*b*) Day 14, (*c*) Day 28, (*d*) Day 42. (From Bending & Read, 1995a).

chemical elements from soil. Increasingly, however, it is recognized that as a result of their location at the interface between root and soil surfaces, mycorrhizal fungi, particularly those of the sheathing kind, have the potential to exert selective influences upon the passage of all chemical elements to the plant. Under some circumstances the ability to exclude those elements which post a threat to physiological equilibrium and thus avoid toxicity, may be as essential as enhancement of nutrient capture.

The acidity typical of most natural substrates of ectomycorrhizal plants, and of many of the man-made substrates such as mine-spoils which they colonize, is conducive to the solubilization of potentially toxic metal ions. Amongst these, aluminium and iron are quantitatively the most significant in natural soil, while mine spoils may be polluted by nickel, lead, zinc and cadmium either in combination or as individual elements, depending upon the nature of the extracting activity. The observation that in nature many spoils polluted with these elements are spontaneously colonized by species of birch (*Betula*) which are routinely ectomycorrhizal has led several investigators to examine the role of colonization in conferring metal resistance on plants of this genus. It has been shown (Brown & Wilkins, 1985; Denny & Wilkins, 1987) that colonization of clonal cuttings of *Betula pubescens* by *Paxillus involutus* provides significant increases of resistance to zinc ions apparently by reducing inflow of the element to the leaves. Jones & Hutchinson (1986) using nickel as the metal element, obtained a similar result in seedlings of *B. papyrifera* which was colonized by a fungus widely present on spoil heaps, *Scleroderma flavidum*. This associate was far more effective in providing resistance than was *Lactarius rufus*. It was subsequently shown that nickel was preferentially concentrated in the roots of the plants where it appeared to be sequestered in polyphosphate bodies (Jones & Hutchinson, 1988). In contrast to the pattern of distribution of zinc, there was some evidence that nickel was also accumulated in senescing leaves so that avoidance of exposure to the metal was largely a feature restricted to stem tissue.

Large differences occur between ectomycorrhizal symbionts in their effectiveness in providing resistance to metal toxicity. There is evidence both in the species and strain levels that exposure to metals in the soil can lead to selection of resistance. Colpaert & Van Assche (1987) isolated strains of *Suillus luteus* from zinc contaminated soil that were able to grow in the presence of 1000 μg g^{-1} of the element. Strains of the same fungus obtained from carpophores growing on uncontaminated soil showed little or no growth above 100 μg g^{-1} Zn. It was subsequently shown (Colpaert & Van Assche, 1992) that zinc-resistant strains of

S. bovinus conferred significantly more zinc tolerance upon plants of *Pinus sylvestris* than did non-resistant strains, and that tolerance was most probably attributable to binding of the metal in the extra-matrical mycelium of the fungus.

Disease suppression by ectomycorrhizal fungi and the epidemiology of colonization

As in the case of plants with VA mycorrhiza, it appears that ectomycorrhizal fungi can provide their hosts with enhanced resistance to attack by fungal pathogens. Both *Pisolithus tinctorius* and *Thelephora terrestris* have been shown to reduce the impacts of the root pathogen *Phytophthora cinnamomi* on *Pinus* spp. (Marx, 1969, 1973), while inoculation with *Laccaria laccata* has been shown to reduce the incidence of disease caused by the pathogen *Fusarium oxysporum* in *Pseudotsuga menziesii* (Sylvia & Sinclair, 1983*a*), *Picea abies* (Sampangi & Perrin, 1985) and *Pinus sylvestris* (Chakravarty & Unestam, 1987*a,b*). It has been hypothesized that protection against fungal pathogens is achieved as a result of the physical barriers imposed by the hyphal mantle (Marx, 1973) or by the production of phenolic compounds in the host tissues in response to the presence of the mycorrhizal fungus (Sylvia & Sinclair, 1983*b*).

While both of these effects may indeed be involved in contributing to defence in the adult plant there is evidence that ectomycorrhizal fungi exert direct antibiotic effects upon would-be pathogens. Duchesne *et al.* (1988*a,b*) observed that inoculation of seedlings of *Pinus resinosa* with the fungus *Paxillus involutus* significantly reduced pathogenicity of *Fusarium oxysporum* before mycorrhizal colonization took place. Increases of seedling survival were associated with a six-fold decrease in sporulation of *F. oxysporum* in the rhizosphere of the host-plant (Duchesne, Peterson & Ellis, 1987). Ethanol extracts of the rhizosphere (Duchesne, Peterson & Ellis, 1989) indicated that fungitoxic effects of *P. involutus* were present within three days of inoculation of seedlings with *P. involutus*. Disease suppression at this critical stage of plant development prior to formation of the ectomycorrhizal symbiosis may be of particular significance in the regeneration niche.

Unfortunately most of the experiments on antibiotic effects of ectomycorrhizal fungi reported to date have been carried out under rather unrealistic conditions. Epidemiological studies under natural conditions are necessary to provide insights comparable with those currently being obtained with VA fungi. To date considerations of epidemiology in nature have been restricted to the processes of colonization of plants by mycor-rhizal fungi. Particular emphasis has been placed upon early symbiotic events and the question of later 'succession' of fungal symbionts.

Rapid colonization of seedlings on the forest floor is facilitated by the high inoculum potential and low host specificity of many ectomycorrhizal fungi, particularly those forming rhizomorphs (Fleming, Deacon & Last, 1986; Newton, 1991). Since the demonstration of the formation by such fungi of linkages between hosts at both the intra- and inter-specific level (Brownlee et al., 1983; Read, 1984; Read et al., 1985) there has been much discussion of the possible significance of the phenomenon in the field (Read et al., 1985; Newman, 1988; Perry et al., 1989; Read, 1991a). While such linkages provide the potential for flow of nutrients between plants, it is, as Newman points out, difficult to obtain unequivocal evidence of net transfer by this route. In fact as Read et al. (1985) point out, it is likely that the main benefits of such linkage will accrue at the establishment phase and be associated with improved access to soil resources. As in the case of the VA system, rapid incorporation into the pre-formed mycorrhizal mycelial network favoured by low specificity, will enable the seedling at little energy cost to itself, potentially to be in contact with the large catch-ment provided by the foraging activities of the heterotroph. Indirect evi-dence for the occurrence of such benefits in the field has been provided by the observation that all Douglas-fir seedlings found under the canopy of maturing 60–75 yr-old stand of the same species were associated with mycelial mats formed by ectomycorrhizal fungi (Griffiths, Castellano & Caldwell, 1991). Newton (1991) has emphasized that benefits of early col-onization may be expressed in the form of improved survivorship rather than growth enhancement but more experimental work is needed to resolve these issues.

The advantages of low host specificity to both host and fungus appear to be so great that the occurrence of the converse phenomenon, strong specificity, widely observed albeit only at the genus level in both hosts and fungi (Molina, Massicotte & Trappe, 1992) is, at first sight, a puzzle. The answer, however, in part may lie in the fact that most genera that are hosts of ectomycorrhizal fungi produce litter of a distinctive quality which can be clearly defined in terms of its nutrient C:N ratio (Read, 1991a,b). There is thus a substrate-induced selection which would be expected to lead to enhanced nutrient exchange and hence greatest compatibility between hosts and those fungi that were most efficient at mobilizing nutri-ents from residues of the particular quality characteristic of the auto-troph. The inter-generic distinctiveness of litter quality produced by ectomycorrhizal hosts and the parallel diversity of physiological attributes seen in their fungal associates contrasts markedly with the homogeneity and high resource quality of most plant litters produced by hosts of vesicular-arbuscular (VA) fungi and the parallel lack of biochemical specialization or specificity shown by the fungi forming the VA type of mycorrhiza.

If, as suggested by this view, selection favouring specificity has been influenced strongly by aspects of resource quality this should be reflected in nature by the appearance, on a given species, of increasing proportions of host specific fungi as its characteristic organic residues accumulate with time. This type of successional pattern was reported by Last, Dighton & Mason (1987) who observed fungal generalists of low host specificity to predominate in the pioneering stages of development of a *Betula* plantation on an agricultural soil, numbers of species of narrow host range increasing with age of the stand, in effect as organic residues accumulated. However, Molina *et al.* (1992) report that in the case of Douglas fir a high proportion of genus specific *Rhizopogon* species is found throughout the life of the plant. In terms of litter quality, that of Douglas fir along with Larch, is amongst the poorest and it may be that in both of these genera the requirements, from an early stage, for fungi effective at mobilizing nutrients from peculiarly recalcitrant substrates was a driving force selecting in favour of narrow specialists. This pattern of selection can be termed 'resource selectivity' and there is a need for further investigation of its role in defining the ecological specificity widely revealed in nature.

The nature and extent of mutualism in the ericoid mycorrhiza

The early history of study of mycorrhiza in ericaceous plants was characterized by extensive debate about the nature of the infection processes, relatively little attention being given to the significance of the association for the partners themselves or for ecosystems in which the partnership occurred. However, in this, as in the previous two categories of mycorrhizal type examined in this paper, emphasis has increasingly shifted in recent times towards consideration of functional aspects of the relationship between host and fungus. The unique structural attributes of the ericoid mycorrhiza and the distinctive nature of the often extreme soil conditions in which plants with this type of colonization occur have both been recognized (Read, 1983; Read & Kerley, 1995), and attention is rightly being directed towards the question of the role of the symbiosis in nature.

It is now acknowledged that the fungi which form the dense hyphal complexes in the epidermal cells of the delicate 'hair roots' of most members of the *Ericaceae, Empetraceae* and *Epacridaceae* are ascomycetes. One in particular, *Hymenoscyphus ericae* (Read) Korf & Kernan (Read, 1974), and its probable anamorph *Scytalidium vaccinii* (Egger & Sigler, 1993), has been the subject of much study. Its involvement in the nutrition of the host plant and in the detoxification of its rooting environment has been extensively documented, over the last twenty years, during which time attention has moved progressively towards analysis of these functions in relation to conditions prevailing in the acidic morhumus which is produced by these plants.

Nutrient relations of plants with ericoid mycorrhiza

Early experiments demonstrated that one of the main advantages accruing to the host plant from colonization was enhancement of its nitrogen supply (Read & Stribley, 1973) and these raised questions as to the chemical nature of the nitrogenous resources of the rooting environment of plants with ericoid mycorrhiza. While small amounts of mineral nitrogen, in the form of ammonium rather than nitrate, are usually detectable in these soils, well over 90%, of the total N fund of the ecosystem is present in organic combinations, the largest organic fraction being of polymers in which the N is labile, at least when exposed to 6N HCl (Stribley & Read, 1974). On the assumption that this fraction must be derived from the proteins and other N containing polymers such as chitin, which were originally constituents of the plant or microbial communities of the heathland, a programme of research was designed to investigate accessibility of N in such polymers when they were presented to *H. ericae* and its host plants as sole sources of the element. The first studies showed that the N contained in amino acids (Stribley & Read, 1980), peptides (Bajwa & Read, 1985) and pure proteins (Bajwa, Abuarghub & Read, 1985) were readily accessible to the fungus and, through it, to the plants. The enzyme primarily involved in proteolysis has been identified and characterized (Leake & Read, 1989). It is a carboxy-proteinase with a strongly acidic optimum pH for both production and activity which is very close to that of the organic substrates in which the fungus proliferates in nature (Fig. 10.10).

It is clear, however, that the half-lives of proteinaceous compounds, especially in soils containing high concentrations of polyphenolic materials, is likely to be short and that much of the N contained in the 'acid-labile' fraction is precipitated with phenolic compounds which will

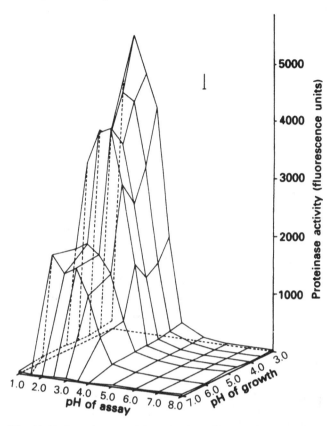

Fig. 10.10. Proteinase activity of culture filtrate of *Hymenoscyphus ericae* over a range of assay pH conditions and when the fungus is grown over the same pH range. Results expressed as fluorescence units released from FITC labelled bovine serum albumen substrate. (From Leake & Read, 1989.)

inevitably be of a more recalcitrant nature. Recent studies have, therefore, recognized that resource quality should reflect as closely as possible that which the mycorrhizal root encounters in nature. While it would be most appropriate to isolate the critical fractions from soil and use them directly as test substrates, the requirement to discriminate between mycorrhizal and general microbial mobilization necessitates sterilization of such natural substrates and this inevitably alters their physico-chemical condition.

An alternative approach is to produce nitrogenous resources of appropriate quality under aseptic conditions and use these as substrates. Mixtures of filter-sterilized soluble proteins and polyphenolics yield dense

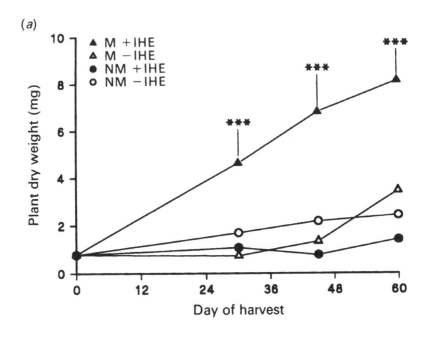

(a)

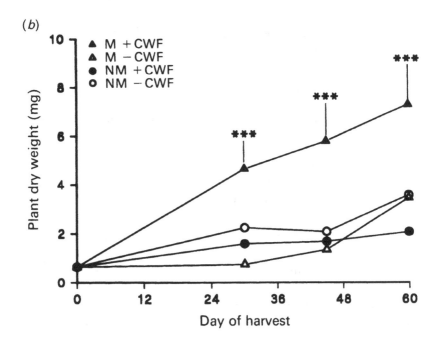

(b)

precipitates which have qualities that are likely to be closely comparable with those produced in return. Studies using substrates of this kind as sole sources of nitrogen (Leake & Read, 1990; Bending & Read, 1995c) have revealed that *H. ericae* retains some ability to mobilize the nitrogen contained in proteins even when they are precipitated in this way.

Another quantitatively significant potential source of nitrogen in heathland soils is that contained in the chitin which constitutes the main structural unit of mycelial cell walls. Calculations have shown that up to 20% of the total soil nitrogen in the rooting environment of a *Pinus-Calluna* heathland soil can be locked in these walls (Bååth & Söderström, 1979). Since these estimates do not include the mycelia intimately associated with the mycorrhizal roots themselves, they probably represent underestimates of the total repository of N in chitin.

In a heathland dominated by *Calluna vulgaris* the mycelium of the ericoid mycorrhizal fungus is likely itself to constitute one of the most important potential sources of chitin-nitrogen. With this in mind, Read and Kerley (1995) designed experiments in which vegetative mycelium of *H. ericae* was grown in liquid culture, before being washed, killed and used, either in the intact condition, or as a purified cell-wall fraction, as sole nitrogenous substrate for aseptically grown mycorrhizal and non-mycorrhizal plants of *Vaccinium macrocarpon*. Whether supplied with intact mycelial necromass in which the N was contained in a mixture of protein and chitin forms (Fig. 10.11(*a*)), or with purified cell-well fraction (Fig. 10.11(*b*)), mycorrhizal plants had sufficient access to the N contained in the necromass to enable significantly more growth to be achieved than when they were grown in the non-mycorrhizal condition. Read and Kerley (1995) pointed out that various natural processes, amongst which melanization is likely to be very important, could be expected to reduce the quality of hyphal walls as potential substrates in nature. However, these results indicate that much of the mycelial nitrogen, including that contained in the larger hyaline walls of the intra-cellular hyphal complexes

Fig. 10.11(*a*) Yield of *Vaccinium macrocarpon* grown in the mycorrhizal (M) or non-mycorrhizal (NM) condition with (+ IHE) or without (- IHE) necromass of the ericoid mycorrhizal endophyte *Hymenoscyphus ericae*, as sole source of nitrogen. *** Indicates significant difference at *p* 0.001 between M + IHE and NM + IHE treatments according to one way ANOVAR.
(*b*) As 10.11(*a*) but using purified cell wall fraction (CWF) of *H. ericae* as sole N source for M and NM plants. *** Indicates significant differences at *p* 0.001 between M + CWF and NM + CWF according to one-way ANOVAR. (From Read & Kerley, 1995.)

of the ericoid mycorrhizal root, may be recycled during senescence, as a consequence of the chitinolytic capability of the living fungus.

Experiments such as these reinforce the view that the considerable saprotrophic potential which has been amply demonstrated in studies of *H. ericae* grown in pure culture, can be expressed under the control of the plant in the mycorrhizal condition, to provide the autotroph with access to a range of chemically complex sources of nitrogen which would otherwise be unavailable to it. The potential benefits to the plant are then clear. Some might argue that in view of such well-developed saprotrophic abilities *H. ericae* could occur in nature quite independently of its host. However, there is some evidence to suggest that it would suffer carbon limitation growing in isolation (Read, 1991*b*) and the observed stimulation of its growth in epidermal cells of its host strongly suggest benefit to the fungus.

Enhancement of resistance to toxicity

Of all plants, those with ericoid mycorrhiza are perhaps the best adapted to extremes of acidity. With the exception of a small number of species which show a preference for organic soils of somewhat higher base status, the majority occur on soils of pH 4.0 or below and by their own biological activity are involved in maintenance of low pH. While the advantages of acidity to plants in which major nutrient mobilizing enzymes such as protease have a pH optimum around 4.0 may be evident, other environmental changes produced by the low pH, most notably the enhancement of availability of metal ions and conversion to the most destructive undissociated condition of aliphatic and phenolic acids, can, in combination, produce extremely hostile soil conditions. The tolerance of ericaceous plants to the syndrome of adversity appears to be attributable to a considerable extent to the activities of their mycorrhizal heterotrophs. Thus, for example, if aseptically grown seedlings of *Calluna vulgaris* are exposed to solutions of phenolic acids at concentration representative of those found in heathland soil solutions (Jalal & Read, 1983,*a*,*b*) extension growth of their hair roots is almost completely inhibited. In contrast, when plants are colonized by *H. ericae*, growth and development of the root system is normal. The alleviation of toxicity is partly a result of assimilation by the fungus of the phenolic compounds, which it has some capacity to utilize as a carbon source (Leake, 1987; Bending & Read, 1995*c*).

Alleviation of toxicity of zinc and copper ions is also provided by *H. ericae*. The roots of plants of major genera such as *Vaccinium* and

Rhododendron in addition to those of *Calluna vulgaris* are extremely sensitive to these ions in the absence of the fungus (Bradley, Burt & Read, 1981; 1982) but grow normally in their presence when mycorrhizal. The main effect of the fungus appears to be exclusion of the metallic ions from the shoot, and to this extent the mechanism of alleviation is comparable with that seen in the ectomycorrhizal symbiosis. However, the levels of tolerance to metals shown by these ericoid fungi are several orders of magnitude greater than those of the latter group.

Discusson and conclusions

It was probably essential in order to ensure some progress that early research on mycorrhizas concentrated upon simplified systems. The reductional approach, typified by studies of the uptake of single elements into mycorrhizal roots, or by analyses of responses to inoculation of single plants grown in pots, has indeed provided a wealth of information about the *potential* of the symbiosis to influence the fitness of plants in the field. The challenge has increasingly been to build upon the established fund of information with experiments designed to evaluate the impacts of the symbiosis upon fitness of its partners in the real world. Such studies inevitably involve questions concerning the nature of substrates being exploited by the heterotroph in the field and the extent to which the root–fungus association can determine the outcome of interspecific interactions between the autotrophs and between the heterotrophs. This paper has selected a few examples of studies which help to provide answers to these questions and hence are leading in the right direction. Collectively they demonstrate that the mycorrhizal relationship can have a broad rather than narrow range of functions, that these can include negative as well as positive impacts upon the components of the association, and that those critical aspects of the relationship which influence the fitness of either partner can be expressed during a very restricted part of the life cycle, or even over a number of cycles. It is increasingly apparent therefore that the blanket application of the term 'mutualism' to the mycorrhizal relationship, wherever it is seen to occur, is inappropriate. The more general notion of 'symbiosis' as used in the sense of de Bary (1879) to indicate, simply, an intimate relationship between partners, is as recommended by Lewis (1985), far preferable. This leaves the onus upon the observer to determine the nature of the relationship. It is then incumbent upon him or her, bearing in mind the nature of the environment in which the 'symbiosis' is observed to occur, first to elevate hypotheses as to possible function, and then to design experiments in which conditions reflect as closely as possible those prevailing in nature. By this means a realistic appraisal of the status of the relationship can be obtained.

If nothing else, this chapter will have served a purpose if it demonstrates that there exists a wide rather than a narrow range of possible hypotheses to be tested. Only by acknowledging the full range of potential attributes of the symbiosis can we expect to meet the challenge of providing an evaluation of the role of mycorrhizas in natural communities and ecosystems.

References

Abuzinadah, R.A. & Read, D.J. (1986a). The role of proteins in the nitrogen nutrition of ectomycorrhizal plants. I. Utilization of peptides and proteins by ectomycorrhizal fungi. *New Phytologist*, **103**, 481–93.

Abuzinadah, R.A. & Read, D.J. (1986b). The role of proteins in the nitrogen nutrition of ectomycorrhizal plants. III. Protein utilization by *Betula*, *Picea* in mycorrhizal association with *Hebeloma crustuliniforme*. *New Phytologist*, **103**, 507–14.

Alexander, I.J. (1989). Mycorrhizas in topical forest. In *Mineral Nutrients in Tropical Forest and Savanna Ecosystems* (ed. J. Proctor), pp. 169–88. Oxford: Blackwell Scientific Publications.

Allen, M.F., Allen, E.B. & Friese, C.F. (1989). Responses of the non-mycotrophic plant *Salsola kali* to invasion by vesicular-arbuscular mycorrhizal fungi. *New Phytologist*, **111**, 45–9.

Bååth, E. & Söderström, B. (1979). Fungal biomass and fungal immobilisation of plant nutrients in Swedish coniferous forest soils. *Revue Ecologie et Biologie des Sols*, **16**, 477–89.

Bajwa, R. & Read, D.J. (1985). The biology of mycorrhiza in the Ericaceae. IX. Peptides as nitrogen sources for the ericoid endophyte and for mycorrhizal and non-mycorrhizal plants. *New Phytologist*, **101**, 459–67.

Bajwa, R., Abuarghub, S. & Read, D.J. (1985). The biology of mycorrhiza in the Ericaceae. X. The utilization of proteins and the production of proteolytic enzymes by the mycorrhizal endophyte and by mycorrhizal plants. *New Phytologist*, **101**, 469–86.

de Bary, A. (1879). *Die Erscheinung der Symbiose*. Strasbourg: Karl J. Trübner.

Baylis, G.T.S. (1975). The magnolioid mycorrhiza and mycotrophy in root systems derived from it. In *Endomycorrhizas* (ed. F.E. Sanders, B. Mosse & P.B. Tinker), pp. 373–89. London: Academic Press.

Bending, G.D. & Read, D.J. (1995a). The structure and function of the vegetative mycelium of ectomycorrhizal plants. V. The foraging behaviour of ectomycorrhizal mycelium and the translocation of nutrients from exploited organic matter. *New Phytologist*, **130**, 401–9.

Bending, G.D. & Read, D.J. (1995b). The structure and function of the vegetative mycelium of ectomycorrhizal plants. VI. Activities of key nutrient mobilising enzymes in birch fermentation horizon organic matter colonized by *Paxillus involutus*. *New Phytologist*, **130**, 411–17.

Bending, G.D. & Read, D.J. (1995c). The impact of the soluble polyphenol tannic acid upon the growth and physiology of ericoid and ectomycorrhizal fungi. *Soil Biology and Biochemistry* (In press).

Bradley, R., Burt, A.J. & Read, D.J. (1981). Mycorrhizal infection and resistance to heavy metal toxicity in *Calluna vulgaris*. *Nature, London*, **292**, 335–7.

Bradley, R., Burt, A.J. & Read, D.J. (1982). The biology of mycorrhiza in the Ericaceae. VIII. The role of mycorrhizal infection in heavy metal resistance. *New Phytologist*, **91**, 197–209.

Brown, M.T. & Wilkins, D.A. (1985). Zinc tolerance of mycorrhizal *Betula*. *New Phytologist*, **99**, 101–6.

Brownleé, C., Duddridge, J.A., Malibari, A. & Read, D.J. (1983). The structure and function of mycelial systems of ecto-mycorrhizal roots with special reference to their role in forming inter-plant connections and providing pathways for assimilate and water transport. *Plant Soil*, **71**, 433–43.

Bryla, D.R. & Koide, R. (1990). Role of mycorrhizal infection in the growth and reproduction of wild vs. cultivated plants. II. Eight wild accessions and two cultivars of *Lycopersicon esculentum* Mill. *Oecologia*, **84**, 82–92.

Burt, A.J., Hashem, A.R., Shaw, G. & Read, D.J. (1986). Comparative analysis of metal tolerance in ericoid and ectomycorrhizal fungi. In *Proceedings of the 1st European Symposium on Mycorrhizas* (ed. V. Gianinazzi-Pearson & S. Gianinazzi), pp. 683–687. Paris: INRA.

Chakravarty, P. & Unestam, T. (1987a). Mycorrhizal fungi prevent disease in stressed pine seedlings. *Journal of Phytopathology*, **188**, 335–40.

Chakravarty, P. & Unestam, T. (1987b). Differential influence of ectomycorrhizae on plant growth and disease resistance of *Pinus sylvestris* seedlings. *Journal of Phytopathology*, **120**, 104–20.

Colpaert, J.V. & Van Assche, J.A. (1987). Heavy metal tolerance in some ectomycorrhizal fungi. *Functional Ecology*, **1**, 415–21.

Colpaert, J.V. & Van Assche, J.A. (1992). Zinc toxicity in ectomycorrhizal *Pinus sylvestris*. *Plant and Soil*, **143**, 201–11.

Denny, H.J. & Wilkins, D.A. (1987). Zinc tolerance in *Betula* spp. IV. The mechanism of ectomycorrhizal amelioration of zinc toxicity. *New Phytologist*, **106**, 545–53.

Duchesne, L.C., Peterson, R.L. & Ellis, B.E. (1987). The accumulation of plant-produced antimicrobial compounds in response to ectomycorrhizal fungi: a review. *New Phytologist*, **68**, 17–27.

Duchesne, L.C., Peterson, R.L. & Ellis, B.E. (1988a). Interaction between the ectomycorrhizal fungus *Paxillus involutus* and *Pinus resinosa* induces resistance to *Fusarium oxysporum*. *Canadian Journal of Botany*, **66**, 558–62.

Duchesne, L.C., Peterson, R.L. & Ellis, B.E. (1988b). Pine root exudate stimulates antibiotic synthesis by the ectomycorrhizal fungus *Paxillus involutus*. *New Phytologist*, **108**, 470–6.

Duchesne, L.C., Peterson, R.L. & Ellis, B.E. (1989). The time-course of disease suppression and antibiosis by the ectomycorrhizal fungus *Paxillus involutus*. *New Phytologist*, **111**, 693–8.

Egger, K.N. & Sigler, L. (1993). Relatedness of the ericoid endophytes *Scytalidium vaccinii* and *Hymenoscyphus ericae* inferred from analysis of ribosomal DNA. *Mycologia*, **85**, 219–30.

Fenner, M. (1978). A comparison of the abilities of colonisers and closed turf species to establish from seed in artificial swards. *Journal of Ecology*, **66**, 953–65.

Finlay, R.D. & Read, D.J. (1986). The structure and function of the vegetative mycelium of ectomycorrhizal plants. I. Translocation of ^{14}C-labelled carbon between plants interconnected by a common mycelium. *New Phytologist*, **103**, 143–56.

Finlay, R.D. & Söderström, B. (1992). Mycorrhiza and carbon flow to the soil. In *Mycorrhiza Functioning* (ed. M.F. Allen), pp. 134–62. Chapman & Hall.

Fitter, A.H. (1985). Functioning of VA mycorrhiza under field conditions. *New Phytologist*, **99**, 257–65.

Fleming, L.C., Deacon, J.W. & Last, F.T. (1986). Ectomycorrhizal succession in a Scottish birch wood. In *Physiological and Genetical Aspects of Mycorrhizae* (ed. V. Gianinazzi-Pearson & S. Gianinazzi), pp. 259–264. Paris: INRA.

Francis, R. & Read, D.J. (1984). Direct transfer of carbon between plants connected by vesicular-arbuscular mycorrhizal mycelium. *Nature, London*, **307**, 53–6.

Francis, R. & Read, D.J. (1994). The contribution of mycorrhizal fungi to the determination of plant community structure. *Plant and Soil*, **159**, 11–25.

Francis, R. & Read, D.J. (1995). Mutualism and antagonism in the mycorrhizal symbiosis with special reference to impacts upon plant community structure. *Canadian Journal of Botany* (in press).

Frank, A.B. (1885). Ueber die Wurzelsymbiose beruhende Ernährung gewisser Bäume durch unterirdische Pilze. *Bericht der Deutschen Botanisch Gesellschaft*, **3**, 128–45.

Frank, A.B. (1894). Die Bedeutung der Mykorrhizapilze für die gemeine Kiefer. *Forstwissenschaftliches Centralblatt*, **16**, 1852–90.

Gadgil, R.L. & Gadgil, P.D. (1975). Suppression of litter decomposition by mycorrhizal roots of *Pinus radiata*. *New Zealand Journal of Forest Science*, **5**, 35–41.

Griffiths, R.P., Castellano, M.A. & Caldwell, B.A. (1991). Ectomycorrhizal mats formed by *Gautieria monticola* and *Hysterangium setchellii* and their association with Douglas-fir seedlings, a case study. *Plant and Soil*, **134**, 255–9.

Grime, P.J., Mackey, J.M.L., Hillier, S.H. & Read, D.J. (1987). Floristic diversity in a model system using experimental microcosms. *Nature, London*, **328**, 420–2.

Grubb, P.J. (1976). A theoretical background to the conservation of ecologically distinct groups of annuals and biennials in the chalk grassland ecosystem. *Biological Conservation*, **76**, 53–76.

Grubb, P.J. (1977). The maintenance of species-richness in plant communities: the importance of the regeneration niche. *Biological Reviews*, **52**, 107–45.

Grubb, P.J. (1986). The ecology of establishment. In *Ecology and Design in Landscape* **24** (ed. A.D. Bradshaw, D.A. Goode and E. Thorpe), pp. 83–97. Symposium of British Ecological Society.

Harley, J.L. (1985). Mycorrhiza: the first 65 years; from the time of Frank till 1950. In *Proceedings of 6th North American Conference on Mycorrhiza* (ed. Molina), pp. 26–33. Corvallis: College of Forestry, Oregon State University.

Harley, J.L. & Harley, E.L. (1987). A check-list of mycorrhiza in the British flora. *New Phytologist*, **105**, 1–102.

Harley, J.L. & Smith, S.E. (1983). *Mycorrhizal Symbiosis*. London: Academic Press.

Jalal, M.A.F. & Read, D.J. (1983*a*). The organic acid composition of *Calluna* heathland soil with special reference to phyto- and fungi-toxicity. I. Isolation and identification of organic acids. *Plant Soil*, **70**, 257–72.

Jalal, M.A.F. & Read, D.J. (1983*b*). The organic acid composition of *Calluna* heathland soil with special reference to phyto- and fungi-toxicity. II. Monthly quantitative determination of the organic acid of *Calluna* and spruce dominated soils. *Plant Soil*, **70**, 273–86.

Jones, M.D. & Hutchinson, T.C. (1986). The effect of mycorrhizal infection on the response of *Betula papyrifera* to nickel and copper. *New Phytologist*, **102**, 429–42.

Jones, M.D. & Hutchinson, T.C. (1988). Nickel toxicity in mycorrhizal birch seedlings infected with *Lactarius rufus* or *Scleroderma flavidum*. I. Effects on growth, photosynthesis, respiration and transpiration. *New Phytologist*, **108**, 451–9.

Koide, R., Li, M., Lewis, J. & Irby, C. (1988). Role of mycorrhizal infection in the growth and reproduction of wild vs. cultivated plants. I. Wild vs. cultivated oats. *Oecologia*, **77**, 537–43.

Last, F.T., Dighton, J. & Mason, P.A. (1987). Successions of sheathing mycorrhizal fungi. *Trends in Ecology and Evolution*, **2**, 157–61.

Le Page, B.A. (1995). Fossil ectomycorrhiza in Eocene *Pinus* roots. In press.

Le Page, B.A., Currah, R.S., Stockey, R.A. & Rothwell, G.W. (1996). Fossil ectomycorrhizae of Eocene *Pinus* roots. *Proceedings of the National Academy of Sciences, USA*. (In press).

Leake, J.R. (1987). Metabolism of phyto- and fungitoxic acids by the ericoid mycorrhizal fungus. In *Proceedings of the seventh North American Mycorrhiza Conference* (ed. D.M. Sylvia, L.L. Hung, & J.H. Graham), pp. 332–3. Gainesville: University of Florida.

Leake, J.R. & Read, D.J. (1989). The biology of mycorrhiza in the Ericaceae. XIII. Some characteristics of the extracellular proteinase activity of the ericoid endophyte *Hymenoscyphus ericae*. *New Phytologist*, **112**, 69–76.

Leake, J.R. & Read, D.J. (1990). The effects of phenolic compounds on nitrogen mobilisation by ericoid mycorrhizal systems. *Agriculture Ecosystem and Environment*, **29**, 225–36.

Lewis, D.H. (1985). Symbiosis and mutualism: crisp concepts and soggy semantics. In *The Biology of Mutualism* (ed. D.H. Boucher).

Linderman, R.G. (1994). Role of VAM fungi in biocontrol. In *Mycorrhizae and Plant Heath* (ed. F.L. Pfleger & R.G. Linderman), pp. 1–25. St Paul, Minnesota: APS Press.

Marx, D.H. (1969). The influence of ectotrophic ectomycorrhizal fungi on the resistance to pathogenic infections. I. Antagonism of mycorrhizal fungi to pathogenic fungi and soil bacteria. *Phytopathology*, **59**, 153–63.

Marx, D.H. (1973). Mycorrhizae and feeder root diseases. In *Ectomycorrhizae: Their Ecology and Physiology* (ed. G.C. Marks & T.T. Kozlowski), pp. 351–82. New York: Academic Press.

Melin, E. (1925). Untersuchungen über die Bedeutung der Baummykorriza. G. Fischer. Jena 152.

Merryweather, J. & Fitter, A. (1995). Phosphorus and carbon budgets: mycorrhizal contributions in *Hyacinthoides non-scripta* (L). Chauard ex Rothm. under natural conditions. *New Phytologist*, **129**, 619–27.

Molina, R., Massicotte, H. & Trappe, J.M. (1992). Specificity phenomena in mycorrhizal symbiosis: Community ecological consequences and practical applications. In *Mycorrhizal Functioning* (ed. M.F. Allen), pp. 357–423. New York: Chapman & Hall.

Morton, J.B. & Benny, G.L. (1990). Revised classification of arbuscular mycorrhizal fungi (Zygomycetes): a new order, Glomales, two new sub-orders Glomineae and Gigasporineae, and two new families, Acaulosporaceae and Gigasporaceae with an emendation of Glomaceae. *Mycotaxon*, **37**, 471–9.

Newman, E.I. (1988). Mycorrhizal links between plants: Functioning and ecological significance. *Advances in Ecological Research*, **18**, 243–70.

Newsham, K.K., Fitter, A.H. & Watkinson, A.R. (1994). Root pathogenic and arbuscular mycorrhizal fungi determine fecundity of asymptomatic plants in the field. *Journal of Ecology*, **82**, 805–14.

Newsham, K.K., Fitter, A.H. & Watkinson, A.R. (1995). Arbuscular mycorrhiza protect an annual grass from root pathogenic fungi, in the field. *Journal of Ecology* (In press).

Newton, A.C. (1991). Mineral nutrition and mycorrhizal infection of seedling oak and birch. III. Epidemiological aspects of ectomycorrhizal infection and the relationship to seedling growth. *New Phytologist*, **117**, 53–60.

Perry, D.A., Amaranthus, M.P., Borchers, J.G., Borchers, S.L. & Brainerd, R.E. (1989). Bootstrapping in ecosystems. *Biological Science*, **39**, 230–7.

Rayner, A.D.M. & Boddy, L. (1988). *Fungal Decomposition of Wood*. Chichester: J. Wiley.

Read, D.J. (1983). The biology of mycorrhiza in the Ericales. *Canadian Journal of Botany*, **61**, 985–1004.

Read, D.J. & Stribley, D.P. (1973). Effect of mycorrhizal infection on nitrogen and phosphorus nutrition of ericaceous plants. *Nature, London*, **244**, 81.

Read, D.J. (1974). *Pezizella ericae* sp. nov., the perfect stage of a typical mycorrhizal endophyte of the Ericaceae. *Transactions of the British Mycological Society*, **63**, 381–3.

Read, D.J. (1984). The structure and function of the vegetative mycelium of mycorrhizal roots. In: *The Ecology and Physiology of the Fungal Mycelium* (ed. D.H. Jennings & A.D.M. Rayner), pp. 215–40. Cambridge: Cambridge University Press.

Read, D.J. (1991*a*). Mycorrhizas in ecosystems. *Experientia*, **47**, 376–91.

Read, D.J. (1991*b*). The mycorrhizal fungal community with special reference to nutrient mobilisation. In *The Fungal Community* (ed. G.C. Carroll and D.T. Wicklow), pp. 631–52. New York: Marcel Dekker.

Read, D.J., Koucheki, H.K. & Hogdom, J. (1976). Vesicular-arbuscular mycorrhiza in natural vegetation systems. I. The occurrence of infection. *New Phytologist*, **77**, 541–655.

Read, D.J., Francis, R. & Finlay, R.D. (1985). Mycorrhizal mycelia and nutrient cycling in plant communities. In *Ecological Interactions in Soil: plants, microbes and animals* (ed. A.H. Fitter, D. Atkinson, D.J. Read & M.B. Usher), pp. 193–217. British Ecological Society Special Publications 4. Oxford: Blackwell Scientific Publications.

Read, D.J., Leake, J.R. & Langdale, A.R. (1989). The nitrogen nutrition of mycorrhizal fungi and their host plants. In *Nitrogen, Phosphorus and Sulphur Utilization by Fungi* (ed. L. Boddy, R. Marchant & D.J. Read), pp. 181–204. Cambridge: Cambridge University Press.

Read, D.J. & Birch, C.P.D. (1988). The effects and implications of disturbance of mycorrhizal mycelial systems. *Proceedings of the Royal Society of Edinburgh*, **94B**, 13–24.

Read, D.J. & Kerley, S. (1995). The status and function of ericoid mycorrhizal systems. In *Mycorrhiza – Structure, Function, Molecular Biology and Biotechnology* (ed. A. Varma & B. Hock), pp. 499–520. Berlin: Springer-Verlag.

Remy, W., Taylor, T.N., Hass, H. & Kerp, H. (1994). Four hundred million year old vesicular-arbuscular mycorrhizae. *Proceedings of the National Academy of Sciences, USA*, **91**, 11841–3.

Sampangi, R. & Perrin, R. (1985). Attempts to elucidate the mechanisms involved in the protective effect of *Laccaria laccata* against *Fusarium oxysporum*. In *Proceedings of the 1st European Symposium on Mycorrhizae* (ed. V. Gianinazzi-Pearson & S. Gianinazzi), pp. 807–10. France: Dijon.

Simon, L., Bousquet, J., Levesque, C. & Lalonde, M. (1993). Origin and diversification of endomycorrhizal fungi and coincidence with vascular land plants. *Nature, London*, **363**, 67–9.

Stribley, D.P. & Read, D.J. (1974). The biology of mycorrhiza in the Ericaceae. IV. The effects of mycorrhizal infection on the uptake of ^{15}N from labelled soil by *Vaccinium macrocarpon* Ait. *New Phytologist*, **73**, 1149–55.

Stribley, D.P. & Read, D.J. (1980). The biology of mycorrhiza in the Ericaceae. VII. The relationship between mycorrhizal infection and the capacity to utilize simple and complex organic nitrogen sources. *New Phytologist*, **86**, 365–71.

Stubblefield, S.P., Taylor, T.N. & Trappe, J.M. (1987). Fossil mycorrhizae: a case for symbiosis. *Science*, **237**, 59–60.

Sylvia, D.M. & Sinclair, W.A. (1983a). Suppressive influence of *Laccaria laccata* on *Fusarium oxysporum* and on Douglas fir seedlings. *Phytopathology*, **73**, 384–9.

Sylvia, D.M. & Sinclair, W.A. (1983b). Phenolic compounds and resistance to fungi pathogens induced in primary roots of Douglas fir seedlings by the ectomycorrhizal fungus *Laccaria laccata*. *Phytopathology*, **73**, 390–7.

Taylor, T.N., Remy, W., Hass, H. & Kerp, H. (1995). Fossil arbuscular mycorrhizae from the early Devonian. *Mycologia*, **87**, 560–73.

Trappe, J.M. (1987). Phylogenetic and ecological aspects of mycotrophy in the angiosperms from an evolutionary standpoint. In *Ecophysiology of VA Mycorrhizal Plants* (ed. G.R. Safir), pp. 2–25. CRC, Boca Raton.

Trappe, J.M. & Berch, S.M. (1985). The prehistory of mycorrhiza: A.B. Frank's predecessors. In *Proceedings of 6th North-American Conference on Mycorrhiza* (ed. R. Molina), pp. 2–11. College of Forestry, Oregon State University: Corvallis.

West, H.M., Fitter, A.H. & Watkinson, A.R. (1993). Response of *Vulpia ciliata* ssp. *ambigua* to removal of mycorrhizal infection and to phosphate application under natural conditions. *Journal of Ecology*, **81**, 351–8.

11

Lichens and the environment

M.R.D. SEAWARD

Lichens are of infinite importance as handmaids of Nature in operating
her changes on the face of our globe, in softening down the pointed
crags of our mountains – in covering with fertile soil alike the bare
surface of the volcanic lava and the coral islet – in a word, that they are
the basis of soil and consequently vegetation.

W. Lauder Lindsey (1856)
A Popular History of British Lichens

Introduction

Thirty-five years ago, the author, like his contemporaries, regarded the
study of lichens, the so-called 'neglected plants', as an esoteric branch of
botany. He never envisaged that much of his professional life would be
devoted to lichenology, since the literature then available, although providing some insight into the uniqueness of lichens, gave little hint of the
major role these apparently insignificant organisms play in the shaping of
the physical and biological environment of our planet and their role in
maintaining its equilibrium.

Because of their symbiotic nature, involving a partnership of one or
more fungi with one or more algae and/or cyanobacteria, lichens cannot
be fitted neatly into conventional systems of classification. For this reason,
lichens were often overlooked in the teaching of botany, being cursorily
touched on by specialists in algology, mycology or bryology, who regarded
these maverick organisms in much the same way as Linnaeus did when he
dubbed them *rustici pauperrimi*, loosely translated by one author as the
'poor trash of vegetation'. As a consequence, lichens were generally
ignored and their importance grossly underestimated in ecological
investigations.

293

Geological context

The biogeography of lichens unquestionably points to their evolutionary antiquity. The oldest known lichen fragments are those preserved in amber (e.g. Garty, Giele & Krumbein, 1982) which can only be dated to the Tertiary period. Triassic fossil material from Lower Franconia, as revealed by detailed SEM studies (Ziegler, 1991), appears to be of a true lichen. There is a fragmentary fossil record prior to the Permo-Triassic period. *Spongiophyton* from Lower Devonian sediments (Stein, Harmon & Hueber, 1993) is possibly a lichen, but the presence of a phycobiont has not been demonstrated. However, the recent interpretation of a pre-Cambrian fossil as a lichen (Retallack, 1994) is very unconvincing, and the identity of lichen-like structures discovered in 1975 in pre-Cambrian deposits in South Africa (Hallbauer, Jahns & Beltmann, 1977) awaits elucidation. However, biogeographical interpretations based on plate tectonics, and employing chemotaxonomy and cladograms, have provided evidence not only for the age of particular genera but also the regions of origin for ancestral species. By means of these techniques, Sheard (1977), for example, hypothesized the radiation of *Dimelaena* species in Triassic times, while Tehler (1983) established that the precursors of *Roccellina* originated more than 225 million years ago, i.e. before the break-up of Pacifica. Similarly, Sipman (1983) has reconstructed a probable evolutionary history of the *Megalosporaceae*, based on the relationship between spore type development and continental break-up, to show that this family originated on eastern Gondwanaland in Palaeozoic times.

Many present-day disjunct distribution patterns of lichen species and chemical races appear to be quite clearly related to plate tectonics and cannot be reliably accounted for on the basis of long-distance transport of propagules (e.g. Kärnefelt, 1990). However, some widespread distribution patterns have been explained in terms of dispersal by propagules on driftwood via sea currents (Westman, 1973), although propagule viability after continuous immersion in seawater has been questioned (Lindsay, 1973); transportation by migratory birds may be another likely agency (Bailey & James, 1979). The increasing mobility of humans over the past two centuries has contributed to dispersal (Lindsay, 1973; Alstrup, 1977), increasing the number of cosmopolitan taxa. Interpretation of worldwide lichen distribution patterns will take many years to accomplish due to lacunae in collecting, recording and taxonomy, but important advances in respect of major geographical regions have already been made; thus, for example, areas as vast as the Pacific can be considered in this context.

Algae, cyanobacteria and fungi certainly do have ancient origins (Hawksworth, 1988*a*; Taylor 1994), and there is no reason to doubt that lichens were formed relatively soon after their appearance, especially since these symbiotic organisms are apparently better equipped to cope with extreme environments (Smith & Douglas, 1987). Lichens are undoubtedly one of the most successful of forms of symbiosis (Galloway, 1994), with partnerships based on a wide variety of symbiotic interactions (Hawksworth, 1988*b*) which are manifested in many, often unique, morphological and physiological adaptations to widely differing environmental conditions throughout the world. The ecological success of lichens over geologic time would infer slow environmental change, since, as will be seen later, most lichens are highly sensitive to rapid environmental disturbance, particularly that brought about by human intervention.

Quite clearly the earliest forms of lichens must have been saxicolous; terricolous forms evolving relatively quickly after these, followed by muscicolous and lichenicolous forms. The evolution of foliicolous, epiphytic and lignicolous species awaited the advent of phorophytes. Foliicolous lichens, already confirmed in Tertiary deposits, have widely distributed genera today indicative of an origin which predates the break-up of Gondwanaland (Sherwood-Pike, 1985).

Pedogenesis and biodeterioration

It is probable that lichens were early colonizers of terrestrial habitats on our planet, coping with harsh environments and relatively low atmospheric oxygen levels. They would have contributed in evolving atmospheres more suited to a much wider variety of life forms. However, the major long-term role of lichens has been as biological weathering agents, with a pedogenic action which is both physical and chemical in nature (Syers & Iskander, 1973; Jones, 1988). Despite early controversy concerning the pedogenic significance of lichens, their effectiveness in the biodeterioration of rocks has been clearly demonstrated by recent research, which has revealed that substantial quantities of substratum can be degraded even over relatively short periods of time (see below). Furthermore, lichens have the capacity to accumulate elements, such as N, P and S, thereby increasing the latters' potential bioavailability to successive life forms which may replace lichens during soil development. Organic material derived from lichen decomposition, together with detached particles of the substratum, and atmospherically derived dusts trapped by thalli all contribute to the development of primitive soils.

Fig. 11.1. Fissures in hornfels rock exposed to reveal penetration by pseudopodetia of *Stereocaulon vesuvianum*.

The weathering action of saxicolous lichens can be physical (Fig. 11.1), due to penetration by rhizinae and expansion and contraction of thalli, and/or chemical, due to carbon dioxide, oxalic acid and the complexing action of lichen substances. The latter have a low but significant solubility in water, forming soluble metal complexes under laboratory conditions when they react with minerals and rocks, particularly limestone (Ascaso, Sancho & Rodriguez-Pascual, 1990). Oxalic acid, formerly considered of minor importance in the biodeterioration process (Syers & Iskander, 1973), has been proved otherwise (see below). The widespread occurrence of metal oxalates, particularly calcium oxalate, in lichens (and in nature generally), the nature of the thallus–substratum interface, and the chemical disruption of the substratum are significant components of the weathering process.

Detailed Raman spectroscopy studies (Edwards, Farwell & Seaward, 1991; Edwards & Seaward, 1993; Seaward & Edwards, 1995) have demonstrated the highly destructive properties of calcium oxalate produced by lichen thalli. Particular attention has been directed in this work towards the dramatic effects caused by the action of certain aggressive lichen species on historic monuments, frescoes and other works of art, where biodeterioration processes have been shown to be devastatingly destructive within a surprisingly short time-scale. Such action on natural substrata is clearly of significance in a pedogenetic context, since lichens are usually regarded as weathering agents on a geological time-scale. Biodeterioration

Fig. 11.2. Italian Renaissance fresco attacked by *Dirina massiliensis* forma *sorediata* showing (centre) further injurious effects caused by its removal.

studies of the lichen *Dirina massiliensis* forma *sorediata* (Fig. 11.2) have revealed that calcium oxalate encrustations can be produced at the thallus–substratum interface up to depths of almost 2 mm within a period of less than 12 years. For a typical thallus of this species with a diameter of 1 cm and an oxalate encrustation thickness of 1288 ± 64 μm in the central area of the thallus (Seaward & Giacobini, 1989), it has been calculated that 135 mg of calcium carbonate is converted into calcium oxalate monohydrate at the interface. On some Italian Renaissance frescoes, this lichen has in many places totally obliterated more than 60% of the surface area and in such cases, 1 square metre of fresco and underlying plaster has probably been converted into more than 1 kg of calcium oxalate. Furthermore, with the incorporation of calcite and gypsum into the thallus encrustation, it is likely that more than four times this amount (*ca* 4.2 kg) of the underlying substratum has been chemically and physically disturbed.

The impact of lichen weathering of rocks on a global scale has been, and continues to be, important in terms of climatic consequences and the habitability of our planet: their disappearance from particular ecosystems would be critical over major areas (see below). Indeed, according to Schwartzman & Volk (1989), if today's weathering were to take place under completely abiotic conditions, dramatic increases in global temperature would result.

Ecological dynamics

Ecological work in recent years has been directed mainly towards the study of growth rates and colonization, ecophysiology, and air pollution monitoring. Little is known of the part played by lichens in ecological dynamics and nutrient and energy budgets. In any study of ecological dynamics, it is necessary to determine the key components in order to reveal the relative importance of lichens for a particular ecosystem. Boreal coniferous forests (Ahti, 1977), cold deserts (Lindsay, 1977; Williams *et al.*, 1978), dune systems (Oksanen, 1984), hot arid and semi-arid lands (Rogers, 1977), maritime rocks (Fletcher, 1980), high altitudes (Kalb, 1970) and tropical rain forests (Sipman & Harris, 1989) contain examples of ecosystems where lichens often contribute a significant proportion of the biomass (and the biodiversity). A variety of ecosystems also exist where a relatively low lichen biomass is of importance due, for example, to a rapid nutrient turnover as a result of intensive herbivore grazing or where there is an unusually fast rate of decomposition of the lichen component.

Bioassay work is an essential complement to autecological studies of each of the key lichen components, particularly in terms of growth, biomass, life expectancy, and rate of decomposition. Such work enables the determination, for example, of the overall pattern of uptake, storage and release of minerals for any ecosystem. Studies by Pike (1978) of some North American forest systems showed a wide variation in lichen biomass, and consequently mineral capital, which seldom accounted for more than 10% of the annual above-ground turnover of a particular element. Biomass studies of epiphytes on subalpine firs on a lava flow of Mount Baker, Washington (Rhoades, 1981) showed there to be a total biomass of 3500 kg ha^{-1}, 50% of which was composed of lichens. Biomass values have also been calculated for lichen epiphytes in a variety of other forests (Wein & Speer, 1975; Lang, Reiners & Pike, 1980; Pike, 1981; Rhoades, 1983; Boucher & Nash, 1990) and for treeless shrublands and bogs (Wein & Speer, 1975), hot deserts (Nash, White & Marsh, 1977; Nash & Moser, 1982) and tundra zones, particularly in respect of reindeer and caribou studies (see below).

Details of the biomass of nitrogen-fixing lichens and their fixation rates have been determined from different habitats including glacial drift in

Iceland (Crittenden, 1975), tall Douglas fir trees in Oregon (Denison, 1973), Columbian montane rain forest (Forman, 1975), and Piedmont oak forest (Becker, Reeder & Stetler, 1977). It has proved difficult to estimate lichen biomass of tall forests, but it is possible to do this by sampling litter-fall (McCune, 1994). His work on three forests of different ages in the Cascade Range showed that the biomass of the lichen litter was about one-hundredth of that remaining on the trees, cyanolichens, alectorioid and other lichens accounting for *ca* 500–1000, 100–200 and 0–1200 kg ha^{-1} respectively of the epiphyte biomass.

Modelling population dynamics and production of lichens is highly complex; for example, Rhoades (1983), in his study of *Lobaria oregana* on Douglas fir, showed that weight classes and total biomass vary dramatically over the tree surface: an average of only 13 large thalli accounted for 98% of the biomass found on the trunks, but represented less than 1% of the biomass of this species over the entire tree, most thalli being small and confined to branches. Sampling problems in ecological inventories of epiphytic lichens are highlighted by McCune & Lesica (1992) who demonstrated that, for example, a high species capture can result in a low accuracy of cover estimates for large plots, but when subsampling with many small plots the accuracy of species capture and cover estimates was reversed.

Detailed studies by Nash and his co-workers have shown that the epiphyte *Ramalina menziesii* (Fig. 11.3) plays an important role in the annual turnover of biomass and macronutrients (Boucher & Nash, 1990). Using destructive sampling techniques, they calculated the standing biomass of this lichen to be 706 kg ha^{-1}, which represented 94% of the total epiphytic lichens by weight. Furthermore, they determined from litterfall studies in blue oak (*Quercus douglasii*) forests that *R. menziesii* contributed 13% N, 4% P, 7% K, 1% Ca, 3% Mg and 8% Na of the annual canopy nutrient turnover, and from transplantation studies that the elemental status showed seasonal (and environmental) variation, with evidence of accumulation by dry deposition (Boonpragob & Nash, 1990). Further work by Knops *et al.* (1991) showed that *R. menziesii* not only contributed 26.4% of the biomass but also 9.4% of the total litter biomass when compared with *Q. douglasii*.

Canopies supporting such pendulous epiphytes considerably extend their surface area, nutrient input being further enhanced by the scavenging nature of lichens which have the ability to accumulate aerosols (including particulate trapping) and ions at exchange sites (Nieboer, Richardson & Tomassini, 1978). Photographic, physical and chemical techniques have

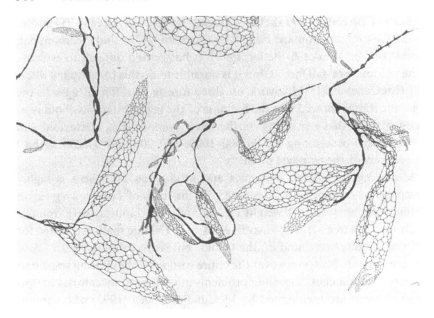

Fig. 11.3. Thalli of *Ramalina menziesii* teased out to show fine reticulations providing an extensive surface area for aerosol accumulation, particle entrapment and ion exchange.

been employed at Bradford University (Seaward *et al.*, unpubl.) to determine the surface area of air-dried samples of *R. menziesii* collected from a blue oak stand in California where this lichen represents *ca* 75% (*ca* 500 kg ha^{-1}) of the total; lichen biomass (Knops & Nash, *in litt.*); the total of *ca* 667 kg ha^{-1} for all pendulous lichen epiphytes at this site is remarkably high when compared with the *ca* 900 kg ha^{-1} of the oak leaves. Our preliminary determinations of the surface area of *R. menziesii* alone for this stand range between 0.15 and 0.7 million square metres per hectare. For the site where this lichen is more abundant (i.e. 706 kg ha^{-1}), a figure between 0.21 and almost 1 million square metres per hectare has been calculated. The lower figures for the two sites are undoubtedly under-estimates as the higher figures were derived from a more sophisticated technique which determined microdetails of thallial surfaces, including pores. Figures, for more realistic comparisons with a leaf area index for the supporting tree (as yet undetermined), are tentatively given as 0.3 and 0.42 million square metres per hectare for the two sites. Quite obviously, loss of such an extensive surface area, with a very high atmospheric scavenging capability, as a result of air pollution (even at moderate levels) and indeed by the felling of the trees themselves is bound to have significant repercussions, not only ecologically but also physiologically since the lichen biomass plays a key role in the water relations of such canopies.

Despite these detailed studies, our knowledge of the total processes involved is still fragmentary (e.g. Wielgolaski, 1975): energy-flow diagrams and budgets are incomplete, and modelling of ecosystem dynamics is therefore frequently based on rudimentary and variable information often derived from a wide variety of sources. It is difficult to determine energy-flow through lichens, since its measurement is dependent on various complex and laborious techniques, such as meticulous sampling, harvesting, age class determination, growth, productivity and litter measurements (Kärenlampi, 1971), analyses of faunal consumption (e.g. Engelmann, 1966) and decomposition (Wetmore, 1982; Biazrov, 1994).

Invertebrate–lichen associations play an important role in soil fertility and ecological energetics, and their value in energy-budgeting, modelling, etc. should therefore not be underestimated (Seaward, 1988). Energetics of lichen faunas are poorly understood due to the wide variety of data from disparate sources. Information on ingestion, assimilation, egestion, respiration, growth and death rates, numbers and biomass (Engelmann, 1966) needs careful coordination and interpretation. Of special interest to our understanding of microenvironmental processes is the pedogenetic study of Wessels & Wessels (1991) on the lichenophagous larvae of a South African bagworm. Clarens sandstone is weathered by endolithic lichens which dissolve the cementing material, thereby loosening quartz crystals which are then used by the larvae to construct the bags in which they live. It is estimated that the larvae contribute 4.4 kg of weathered sandstone per hectare per year to the area; furthermore, the larvae utilize the lichens as a food source, the resulting faeces providing 200 g of organic material per hectare per year and contributing to mineral cycling.

Faunal associations

Quantitative data on numbers and weight of the various faunal groups per unit area of lichen thallus, thalli or community, on which to base ecological energetic studies, are available (e.g. Gerson & Seaward, 1977). Data have been derived mainly from microhabitat studies, but more attention is now being focused on environments where lichens form a major component of the flora, such as littoral zones (Sochting & Gjelstrup, 1985), Arctic and Antarctic ecosystems (Gressitt, 1967), and certain types of woodland (Biazrov, 1994). There is also a lack of factual information relating to food-webs and nutrient recycling (cf. Gerson & Seaward, 1977).

The effects of environmental pollutants need more detailed investigation, since air pollution adversely affects lichens, and hence their primary consumers and in turn their predators (Gilbert, 1971). A significant uptake of heavy metals, pesticides and radioelements by invertebrate grazers via lichens should be expected.

For an integrated model of the dynamics of an ecosystem containing a significant lichen component, response patterns and typical parameter values for a full range of abiotic and biotic factors involved need to be established (e.g. Bunnell, Karenlampi & Russell, 1973). At best, energy-flow and mineral cycling calculations can only be an approximation (Seaward, 1988). Minerals taken up by lichens may subsequently reach other components of the surrounding biosphere via litterfall, leaching, bacterial incorporation, or non-cellular particle formation. Data on lichen decomposition, and consequent release of minerals, are scarce; such studies on oak woodlands, Douglas and balsam fir forests, heathlands and black spruce bogs (Pike, 1978; Gloaguen, Touffet & Forgeard, 1980; Wetmore, 1982; Moore, 1984) employ different techniques and methods of measurement, making comparisons difficult. It would appear, however, that the rate of lichen decomposition is rapid (half-lives of *ca* 90–760 days), but generally less than that of vascular plants. Interestingly, common saprophytic agencies rarely appear to be involved in the decomposition process, perhaps due to the inhibiting effect of lichen substances. The abundance and diversity of those invertebrates involved in the process can be very considerable (Biazrov, 1994).

As well as providing food for invertebrates, lichens provide a milieu that enables them to survive extended periods of desiccation within a thallial microhabitat, where they can also be protected against predation either by physical concealment or by blending with their background through camouflage or mimicry (Seaward, 1988). Camouflage is effected by these animals in various ways, including covering themselves with lichen fragments and soredia, or in some cases even providing a foundation on which an entire lichen will establish itself. Mimicry has been achieved through evolution of adaptive colouration and body shape. Colouration has been well researched, mainly with respect to industrial melanism and more particularly in the study of the ascendancy of melanic morphs of various moth species, the peppered moth (*Biston betularia*) being the best-known exemplar (Kettlewell, 1973). Lichen cover appears to be critical for the maintenance of light-coloured morphs of *B. betularia*: the rise in air pollution levels over the past two centuries throughout Europe has seen not only a decline in lichen diversity (see below), and thereby protective crypsis for light-coloured morphs, but also the recent rise in dominance of a few lichen species creating monotonous verdant backgrounds which appear to be almost equally advantageous/disadvantageous to both light- and dark-coloured morphs; the ratio between the two morphs is being carefully monitored in England (Clarke, Mani & Wynne, 1985; Cook, Rigby & Seaward, 1990). The implementation of clean air policies has frequently seen a significant improvement in the lichen flora in terms of increasing species diversity but the above-mentioned dominating lichens continue to exert a major influence on their associated fauna.

The effects of pollutant accumulation on the palatability of lichens, as well as ecosystem dynamics studies to determine residence times of toxins, and rate of detoxification, decomposition, are but a few of the many aspects of lichen–invertebrate associations in need of further study. Undoubtedly, the biodiversity and abundance of invertebrates which feed on, shelter in, and are camouflaged by lichens are enormous. Calculations of their biomass and further study of their role in mineral cycling and energy-flow are likely to reveal how highly significant they are to many ecosystems.

Many bird species use lichens as nesting material (Richardson & Young, 1977), some showing a definite preference for certain types of lichen not only for nest construction and camouflage, but also, as in the case of bower birds (Tibell & Gibson, 1986), for purely decorative purposes. The disappearance of lichens, mainly due to air pollution, has depleted this source of nest-building material, and, more importantly, the consequent loss of the associated lichenophagous insect fauna. The loss of this important food source as well as egg-laying sites for invertebrates will affect the complex food webs upon which bird populations depend. Only a few studies have highlighted the significant differences in the invertebrate faunas of unpolluted and polluted environments (e.g. Gilbert, 1971; Roberts & Zimmer, 1990).

Lichens are also used as nesting material by flying squirrels. Detailed studies in Idaho (Hayward & Rosentreter, 1994) revealed that these animals were highly selective in their choice of lichens, 96% by volume of their nests being constructed of arboreal lichens, almost entirely composed of three species of *Bryoria*; interestingly, these three species contain only traces of, or no, lichen substances, making them more palatable to these squirrels which are known to feed off lichens (McKeever, 1960).

A large number of mammal species are known to feed on lichens, although their importance in the animal's diet varies considerably. Deer, elk, ibex, gazelle, musk ox, mountain goat, polar bear, lemming, vole, tree mouse, marmot, squirel, and some domestic animals may include lichens in their diet, perhaps fortuitously, but more likely as a means of supplementing their normal diet or as winter feed (Richardson & Young, 1977). By far the most important mammalian lichen feeders are reindeer and caribou, the Eurasian and North American subspecies, respectively, of *Rangifer tarandus*.

The winter diet of *R. tarandus* contains more than 50% of lichens, consumption being *ca* 3 to 5 kg day⁻¹, varying with the geographic location of the herd. Animals not only scrape away the snow cover to expose terricolous lichens on which to feed, but also consume arboreal and saxicolous lichens where woodland and rocks prevail. Reindeer and caribou have been researched extensively in terms of the composition, growth and productivity of the lichen stands on which they feed, intake rates, the nutritive value and chemical composition of lichen and lichen-supplemented diets, their ability to ingest lichens and the effects of grazing and trampling on lichen-dominated communities. From these studies, it is clear that the lichens on which *R. tarandus* feeds have a low nutritive value, being particularly deficient in protein, calcium and phosphorus (Scotter, 1965; Soppela, Nieminen & Saarela, 1992), often leading to overgrazing of this resource and the need for an optimal harvesting policy (Virtala, 1992). A level of *ca* 5% protein, normally regarded as acceptable for the survival of domestic animals, is not attained in the majority of lichens; a decrease in muscle weight over the winter period is therefore inevitable. However, the 7.4% protein content of *Ramalina lacera* may adequately sustain grazing desert gazelles if regularly available to them (Hawksworth *et al.*, 1984). Lichens are well known for their capacity to accumulate elements, particularly from metal enriched substrata and polluted environments, and excessive amounts of certain heavy metals have occasionally been reported as occurring in lichens grazed by *R. tarandus*; in spite of this, many symptoms of malnutrition appear in *R. tarandus* resulting from deficiencies of such elements as K, Mg, Zn and Cu in their winter lichen diet.

Lichens also have the ability to accumulate substantial levels of radionuclides (Richardson, 1992) and can therefore be used as bioindicators to monitor the extent and severity of environmental contamination. The fate of such contaminants through food chains has been followed, particularly where humans are at risk, as in the case of consumption of meat from *R. tarandus* which are reliant upon lichens for a substantial part of their diet. Studies have been extended to other domestic animals, particularly mountain sheep, which normally eat ericaceous plants, but may fortuitously graze lichens which may be rich in radionuclides, as in the case of those contaminated as a consequence of the Chernobyl disaster (see below).

Human impact

Undoubtedly, human beings are currently the paramount agents of lichen destruction, by causing disturbance of ecosystems worldwide, through deforestation, agricultural practices, urbanization, pollution of air, water and soil, and exploitation of natural resources.

In the past it has been difficult to make a convincing case to justify the conservation of lichens when weighed against other pressing needs (Seaward, 1982). However, lichens currently occupy *ca* 8% of the earth's terrestrial surface (Larson, 1987) and as clearly demonstrated above, form such an important component of the complex web of life that their disappearance affects the balance of nature to a surprising degree. This is particularly the case in tundra zones, high altitudes, cold deserts, dune systems, semi-arid lands and deserts, and even urban areas, where they provide vital links in food chains and are important in community development and succession on rocks and soils.

Even more dramatic, there are areas of the globe where the results of lichen denudation are now being detected by means of remote sensing. Such losses may well have climatic repercussions and exert a measurable influence on global warming. Of particular interest in this respect is the disappearance of epilithic lichens over very large areas of the Canadian shield as a direct consequence of atmospheric pollution formerly (but less so now) emanating from the smelting operations at Sudbury, Ontario. Here the barren rock surfaces now have different reflectance characteristics (Fig. 11.4), as their essentially light-absorbing lichen cover no longer exists (cf. Petzold & Goward, 1988; Rollin, Milton & Roche, 1994).

Similar situations prevail in terricolous lichen dominated areas where human disturbance has significantly reduced lichen cover. This is particularly apparent in mountainous regions where skiing, once the pastime of a select few, is now enjoyed by thousands, resulting in the wholesale erosion of lichen dominated vegetational cover; in some cases the entire ecosystem has been buried beneath alien imported materials such as bitumen.

In hot and cold deserts sensitive lichen dominated ecosystems have been similarly destroyed through human activities (Klopatek, 1992; Eldridge & Greene, 1994). It has been shown, for example, that the microphytic crusts of lichens, mosses and cyanobacteria in semi-arid regions of eastern Australia contribute up to 27% of the ground cover and decrease reflectance (O'Neill, 1994). Some gypsiferous soils in the intermountain area of the western United States support soil crust communities with a high species diversity, often with a 100% ground cover of lichens (St Clair,

Fig. 11.4. Rocks of the Canadian shield near Sudbury, Ontario, showing total denudation of former cryptogamic cover as a consequence of air pollution.

Fig. 11.5. Virtually complete cover of hot desert soil provided by cryptogamic crusts, mainly composed of *Psora decipiens*.

Johansen & Rushforth, 1993). Over a long period of time, lichen crust communities have changed the soil's physico-chemical properties, enhancing their stability and fertility (Fig. 11.5). These benefits are lost when human interference affects the natural equilibrium by, for example, the introduction of domestic animals (Eldridge & Greene, 1994).

Changes in reflectance and in ecophysiological responses, such as photosynthetic activity, chlorophyll levels, gaseous exchange and water absorbance (Lange, Schulze & Koch, 1970; Lange *et al.*, 1975, 1992; Lindsay, 1977), brought about by anthropogenic disturbances to lichen dominated communities are being detected in remotely sensed images (cf. Eldridge & Greene, 1994). As a consequence, the environmental significance of variations in these activities is increasingly being recognized. In arctic and alpine areas, lichens possess specialized physiological mechanisms enabling them to photosynthesize and take up water at low temperatures (e.g. Kappen, 1989). Field studies have established the importance of lichens for contributing to environmental modification. Lichens may also be effective ice nucleating agents (Kieft, 1988) and can therefore initiate freezing of supercooled water at relatively warm temperatures. Such phenomena may be used to interpret data derived from remote sensing. In the case of global warming, temperature differences in remote areas such as the polar regions could be dictated by the presence or absence of lichen cover.

Environmental monitoring

As has been shown above, lichens play a major part in shaping the natural world, both physically and biologically. Lichens are also important in terms of human needs, not only as sources of medicines, dyes, foods, perfumes and decorations (Richardson, 1988, 1991; Crittenden & Porter, 1991; Yamamoto *et al.*, 1993; Ramelow *et al.*, 1993; Hawksworth, 1994), but also as natural sensors of our changing environment. The sensitivity of particular lichen species and assemblages to a very broad spectrum of environmental conditions, both natural and unnatural, is widely appreciated. Lichens are therefore used increasingly in evaluating threatened habitats, in environmental impact assessments, and in monitoring environmental perturbations due to a disturbingly large and growing number of chemical pollutants.

Environmental interpretation by means of lichens can be based on the presence and/or absence of particular species. The biodiversity of lichen assemblages has proved to be strongly indicative of one or more identifiable factors. It is therefore possible to use the composition of the lichen flora to evaluate habitat and ecosystem stability, often in terms of ecological continuity over time; for example, the woodland continuity index devised by Rose (1976) can be used to determine past management and approximate age of British deciduous woodlands, and similar indices are being devised for use elsewhere, such as Scandinavian boreal coniferous forests (Tibell, 1992).

Information gained from our knowledge of how lichens respond to long-term perturbations and short-term upheavals in nature can be applied to the interpretation and monitoring of environmental changes and disasters brought about through a wide range of human activities. The reaction of lichens to sudden natural events such as fire, volcanic eruptions and earthquakes on the one hand and to the long-term effects of glaciers, snow and water on the other can be effectively employed to determine those human impacts which destabilize soil, rock and water systems. Thus, lichens can often be used as an early warning system for other biota which without remedial action would subsequently suffer stress or indeed extinction through forest and agricultural mismanagement, desertification, urbanization, industrialization and a whole host of other problems arising from world overpopulation.

Baseline information on lichen assemblages and ecosystems which are ecologically or geographically zoned on the basis of particular natural phenomena have proved invaluable in assessing widespread increases in various pollutants and climatic changes resulting from global warming (Galloway, 1992); for example, distinctive lichen zonations at freshwater and marine water-lines can be affected by acidification, hypertrophication and other polluting agencies. Delimitation will change as water levels rise as a consequence of increases in global temperature. From detailed field studies and remote sensing, it should be possible to monitor changes in the lichen flora of terrestrial environments resulting from displaced snow-lines (cf. Sonesson, Osborne & Sandberg, 1994), episodic snow-kill (Benedict, 1991, 1993), avalanches (McCarroll, 1993), seismic landslides (Bull *et al.*, 1994) and other unstable debris flows (e.g. Innes, 1983), and retreating glaciers due to climatic shifts. Such time-space investigations can fully exploit lichenometrical techniques widely developed over the past few decades (Innes, 1985).

The greatest use of evaluating changes in the lichen flora, based on diversity counts and phytosociological analyses to generate scales and indices, is in the field of pollution monitoring. Such investigations have been applied mainly to the assessment of air quality, but have also proved useful in evaluating the extent and nature of soil contamination. The generally detrimental effects of air pollution on lichens have been known

Fig. 11.6. 1993 distributions of *Hypogymnia physodes*, *Parmelia sulcata* and *Usnea subfloridana* derived from the British Lichen Society's database (currently housed at the University of Bradford). Their relative sensitivity to air pollution is shown not only in terms of the size of those areas from which they have disappeared but also of their ability to re-establish themselves over the three decades following implementation of the Clean Air Acts.

since the mid-nineteenth century. The role of specific components of such pollution as reflected by lichen performance is still in the process of elucidation. However, sulphur dioxide is undoubtedly implicated and the worldwide use of scales, based on specific lichens, for estimating its level, stemming mainly from the work on the British lichen flora which culminated in the publication of the scale by Hawksworth & Rose (1970), has proved invaluable in air pollution monitoring.

In the past, air pollution has undoubtedly devastated the lichen flora over relatively small areas, usually within the immediate vicinity of urban and/or industrial complexes, but there is also a wealth of evidence to demonstrate the selective decline of lichen species over major geographic regions, attributable to the rise in sulphur dioxide and other pollutants (Fig. 11.6), as well as changing forestry and agricultural practices. In

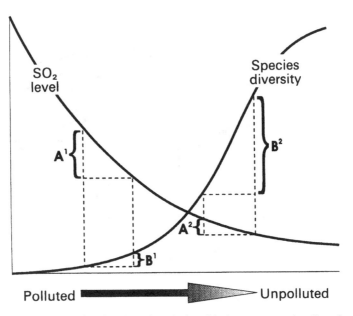

Fig. 11.7. Model showing the relationship between species diversity and sulphur dioxide level (Seaward, 1993, Fig. 3). Biological scales are logarithmic in nature: major reductions in the level of the sulphur dioxide (A¹) at the polluted end of the scale may have little or no effect on lichen species diversity (B¹); however, moving from a polluted to a less polluted situation, the effect increases exponentially, such that only a minor increase in sulphur dioxide (A²) would dramatically reduce species diversity (B²).

conditions of reasonably stable pollution there is a clearly defined negative relationship between species diversity of lichens and sulphur dioxide level (Fig. 11.7). Lichens respond relatively quickly to a rise in its concentration and can therefore be satisfactorily employed to monitor ambient levels; however, in areas experiencing a fall in sulphur dioxide concentration, the situation is less straightforward. The rate at which the different lichen species reinvade cleaned-up areas is enormously variable, but bioindicational scales which take account of these hierarchical changes are currently being developed (cf. Hawksworth & McManus, 1989).

Long-term field investigations involving stringent ecological and phytogeographical criteria, such as those made possible in the British Isles through a comprehensive on-going programme of detailed mapping by the British Lichen Society since 1963 (Hawksworth & Seaward, 1990; Seaward, 1995), can provide the basis for large-scale monitoring of quantitative and qualitative changes in air pollution regimes (Seaward, 1992). Steps are being taken to establish an international network of lichen mapping recorders, primarily for Europe, to monitor air pollution over wide geographical areas by species diversity counts, the presence or absence of particular pollution sensitive species, and by means of bioassays of selected common species with widespread distributions in order to determine spatially levels of elements (particularly nitrogen), heavy metals, radionuclides, etc. (see below). The determination of background and/or baseline levels of these chemicals is of particular importance to gauge their environmental impact.

Intensive lichen monitoring is a necessary component of any programme aimed at effective long-term observation of air pollution (Fig. 11.8). Numerous difficulties can arise over taxonomy, thallial chronology, genetic plasticity, site differences, etc., but a fundamental advantage of using lichens (or other organisms) as indicators is that they show the results of the action of pollutants on living material – a relevant, if at times rather emotive approach to determining the many facets of the technological impact on the biosphere. It will never be possible to replace completely direct physical and chemical measurements of air pollutant concentrations by bioindicators; nevertheless, both approaches are necessary for making detailed or large-scale surveys of the distribution of air pollutants, where the extensive use of technical equipment is costly or impractical.

Lichens display an exceptional ability to accumulate chemicals, which usually reflect ambient pollution loads, often with considerable accuracy. Most of the work in this field has involved studies on heavy metals and radionuclides, with bioassays being used to evaluate the degree of natural and unnatural environmental contamination, occasionally for geochemical prospecting. The increasing burden of these elements in ecosystems requires continual monitoring, and the establishment of background and baseline levels is of prime importance in the light of disasters such as that at Chernobyl (Seaward, 1991; Hofmann *et al.*, 1993), the effects of which will be detectable for many years to come, particularly in the subarctic where the lichen–reindeer–human food-chain is involved (Rissanen & Rahola, 1989).

Recently, attention has been focused on the use of lichens to monitor

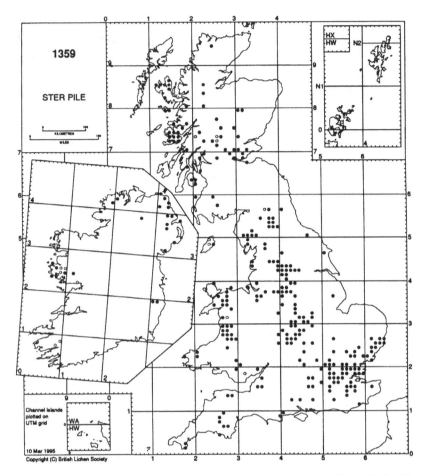

Fig. 11.8. Distribution of *Stereocaulon pileatum* in the British Isles (March 1995): normally found on metal-enriched substrata, mainly to the north and west, its present distribution, corresponding to urban and motorway developments, is attributed to the increase in lead as an environmental contaminant.

atmospheric nitrogen (e.g. Bruteig, 1993); while sulphur dioxide levels are falling in many regions, nitrogen is on the increase, mainly attributable to automobile emissions, agrochemicals and intensive animal husbandry. A very wide range of other pollutants and toxins go largely undetected, but with advances in analytical equipment, lichen monitoring has been extended to redress this, as in the case of organochlorine contaminants (Muir *et al.*, 1993).

The proper use of lichens as indicators and samplers of ambient conditions is a valuable resource for the environmentalist. In theory, the techniques which have been developed by the ecologist could be employed on a large scale for 'low technology' environmental appraisals and impact studies, particularly where on-site instrumentation would be expensive to install and maintain. Unfortunately, most of the methodologies require a fairly detailed understanding of the taxonomy of lichens; furthermore, techniques based on bioassays necessitate depletion of the resource material, rigorous protocols for its collection, preparation and analysis, and sophisticated analytical equipment.

Undoubtedly, pollution monitoring by means of lichens has captured the attention not only of lichenologists but also of biologists and environmental scientists in general. There has been an endless flood of publications on this topic, a review of which is beyond the scope of this paper. Information and detailed bibliographies on this aspect of lichenology are to be found in Hawksworth (1971, 1990), Ferry, Baddeley & Hawksworth (1973), Bates & Farmer (1992), Richardson (1992), Seaward (1993) and the on-going series 'Literature on air pollution and lichens' in *The Lichenologist*.

Postscript

At the outset of this chapter I quoted from the work of the far-sighted British lichenologist William Lauder Lindsay (1829–1880) who fully appreciated the major role played by lichens in shaping our environment, a role which it has only been possible to substantiate fully in recent years by advancements in a wide range of high technology. It is fitting that Lindsay (1856) should also provide me with my closing words:

> We may now be said to be entering on a new era in Lichenology; it is now being studied in a more philosophic spirit, and with all the aids which modern discoveries in science – which the microscope and chemistry – can furnish. Facts are being earnestly and patiently sought after; generalization and theory avoided until a sufficiency of data be accumulated . . . labourers are increasing and volunteers are coming forward who esteem it an honour to join this forlorn hope of Cryptogamic Botany, who are eager for the work solely on account of its difficulty . . . as the Lichens are more fully studied by the reflected light of modern science . . . so will the study of Lichenology become more simple and attractive . . . But the labours of the student must equally begin and terminate on the spot where the Lichens grow . . . there he must watch patiently and note accurately – it may be for a series of years – the stages of origin, growth, and decay of species under all the influences, terrestrial and aerial, by which these are so liable to be affected.

Acknowledgements

I am most grateful to Dr Jean Knops and Professor Tom Nash of Arizona State University for providing field data and lichen material on which to experiment, and to my colleagues Dr R.I. Bickley, Dr H.G.M. Edwards and Mr R. Wilkes at Bradford University for their continuing researches into the measurement of lichen surface area. I also wish to thank Professor David H.S. Richardson for his helpful comments on a draft of this paper.

References

Ahti, T. (1977). Lichens of the boreal coniferous zone. In *Lichen Ecology* (ed. M.R.D. Seaward), pp. 145–81. London, Academic Press.

Alstrup, V. (1977). Cryptogams on imported timber in west Greenland. *The Lichenologist*, **9**, 113–17.

Ascaso, C., Sancho, L.G. & Rodriguez-Pascual, C. (1990). The weathering action of saxicolous lichens in maritime Antarctica. *Polar Biology*, **11**, 33–9.

Bailey, R.H. & James, P.W. (1979). Birds and the dispersal of lichen propagules. *The Lichenologist*, **11**, 105–6.

Bates, J.W. & Farmer, A.W. (ed.) (1992). *Bryophytes and Lichens in a Changing Environment*. Oxford, Clarendon Press.

Becker, V.E., Reeder, J. & Stetler, R. (1977). Biomass and habitat of nitrogen fixing lichens in an oak forest in the North Carolina Piedmont. *The Bryologist*, **80**, 93–9.

Benedict, J.B. (1991). Experiments on lichen growth. II. Effects of a seasonal snow cover. *Arctic and Alpine Research*, **23**, 189–99.

Benedict, J.B. (1993). A 2000-year lichen-snowkill chronology for the Colorado Front Range, USA. *The Holocene*, **3**, 27–33.

Biazrov, L.G. (1994). Decomposition of epiphytic lichen litter and the involvement of invertebrates in forests near Moscow. *Cryptogamic Botany*, **4**, 130–4.

Boonpragob, K. & Nash, T.H. (1990). Seasonal variation of elemental status in the lichen *Ramalina menziesii* Tayl. from two sites in southern California: evidence for dry deposition accumulation. *Environmental and Experimental Botany*, **30**, 415–28.

Boucher, V.L. & Nash, T.H. (1990). The role of the fruticose lichen *Ramalina menziesii* in the annual turnover of biomass and macronutrients in a blue oak woodland. *Botanical Gazette*, **151**, 114–18.

Bruteig, I.E. (1993). The epiphytic lichen *Hypogymnia physodes* as a biomonitor of atmospheric nitrogen and sulphur deposition in Norway. *Environmental Monitoring and Assessment*, **26**, 27–47.

Bull, W.B., King, J., Kong, F., Moutoux, T. & Phillips, W.M. (1994). Lichen dating of coseismic landslide hazards in alpine mountains. *Geomorphology*, **10**, 253–64.

Bunnell, F.L., Karenlampi, L. & Russell, D.E. (1973). A simulation model of lichen–*Rangifer* interactions in northern Finland. *Report of the Kevo Subarctic Research Station*, **10**, 1–8.

Clarke, C.A., Mani, G.S. & Wynne, G. (1985). Evolution in reverse: clean air and the peppered moth. *Biological Journal of the Linnean Society*, **26**, 189–99.

Cook, L.M., Rigby, K.D. & Seaward, M.R.D. (1990). Melanic moths and changes in epiphytic vegetation in north-west England and north Wales. *Biological Journal of the Linnean Society*, **39**, 343–54.

Crittenden, P.D. (1975). Nitrogen fixation on the glacial drift of Iceland. *New Phytologist*, **74**, 41–9.

Crittenden, P.D. & Porter, N. (1991). Lichen-forming fungi: potential sources of novel metabolites. *Trends in Biotechnology*, **9**, 409–14.

Denison, W.C. (1973). Life in tall trees. *Scientific American*, **228**, 74–80.

Edwards, H.G.M., Farwell, D.W. & Seaward, M.R.D. (1991). Raman spectra of oxalates in lichen encrustations on Renaissance frescoes. *Spectrochimica Acta*, **47A**, 1531–9.

Edwards, H.G.M. & Seaward, M.R.D. (1993). Raman microscopy of lichen–substratum interfaces. *Journal of the Hattori Botanical Laboratory*, **74**, 303–16.

Eldridge, D.J. & Greene, R.S.B. (1994). Microbiotic soil crusts: a review of their roles in soil and ecological processes in the rangelands of Australia. *Australian Journal of Soil Research*, **32**, 389–415.

Engelmann, M.D. (1966). Energetics, terrestrial field studies, and animal productivity. *Advances in Ecological Research*, **3**, 73–115.

Ferry, B.W., Baddeley, M.S. & Hawksworth, D.L. (ed.) (1973). *Air Pollution and Lichens*. London: Athlone Press.

Fletcher, A. (1980). Marine and maritime lichens of rocky shores: their ecology, physiology and biological interactions. In *The Shore Environment* 2 (ed. J.H. Price, D.E.G. Irvine & W.F. Farnham), pp. 789–842. New York: Academic Press.

Forman, R.T.T. (1975). Canopy lichens with blue-green algae: a nitrogen source in a Columbian rain forest. *Ecology*, **56**, 1176–84.

Galloway, D.J. (1992). Biodiversity: a lichenological perspective. *Biodiversity and Conservation*, **1**, 312–23.

Galloway, D.J. (1994). Biogeography and ancestry of lichens and other Ascomycetes. In *Ascomycete Systematics: Problems and Perspectives in the Nineties* (ed. D.L. Hawksworth), pp. 175–184. New York: Plenum Press.

Garty, J., Giele, C. & Krumbein, W.E. (1982). On the occurrence of pyrite in a lichen-like inclusion in Eocene amber (Baltic). *Palaeogeography, Palaeoclimatology and Palaeoecology*, **39**, 139–47.

Gerson, U. & Seaward, M.R.D. (1977). Lichen–invertebrate associations. In *Lichen Ecology* (ed. M.R.D. Seaward), pp. 69–119. London: Academic Press.

Gilbert, O.L. (1971). Some indirect effects of air pollution on bark-living invertebrates. *Journal of Applied Ecology*, **8**, 77–84.

Gloaguen, J.C., Touffet, J. & Forgeard, F. (1980). Vitesse de décomposition et évolution minérale des litières sous climat atlantique. *Oecologica, Plantarum*, **1**, 257–73.

Gressitt, J.L. (1967). Notes on arthropod populations in the Antarctic Peninsula–South Shetland Islands–South Orkney Islands area. In *Entomology in Antarctica* (ed. J.L. Gressitt), pp. 373–391. American Geophysical Union, Washington.

Hallbauer, D.K., Jahns, H.M. & Beltmann, H.A. (1977). Morphological and anatomical observations on some Precambrian plants from the Witwatersrand, South Africa. *Geologische Rundschau*, **66**, 477–91.

Hawksworth, D.L. (1971). Lichens as litmus for air pollution: a historical review. *International Journal of Environmental Studies*, **1**, 281–96.

Hawksworth, D.L. (1988*a*). Co-evolution of fungi and algae in lichen associations. In *Co-evolution of Fungi with Plants and Animals* (ed. K.A. Pirozynski & D.L. Hawksworth), pp. 125–148. London: Academic Press.

Hawksworth, D.L. (1988*b*). The variety of fungal–algal symbioses, their evolutionary significance, and the nature of lichens. *Botanical Journal of the Linnean Society*, **96**, 3–20.

Hawksworth, D.L. (1990). The long-term effects of air pollutants on lichen communities in Europe and North America. In *The Earth in Transition: Patterns and Processes of Biotic Impoverishment* (ed. G.M. Woodwell), pp. 45–64. Cambridge: Cambridge University Press.

Hawksworth, D.L. (1994). The recent evolution of lichenology: a science for our times. *Cryptogamic Botany*, **4**, 117–29.

Hawksworth, D.L., Lawton, R.M., Martin, P.G. & Stanley-Price, K. (1984). Nutritive value of *Ramalina duriaei* grazed by gazelles in Oman. *The Lichenologist*, **16**, 93–4.

Hawksworth, D.L. & McManus, P.M. (1989). Lichen recolonization in London under conditions of rapidly falling sulphur dioxide levels, and the concept of zone skipping. *Botanical Journal of the Linnean Society*, **100**, 99–109.

Hawksworth, D.L. & Rose, F. (1970). Qualitative scale for estimating sulphur dioxide air pollution in England and Wales using epiphytic lichens. *Nature, London*, **227**, 145–8.

Hawksworth, D.L. & Seaward, M.R.D. (1990). Twenty-five years of lichen mapping in Great Britain and Ireland. *Stuttgarter Beitrage zur Naturkunde, ser. A (Biol.)*, **456**, 5–10.

Hayward, G.D. & Rosentreter, R. (1994). Lichens as nesting material for northern flying squirrels in the northern Rocky Mountains. *Journal of Mammalogy*, **75**, 663–73.

Hofmann, W., Attarpour, N., Lettner, H. & Türk, R. (1993). [137]Cs concentrations in lichens before and after the Chernobyl accident. *Health Physics*, **64**, 70–3.

Innes, J.L. (1983). Lichenometric dating of debris-flow deposits in the Scottish Highlands. *Earth Surface Processes and Landforms*, **8**, 579–88.

Innes, J.L. (1985). Lichenometry. *Progress in Physical Geography*, **9**, 187–254.

Jones, D. (1988). Lichens and pedogenesis. In *Handbook of Lichenology* 3 (ed. M. Galun), pp. 109–24. Boca Raton, CRC Press.

Kalb, K. (1970). Flechtengesellschaften der Vorderen Ötztaler Alpen. *Dissertationes Botanicae*, **9**, 1–120.

Kappen, L. (1989). Field measurements of carbon dioxide exchange of the Antarctic lichen *Usnea sphacelata* in the frozen state. *Antarctic Science*, **1**, 31–4.

Kärenlampi, L. (1971). On methods for measuring and calculating the energy flow through lichens. *Report of the Kevo Subarctic Research Station*, **7**, 40–6.

Kärnefelt, I. (1990). Evidence of a slow evolutionary change in the speciation of lichens. *Bibliotheca Lichenologica*, **38**, 291–306.

Kettlewell, H.B.D. (1973). *The Evolution of Melanism*. Oxford: Oxford University Press.

Kieft, T.L. (1988). Ice nucleation activity in lichens. *Applied Environmental Microbiology*, **54**, 1678–81.

Klopatek, J.M. (1992). Cryptogamic crusts as potential indicators of disturbance in semi-arid landscapes. In *Ecological Inidicators* 1 (ed. D.H. McKenzie, D.E. Hyatt & V.J. McDonald), pp. 773–86. London: Elsevier Applied Science.

Knops, J.M.H., Nash, T.H., Boucher, V.L. & Schlesinger, W.H. (1991). Mineral cycling and epiphytic lichens: implications at the ecosystem level. *The Lichenologist*, **23**, 309–21.

Lang, G.E., Reiners, W.A. & Pike, L.H. (1980). Structure and biomass of epiphytic lichen communities of balsam fir forests in New Hampshire. *Ecology*, **61**, 541–50.

Lange, O.L., Kidron, G.J., Büdel, B., Meyer, A., Kilian, E. & Abeliovich, A. (1992). Taxonomic composition and photosynthetic characteristics of the 'biological soil crusts' covering sand dunes in the western Negev Desert. *Functional Ecology*, **6**, 519–27.

Lange, O.L., Schulze, E., Kappen, L., Buschbom, V. & Evenari, M. (1975). Adaptations of desert lichens to drought and extreme temperatures. In *Environmental Physiology of Desert Organisms* (ed. N.P. Hadley), pp. 20–37. Dowden, Hutchinson & Ross; Stroudsburg.

Lange, O.L., Schulze, E. & Koch, W. (1970). Experimentellokologische Untersuchungen an Flechten der Negev-Wuste. II. CO_2-Gaswechsel und Wasserhaushalt von *Ramalina maciformis* (Del.) Bory am naturlichen Standort wahrend der sommerlichen Trockenperiode. *Flora, Jena*, **159**, 38–62.

Larson, D.W. (1987). The absorption and release of water by lichens. *Bibliotheca Lichenologica*, **25**, 351–60.

Lindsay, D.C. (1973). Probable introductions of lichens to South Georgia. *British Antarctic Survey Bulletin*, **33/34**, 169–72.

Lindsay, D.C. (1977). Lichens of cold deserts. In *Lichen Ecology* (ed. M.R.D. Seaward), pp. 183–209. London, Academic Press.

Lindsay, W.L. (1856). *A Popular History of British Lichens*. London: Lovell Reeve.

McCarroll, D. (1993). Modelling late-Holocene snow-avalanche activity: incorporating a new approach to lichenometry. *Earth Surface Processes and Landforms*, **18**, 527–39.

McCune, B. (1994). Using epiphyte litter to estimate epiphyte biomass. *The Bryologist*, **97**, 396–401.

McCune, B. & Lesica, P. (1992). The trade-off between species capture and quantitative accuracy in ecological inventory of lichens and bryophytes in forest in Montana. *The Bryologist*, **95**, 296–304.

McKeever, S. (1960). Food of the northern flying squirrel in northeastern California. *Journal of Mammology*, **41**, 270–1.

Moore, T.R. (1984). Litter decomposition in a subarctic spruce-lichen woodland, eastern Canada. *Ecology*, **65**, 299–308.

Muir, D.C.G., Segstro, M.D., Welbourn, P.M., Toom, D., Lsenreich, S.J., Macdonald, C.R. & Whelpdale, D.M. (1993). Patterns of accumulation of airborne organochlorine contaminants in the lichens from the Upper Great Lakes region of Ontario. *Environmental Science and Technology*, **27**, 1201–10.

Nash, T.H. & Moser, T.J. (1982). Vegetational and physiological patterns of lichens in North American deserts. *Journal of the Hattori Botanical Laboratory*, **53**, 331–6.

Nash, T.H., White, S.L. & Marsh, J.E. (1977). Lichen and moss distribution and biomass in hot desert ecosystems. *The Bryologist*, **80**, 470–9.

Nieboer, E., Richardson, D.H.S. & Tomassini, F.D. (1978). Mineral uptake and release by lichens: an overview. *The Bryologist*, **81**, 226–46.

Oksanen, J. (1984). Interspecific contact and association in sand dune vegetation dominated by bryophytes and lichens. *Annales Botanici Fennici*, **21**, 189–99.

O'Neill, A.L. (1994). Reflectance spectra of microphytic soil crusts in semi-arid Australia. *International Journal of Remote Sensing*, **15**, 675–81.

Petzold, D.E. & Goward, S.N. (1988). Reflectance spectra of subarctic lichens. *Remote Sensing of Environment*, **24**, 481–92.

Pike, L.H. (1978). The importance of epiphytic lichens in mineral cycling. *The Bryologist*, **81**, 247–57.

Pike, L.H. (1981). Estimation of lichen biomass and production with special reference to the use of ratios. In *The Fungal Community: Its Organization and Role in the Ecosystem* (ed. D. Wicklow & G. Carroll), pp. 533–52. New York; Marcel Dekker.

Ramelow, G.J., Liu, L., Himel, C., Fralick, D., Zhao, Y. & Tong, C. (1993). The analysis of dissolved metals in natural waters after preconcentration on biosorbents of immobilized lichen and seaweed biomass in silica. *International Journal of Analytical Chemistry*, **53**, 219–32.

Retallack, G.J. (1994). Were the Ediacaran fossils lichens? *Paleobiology*, **20**, 523–44.

Rhoades, F.M. (1981). Biomass of epiphytic lichens and bryophytes on *Abies lasiocarpa* on a Mt. Baker lava flow. *The Bryologist*, **84**, 39–47.

Rhoades, F.M. (1983). Distribution of thalli in a population of the epiphytic lichen *Lobaria oregana* and a model of population dynamics and production. *The Bryologist*, **86**, 309–31.

Richardson, D.H.S. (1988). Medicinal and other economic aspects of lichens. In *Handbook of Lichenology* 3 (ed. M. Galun), pp. 93–108. Boca Raton, CRC Press.

Richardson, D.H.S. (1991). Lichens and man. In *Frontiers in Mycology* (ed. D.L. Hawksworth), pp. 187–210. Wallingford; CAB International.

Richardson, D.H.S. (1992). *Pollution Monitoring with Lichens*. Slough; Richmond Publishing.

Richardson, D.H.S. & Young, C.M. (1977). Lichens and vertebrates. In *Lichen Ecology* (ed. M.R.D. Seaward), pp. 121–44. London, Academic Press.

Rissanen, K. & Rahola, T. (1989). Cs-137 concentration in reindeer and its fodder plants. *The Science of the Total Environment*, **85**, 199–206.

Roberts, D. & Zimmer, D. (1990). Microfaunal communities associated with epiphytic lichens in Belfast. *The Lichenologist*, **22**, 163–71.

Rogers, R.W. (1977). Lichens of hot arid and semi-arid lands. In *Lichen Ecology* (ed. M.R.D. Seaward), pp. 211–52. London; Academic Press.

Rollin, E.M., Milton, E.J. & Roche, P. (1994). The influence of weathering and lichen cover on the reflectance spectra of granitic rocks. *Remote Sensing of Environment*, **50**, 194–9.

Rose, F. (1976). Lichenological indicators of age and environmental quality in woodlands. In *Lichenology: Progress and Problems* (ed. D.H. Brown, D.L. Hawksworth & R.H. Bailey), pp. 279–307. London; Academic Press.

St Clair, L.L., Johansen, J.R. & Rushforth, S.R. (1993). Lichens of soil crust communities in the intermountain area of the western United States. *Great Basin Naturalist*, **53**, 5–12.

Schwartzman, D.W. & Volk, T. (1989). Biotic enhancement of weathering and the habitability of Earth. *Nature, London*, **340**, 457–60.

Scotter, G.W. (1965). Chemical composition of forage lichens from northern Saskatchewan as related to use by barren-ground caribou in the taiga of northern Canada. *Canadian Journal of Plant Sciences*, **45**, 246–50.

Seaward, M.R.D. (1982). Principles and priorities of lichen conservation. *Journal of the Hattori Botanical Laboratory*, **52**, 401–6.

Seaward, M.R.D. (1988). Contribution of lichens to ecosystems. In *Handbook of Lichenology* 2 (ed. M. Galun), pp. 107–129. Boca Raton; CRC Press.

Seaward, M.R.D. (1991). Biomonitoring radionuclides in eastern Europe, pre- and post-Chernobyl. In *Environmental Pollution and Control* (ed. Z. Ayvaz), pp. 80–9. Izmir: Ege University.

Seaward, M.R.D. (1992). Large-scale air pollution monitoring using lichens. *GeoJournal*, **28**, 403–11.

Seaward, M.R.D. (1993). Lichens and sulphur dioxide air pollution: field studies. *Environmental Reviews*, **1**, 73–91.

Seaward, M.R.D. (ed.) (1995). *Lichen Atlas of the British Isles*. London, British Lichen Society.

Seaward, M.R.D. & Edwards, H.G.M. (1995). Lichen – Substratum interface studies, with particular reference to Roman microscopic analysis. I. Deterioration of works of arts by *Dirina massiliensis* forma *soredita*. *Cryptogamic Botany*, **5**, 282–7.

Seaward, M.R.D. & Giacobini, C. (1989). Oxalate encrustation by the lichen *Dirina massiliensis* forma *sorediata* and its role in the deterioration of works of art. In *Le Pellicole ad Ossalati: origini e significato nella conservazione della opere d'arte*. pp. 409–14. Milan: Centro CNR.

Sheard, J.W. (1977). Palaeogeography, chemistry and taxonomy of the lichenized Ascomycetes *Dimelaena* and *Thamnolia*. *The Bryologist*, **80**, 100–18.

Sherwood-Pike, M.A. (1985). *Pelicothallus* Dilcher, an overlooked fossil lichen. *The Lichenologist*, **17**, 114–15.

Sipman, H.J.M. (1983). A monograph of the lichen family Megalosporaceae. *Bibliotheca Lichenologica*, **18**, 1–241.

Sipman, H.J.M. & Harris, R.C. (1989). Lichens. In *Tropical Rain Forest Ecosystems* (ed. H. Leith & M.J.A. Werger), pp. 303–9. Amsterdam; Elsevier.

Smith, D.C. & Douglas, A.E. (1987). *The Biology of Symbiosis*. London; Edward Arnold.

Sochting, U. & Gjelstrup, P. (1985). Lichen communities and associated fauna on a rocky sea shore on Bornholm in the Baltic. *Holarctic Ecology*, **8**, 66–75.

Sonesson, M., Osborne, C. & Sandberg, G. (1994). Epiphytic lichens as indicators of snow depth. *Arctic and Alpine Research*, **26**, 159–65.

Soppela, P., Nieminen, M. & Saarela, S. (1992). Water intake and its thermal energy cost in reindeer fed lichen or various protein rations during winter. *Acta Physiologica Scandinavica*, **145**, 65–73.

Stein, W.E., Harmon, G.D. & Hueber, F.M. (1993). *Spongiophyton* from the Lower Devonian of North America reinterpreted as a lichen. *American Journal of Botany*, suppl., **80**, 93.

Syers, J.K. & Iskander, I.K. (1973). Pedogenetic significance of lichens. In *The Lichens* (ed. V. Ahmadjian & M.E. Hale), pp. 225–248. New York; Academic Press.

Taylor, T.N. (1994). The fossil history of Ascomycetes. In *Ascomycete Systematics: Problems and Perspectives in the Nineties* (ed. D.L. Hawksworth), pp. 167–74. New York; Plenum Press.

Tehler, A. (1983). The genera *Dirina* and *Roccellina* (Roccellaceae). *Opera Botanica*, **70**, 1–86.

Tibell, L. (1992). Crustose lichens as indicators of forest continuity in boreal coniferous forests. *Nordic Journal of Botany*, **12**, 427–50.

Tibell, L. & Gibson, C.J. (1986). Bower decoration with *Usnea* species in the golden bower bird. *The Lichenologist*, **18**, 95–6.

Virtala, M. (1992). Optimal harvesting of a plant–herbivore system: lichen and reindeer in northern Finland. *Ecological Modelling*, **60**, 233–55.

Wein, R.W. & Speer, J.E. (1975). Lichen biomass in Acadian and boreal forests of Cape Breton Island, Nova Scotia. *The Bryologist*, **78**, 328–33.

Wessels, D.C.J. & Wessels, L.A. (1991). Erosion of biogenically weathered Clarens sandstone by lichenophagous bagworm larvae (Lepidoptera; Psychidae). *The Lichenologist*, **23**, 283–91.

Westman, L. (1973). Notes on the taxonomy and ecology of an arctic lichen: *Lecanora symmicta* var. *sorediosa* Westm. *The Lichenologist*, **5**, 457–60.

Wetmore, C.M. (1982). Lichen decomposition in a black spruce bog. *The Lichenologist*, **14**, 267–71.

Wielgolaski, F.E. (1975). Functioning of Fennoscandian tundra ecosystems. In *Fennoscandian Tundra Ecosystems. Part 2. Animals and Systems Analysis* (ed. F.E. Wielgolaski), pp. 300–26. Berlin; Springer-Verlag.

Williams, M.E., Rudolph, E.D., Schofield, E.A. & Prasher, D.C. (1978). The role of lichens in the structure, productivity, and mineral cycling of the wet coastal Alaskan tundra. In *Vegetation and Production Ecology of an Alaskan Arctic Tundra* (ed. L.L. Tieszen), pp. 185–206. New York; Springer-Verlag.

Yamamoto, Y., Miura, Y., Higuchi, M., Kinoshita, Y. & Yoshimura, I. (1993). Using lichen tissue cultures in modern biology. *The Bryologist*, **96**, 384–93.

Ziegler, R. (1991). Komplex-thallose, fossile Organismen mit blattflechtenartigem Bau aus dem mittleren Keuper (Trias, Karn) Unterfrankens. In *Palaeovegetational Development in Europe* (ed. J. Kovar-Eder), pp. 341–9. Vienna; PEPC.

12
Recording and mapping fungi

D.W. MINTER

Introduction

Over the last century, the BMS has held over 1000 residential forays and one-day field meetings during which many hundred thousand individual observations have been made of fungi, of when, where and on what they occur. No one knows the full extent of those observations: humans have a wonderful capacity to take in information, and a single walk in the countryside may provide vast amounts of data of which the proportion transmitted to storage outside of the human mind, in earlier days to paper, and now to computer, is inevitably only a small part of the whole: the collective consciousness of our Society's membership forms one of its great assets. Nevertheless, even this small proportion stored outside of human memory forms an important contribution to the international effort to describe fungi and note their occurrence and distribution. The Society's centenary provides an opportunity to reflect on that contribution, to assess it within the context of the present interest in biodiversity, to evaluate its importance, and to make plans to ensure that it not only continues through the Society's next century, but is also refined and kept up to date.

Up to about 1970, virtually all recording of living organisms was done on paper. Since then, and particularly since the early 1980s, there has been a huge trend towards using computers. At the same time, the amount of recording being carried out has soared: mycology has been no exception, and there the BMS has led. By the end of the century, it seems likely that virtually all biological recording will use computer technology. During the last 15 years there has been a great debate on how to cope with all this machine-readable information: how to generate, receive, store, output and otherwise share it.

This chapter looks at some of these questions from the point of view of mycologists and their fungi. In doing so, it is important to recognize that data handling technologies are currently evolving at a phenomenal rate (it has been said that you should throw out everything you've learned about databases every two years and start again), so questions of what computer hardware and commercial software should be used, being ephemeral, will rarely be addressed in this chapter, nor will back-up regimes or computer virus protection systems, nor the wonderful but rapidly changing possibilities being opened up by information superhighways.

The great benefits of having computerized information are rapid searching, mechanical manipulation, and easy duplication of data. These benefits, however, are by no means guaranteed (nor are they always necessarily benefits). The adage 'garbage in, garbage out' will always be true: if the information on disc is incorrect, or has been stored in a poor structure, it may be even less accessible for searching than its paper equivalent, mechanical manipulation may be impossible, and the ease with which it is duplicated may turn out to be a two-edged and very sharp sword! Until artificial intelligence truly arrives and computers can make allowances for such defects, we have to rely on man's limited albeit natural abilities to avoid them.

It is thus very important to recognize the constraints under which late twentieth century computers work and make sure our recording systems function correctly within those constraints. The chapter will accordingly concentrate on the rather less rapidly changing practicalities of gathering mycological information, editing it and using it for output. Practical problems in setting up feeder databases for gathering data will be briefly examined, and when output is considered, it should be obvious that creation of distribution maps, while fascinating and useful, constitutes only a small facet of what it is possible to produce from a properly-structured system.

In particular, though, the first, and largest part of the chapter will be devoted to the question of what information we are collecting, and how it should be structured. There has been a lot of work published by various groups and individuals, including TADWG (the Taxonomic Databases Working Group), IOPI (the International Organization for Plant Information), ERIN (the Environmental Resources Information Network), MINE (the Microbial Information Network for Europe), but much of this work is in a form not readily accessible to, or easily assimilated by the field biologist, and almost none of it is presented from the point of view of mycology.

The meat of this first part comes from hitherto unpublished work accumulated over the last ten years at the IMI, designing, constructing and using databases for recording fungi and data associated with them. The need for scores of tables and hundreds of fields has been recognized at IMI and elsewhere, and the task of gathering data and allocating its elements into such a structure is daunting. At IMI, this work is in progress, but by no means complete: many fields, and sometimes even whole tables of data are not in use, or are only in use by one or two individuals within the Institute, some still experimentally. Many others, though, are in daily and successful operation.

In this review, only tables and fields of interest in biological recording (in a broad sense) will be considered in detail. For each a comparison will be made with the fields used by the BMS, and possible improvements suggested by recent experience will be discussed. In this way, some assessment will be made of the BMS's contribution to recording fungi. Lest this assessment should seem at times critical, I would like to point out that almost all the BMS database fields are virtually unchanged in design since I set them up as Foray Secretary in the mid-1980s. Any criticisms therefore reflect purely on myself and on the rate of change of my own perception of problems in recording by computer. A fundamental familiarity with relational database concepts of tables, fields and indexes, and a certain basic knowledge of computers is assumed.

General considerations

Major data groups

When an analysis is made of information used in recording fungi, seven major groups of data stand out. In this chapter, each of these groups is treated as a separate database. These databases are not exhaustive, and each one is actually a suite of different but inter-connected tables. The seven databases, all identified in the following paragraph, are each of roughly similar size and complexity. In the present chapter there is not room to review them all. Only the first database, being the most important from the point of view of 'field' recording, will be reviewed, and even then the review will concentrate only on those parts which are most relevant to foraying. The ways in which this database links with and relates to the other databases will, however, be indicated wherever appropriate.

The first database comprises information about observations on the occurrence of fungi and of other organisms associated with them in time and space, and about living or dried reference collections of these organisms. Although this database deals with not only fungi, but also plants and animals, and not just with collections, but with field observations not backed by voucher material, it will for convenience be referred to as the *Collections Database*. The second database comprises information about names of organisms and their taxonomic position. This will be referred to as the *Nomenclature and Taxonomy Database*. The third, the *Bibliography Database*, deals with relevant publications: books, journals, pamphlets and other printed material. The fourth, the *People Database*, contains information about relevant individuals. The fifth, the *Descriptions Database*, stores descriptions of fungi, while the sixth, the *Illustrations Database*, stores information about illustrations of fungi, and links to digitized illustrations. Lastly, there is the *Geography Database*, which stores data relating to different locations.

Of those seven databases, the BMS Foray Records Database and BMS/JNCC Database (the paper-based foray lists found in back-numbers of its publications and now computerized, and all the records of British fungi referred to in BMS publications, now being computerized) correspond roughly to the *Collections Database*. Through publication of various checklists over the years, the BMS has built up a significant but often ageing, and mainly paper-based collection of records on the nomenclature and taxonomy of British fungi. It also holds considerable bibliographic information, again mostly on paper. Its computerized membership records form a potentially valuable start (but only a start) to a database of its human resources for field mycology. Through the present joint project on the ascomycete flora of Britain, it is building up a computerized database of descriptions, to which must be added a vast but disorganized paper database of descriptions and illustrations within the back-numbers of its publications. Specialist geographical information is not held by the BMS, but can be imported.

Field names

All field names used in this chapter are identified by being placed within square brackets (e.g. [CloxAccouA]). To assist in keeping track of all of the fields of these tables, a convention used in IMI to name fields will be adopted: each field has a name which is unique ten characters long, and a standard unique short description less than one hundred characters long. In the present chapter, each field will be identified by its name in the heading preceding discussion of that field. The short description which accompanies that name in the heading is not the standard unique short description used at IMI, but is a short description tailored for the needs of the current chapter. The BMS employs no conventions for naming fields.

Field names in BMS databases vary in length, and tend to be a short description of the main contents.

The IMI convention for field names is structured: the first character identifies the database to which the field belongs ('C', collections; 'N', nomenclature and taxonomy; 'B', bibliography; 'P', people; 'D', descriptions; 'I', illustrations; 'G', geography), and is always upper case; the next three characters, all lower case letters or digits, identify the individual table containing that field within the database (e.g. 'Clox' identifies the *Locality Cross-reference Table* within the *Collections Database*); the next five characters (the first an upper case letter, the remaining four lower case letters, digits or the underline character '_') constitute a mnemonic of the function of the field; the last character, which is also upper case, indicates the data type to which the field belongs ('N', numeric; 'A', alphanumeric; 'D', date).

General data conventions

A note on some general conventions used for storage of data in the text fields of these databases may be useful. The first character of a text field is never upper case, unless it relates to a proper noun or would for some other reason be upper case if it were to appear in the middle of a normal text sentence. The last character of a text field is never a full stop, unless it relates to an abbreviation. In alphanumeric some fields which contain a text version of data which could alternatively have been expressed in numeric form, inadequate information is expressed using the character 'x' (thus, for example, the field [CpexSyearA] (year in which action began) should contain '18xx' for a collection known to have been made in the 19th century): for such fields, an alphanumeric format has the advantage that uncertainties and imprecisions of this type can be expressed. The standard shorter ASCII code values 32, 33 and 35–127 are used for text entry. ASCII value 34 (double inverted commas), not permitted because its presence in data disrupts output of comma-delimited ASCII files, is represented as the embedded typesetting command '<ic>'. ASCII values higher than 127 for accented characters are also used where available in the Hewlett Packard 'PC 8' symbol set.

Where not available, or where the keyboarder is unsure, these accented characters and other typesetting commands are embedded in the text within single chevrons (examples: 'a<breve>', 'u<circle over>', 'o<double acute>', 'e<hacek>', 'e<hook>', 'l<line through>', 'i<macron>', 'z<dot over>', '<italic>', '<roman>', '<bold>', '<light>', '<new para>' etc.).

Leading, trailing and duplicate spaces are automatically eliminated. Where information is in other alphabets, the national standards tend to use ASCII values higher than 127 for those alphabet characters, and those standards are followed, with switches to indicate the change in character set from and back to the Hewlett Packard 'PC 8' series (e.g. '<cyrillic>', '<greek>', '<latin>'). Although its practice is probably similar to what has just been described, the BMS appears not to have any clearly defined general data conventions.

Data problems

At first glance, the information involved in recording fungi does not seem too complex: like the lady from Khartoum in the limerick, all we seem to need to know is 'on what, where, when, and by whom?' A typical record might read: '*Ascodichaena rugosa*, on dead twig of beech, Chobham Common, Surrey, England, 8 October 1994, identified by M. Cooke'. Within that record, we can identify the different main data elements: the fungus name, the substratum, an associated organism, the exact locality, county and country, the data of observation, and name of the person identifying the fungus. The trouble is that experience quickly shows how each of these apparently simple elements can contain problems.

Fungi are organisms with unstable names and an unstable taxonomy: what is an acceptable name for a distinct species may, as opinion changes, become a synonym of a different species (quite possibly in a very different taxonomic group). It is not clear whether the substratum represents the material on which the fungus was growing, or its source of nutrition or both. The associated organism is referred to by a vernacular rather than a scientific name. Exact localities can change their names: the city of Milton Keynes didn't exist fifty years ago. So can counties: Avon, Cleveland and Gwynedd are all examples of county names created through local government re-arrangements over the last fifty years. Even countries can change: the USSR ceased to exist as a country in the early 1990s. The data of observation are usually straightforward, until, for example, records from Tsarist Russia (there are an awful lot!) are accessed: these were all generated using the unreformed calendar. Finally, the person making the observation may change their name following marriage. They may even have transsexual surgery, or an alter ego: examples of both are known in mycology!

It is clear from this brief analysis that information, even apparently simple and factual information is inherently unstable. As a result, a most important principle adopted in the design of all the databases described in this chapter is that, wherever possible, the original information for each record is preserved in the form it was received into the system. The corollary of this is that each database has to be designed with additional fields, and sometimes even whole tables to permit the expression of different opinions as to what that original information means, and to record the dates on which and the people by whom those opinions were expressed. In the *BMS Foray Records Database*, there is little protection of original data: only the name of the fungus can be stored in both an original and current form.

The Collections Database

This group of databases stores information about individual observations of organisms. The term 'observation' is interpreted in a wide sense. In addition to observations backed by physical material in a living or preserved reference collection, the term may also cover field observations and records derived from literature for which no physical material is available for examination. It is important to note that observations of unsuccessful searches are also stored in the *Collections Database*. The tables are structured so that observations of any type of living organism can be stored, and the ecological relationship between any individual organism and any other individual organism observed together can be noted. In the notes in the following two paragraphs, tables which contain fields of particular interest for foraying will be italicized, and the fields from those tables relevant to foraying will then be commented on.

Core information, including information for which indexed one-to-one links are required, is stored in the *Collections Core Table* ([Ccor......]). This table has fixed-length fields which are virtually all numeric pointers to other databases. Each record in the *Collections Core Table* represents a single observation of a single organism. Where several organisms are observed simultaneously in association with one another (for example a fungus on its host), a separate record is made in the *Collections Core Table* for each organism observed, but the records are linked through a common number in [CcorColnoN] (the *Collections Core Table* individual collection identifier). The *Collections Core Supplementary Table* ([Cco0......]) stores the remaining core information, particularly the longer textual information for which a variable-length field structure is more appropriate.

Information which requires indexing, and for which many-to-one links between the *Collections Core Table* and other tables are needed may be stored in the following series of cross-reference tables: *Collections Administration Table* ([Cadm......]); *Collections Bibliography Cross-reference Table*

([Cbix......]); Collections Description Cross-reference Table ([Cdex......]); Collections Illustration Cross-reference Table ([Cilx......]); Collections Culture Collection Table ([Ciso......]); Collections Culture Collection Maintenance Table ([Cisx......]); *Collections Locality Cross-reference Table* ([Clox......]); *Collections Other Collections Cross-reference Table* ([Coth......]); *Collections People Cross-reference Table* ([Cpex......]); Collections Properties Table ([Cpro......]); *Collections Substratum Cross-reference Table* ([Csux......]); Collections Taxonomists' Table ([Ctax......]); Collections Technicians' Table ([Ctec......]).

The principal uses of the *Collections Database* are to store information about fungi in living and dried reference collections, about occurrences of fungi reported in the literature and elsewhere, about observations of fungi backed by neither a literature reference nor a specimen, and about organisms associated with the fungi in all of those records. Information about unsuccessful searches is also stored here. This information can be used during curation of living and dried reference collections, administration of identification services, and can form part of standard catalogues and other publications, such as the *Index of Fungi, Bibliography of Systematic Mycology* and *Distribution Maps of Plant Diseases*. The information can also be incorporated in many research projects.

Because its design permits storage of data on collections of any organisms in any reference collection, and on observations arising from the literature and from foraying, and because its design permits these sources of data to be distinguished, this group of tables has important further uses in the production of a wide range of other scientific documents, for example host and country check lists. Examples of mechanically-produced publications deriving a large amount of their data from this group of tables are the checklists of fungi described by Batista and co-workers (Da Silva & Minter, 1995), and of fungi on *Eucalyptus* (Sankaran, Sutton & Minter, 1995). Furthermore, the tables have the potential for expansion to cover identification work on any group of living organisms.

A commentary on selected fields of the Collections Database

The Collections Core Tables (Ccor and Cco0)

A unique observation identifier, 8 characters, indexed
[CcorLink_N]

Each record in the *Collections Database* represents a single observation about a single taxon within one space of time and in one place. To ensure that each record can be identified separately, every record should be issued with a record number which is unique, stored in [CcorLink_N]. It is worth noting that, all over the world, many people are setting up lots of 'collections databases', and they are all starting their 'unique numbering' at the same point: number one. Sooner or later, there is going to be a need to amalgamate data from different sources, and when that happens, this multiplicity of so-called unique numbers is probably going to be a real problem. There are plenty of forward-lookers expressing concern about this. The BMS Databases at present do not have a properly-designated unique number of this type in any of their records. The BMS should consider restructuring its database to accommodate such numbers, and should participate in the setting up of a global observation numbering system.

A unique collection identifier, 8 characters, indexed [CcorColnoN]

The recording of higher plants, and of many animals is very frequently carried out with no regard for other organisms with which they might be associated: those who map the distribution of forest trees rarely bother to note in their databases the many fungi, insects or other smaller living things which live on, in or under these huge plants. Mycologists can be proud that they have done rather better for many years. Because fungi are heterotrophs, when observing a fungus it is normal to note at least one, and often several other organisms associated with it: phrases such as '*Amanita muscaria*, on soil beneath birch, with hazel and oak nearby', or '*Beauveria bassiana*, on a lepidopteran larva on *Morus* sp.' are common in the history of recording fungi.

The first computerized 'collections databases' designed for recording fungi simply copied the practice of old printed records. Each observation of a fungus was regarded as being a separate record, and information about associated organisms was treated as subsidiary data. There are undoubtedly a lot of advantages to this procedure from the point of view of easy gathering of data: mycologists tend to think in terms of the fungi they observe, and the associated organisms, while interesting, and even quite often important, are nevertheless merely an adjunct to the main fact that a fungus has been observed. Not surprisingly, the present BMS Databases reflect this viewpoint: within one record, there are fields for the fungus name and information associated with that fungus, and there are other fields for the name of an associated organism, with its associated information.

A little reflexion shows, however, that such a structure is very problematic for long-term storage of the data. In the first place, there is provision for information about only one associated organism, yet it is the norm that many organisms can live together in associations of many different kinds. During recent compilation at IMI of the records of microfungi from Brazil made by Batista and co-workers, a very large proportion of observations related to multiple groups of organisms, to fungi, algae, higher plants and animals all associated with each other in various ways. A structure which permits each record to have only one fungus and one associated organism simply cannot express such complexities.

The second problem is that in a significant number of records, those dealing with fungi parasitic on other fungi, one of the two fungi must be an associated organism if the structures currently used by the BMS are employed. The person generating a record of *Eudarluca* on a rust, for example, will put the rust as the fungus and the *Eudarluca* as the associated organism, or the other way round, depending on their particular mycological point of view. The result is that, to find all records of *Eudarluca* by a mechanical search, not only the fungus fields, but also the associated organism fields have to be scanned.

The third problem is that this simple structure encourages those generating records to treat information relating to associated organisms as, in some way, second-class data. Thus there is no provision within the BMS Foray Records Database to note down who identified the oak, or when they identified it, and you cannot record how many oaks were observed, nor in what condition they were. This objection may seem trivial to recorders in Britain ('all British field biologists know what an oak looks like'), but on a global scale it starts to look important. If British records are to be used by a researcher in the tropics who is unfamiliar with the genus *Quercus*, information about who identified this host could be important, and British researchers would similarly appreciate knowing who identified the *Acacia* in mycological data coming from Africa.

The fourth difficulty is that, by treating associated organism information as second class data, our databases are not functioning efficiently. During a year of collecting data on fungi of Ukraine, of the 24 000 records gathered, over 23 000 contained floristic information about higher plants of Ukraine. Properly structured, our mycological records are also a vast resource of data on the occurrence, distribution, ecological preferences and associations of many other organisms.

The solution to this problem is to recognize that, for long-term storage of data, each fungus and each associated organism observed actually represent separate floristic records. Thus, if *Polyporus squamosus* is observed on sycamore, two records should be generated in the *Collections*

Database, one for *Polyporus squamosus*, and the other for sycamore. If *Eudarluca* is recorded on a rust on *Picea*, then three records should similarly be generated. If 38 different insects, twelve fungi, three nematodes, and two spiders are observed on the leaf of one plant, 56 records should be generated: with such a solution, possibilities for recording relationships become open-ended. For the BMS Databases to achieve this, major restructuring will be necessary.

To link this cluster of different related observations made at the same time, a unique collection identifier is needed ('collection' is not used here in the sense of 'herbarium collection'). This should be a number, stored in [CcorColnoN]. The BMS Foray Records Database and the BMS/JNCC Database do issue such numbers ([BMS Accession Number] and [BMSFRD Record number], respectively), since each record in their structures usually comprises a collection of two different observations made at the same time, one for the fungus, the other for the associated organism, but at present that database makes no distinction between the unique observation identifier and the unique collection identifier.

Treating different observations made at the same time and place as different records in the *Collections Database* has implications for information about substratum and ecological relationships. That is another area where the structures of the BMS Databases are inadequate for properly structured long-term storage of information. These implications will be considered later.

A unique individual identifier, 8 characters [CcorIndivN]

According to the *Guinness Book of Records*, the largest individual fungus fruitbody in the world is to be found in Kew Gardens, in what used to be the grounds of IMI. It is regularly and proudly but anxiously measured. If that information were to be recorded in database format, its occurrence in time and space would be floristic information, and its size would be descriptional, but for the sequence of observations to be meaningful, there would have to be some way of noting that they all came from the same individual. The present field provides the opportunity to issue an individual organism with a number which is unique, so that linked sequential observations on that individual can be stored in the *Collections Database* (and by linkage in the *Descriptions Database* and the *Illustrations Database*).

What constitutes an individual in the fungal world is rarely so straightforward as the case just considered, and doubtless some of us would quite reasonably argue that on a genetic basis even that huge fruitbody represented a population rather than an individual. But the *Collections Database* deals not just with fungi, but with any organism, and in particular with organisms associated with fungi. It means that this field can be used, for example, to define a particular tree from which regular samples are being taken, and the presence of this field opens the door to researchers wishing to use the *Collections Database* as a vehicle for sequential observations of many different sorts. The BMS Foray Records Database and the BMS/JNCC Database do not have this field, but could easily have a need for it if, for example, the database were to be used as part of a strategy for conserving rare polypores.

A flag to mark an unsuccessful search, 1 character [CcorAbestA]

As already noted, each record in the *Collections Database* represents a single observation about a single taxon at one time and in one place. It is human nature to remember discoveries, and quietly forget unsuccessful searches: not surprisingly, the default condition of records in the *Collections Database* indicates a successful discovery. In recording fungi, this seems particularly reasonable when it is remembered that most records relate to larger basidiomycetes and, in particular the Agaricales, where owing to their often ephemeral and unpredictable fruiting, serendipity is often the largest single factor in generating a record: 'when I go on a foray, there are thousands of toadstools I don't see! You don't seriously expect me to record them?' Of course not.

On the other hand, an unsuccessful search at the right season of a site where *Cantharellus* has regularly been recorded for many years, but has been becoming less abundant could be significant. There is growing concern in European mycology about the decline of larger basidiomycetes as a result of pollution. How can the unwelcome but very real extinction aspects of that decline be recorded except through unsuccessful searches? And how is such information to be stored?

Besides, larger basidiomycetes are not the only fungi, and even of them not many produce long-lived fruitbodies. Among the numerically far larger ascomycetes, very many produce identifiable colonies, stromata, thalli and fruitbodies which are long-persisting. For these, serendipity is a far smaller factor: if the correct associated organism or other substratum is present, the recording of an unsuccessful search becomes all the more meaningful.

Without an ability to record unsuccessful searches, our recording system remains open to the criticism that its distribution maps merely show where mycologists have been. Without it, there are no means to record the often considerable time spent looking for fungi in marginal habitats. Without it, there is no evidence that a fungus which appears to be increasing its range was not simply unnoticed before. Without it, there is no evidence that an apparent decline is not merely due to lack of attention from present-day mycologists. To record an unsuccessful search is not the same as saying the fungus is absent, but it is meaningful, and often important. IMI holds several hundred records of unsuccessful searches. Neither BMS database has provision for recording unsuccessful searches.

The original name of the organism, 100 characters [CcoOOrinaA]

At the heart of any *Collections Database* record is the statement 'an organism was observed'. For each record generated, the *Collections Database* needs to store information about the identity of that organism. At first sight, the information might seem to comprise a name and nothing else, but things are not that simple. One organism can have many different names, some vernacular and others scientific (e.g. 'raspberry' and '*Rubus idaeus*'). In English there are many vernacular names for the same organism ('lapwing', 'peewit', 'green plover'), and English is not the only language in the world nor is the Latin alphabet unique! Scientific or quasi-scientific names may be used (e.g. '*Fuchsia*' or 'fuchsia'), and it may be difficult to distinguish which is meant. Often the person generating the information will have used such a name unaware of the potential problem.

Even if the name received is scientific, it may not be the only name in use for the organism (e.g. '*Rhytisma acerinum*' and '*Melasmia acerina*' both quite correctly refer to the same fungus), or its meaning may not be clear ('*Euphrasia officinalis*', '*Rubus fruticosus*' and '*Taraxacum officinale*' are three examples of binomials used for species aggregates), or the name may be used in different senses ('*sensu lato*', '*sensu stricto*' etc.), and again the person supplying the information may be unaware of such subtleties.

Schemes designed to collect fresh records try to get round this problem by encouraging participants to supply only correct scientific names. This certainly helps to ensure that incoming records meet certain basic standards, but even such schemes can only compel by imposing the artificial constraint of a monolithic taxonomy on their participants. This is undesirable in that the nuances of different opinions themselves comprise a valuable part of many records. In any case, sooner or later all schemes start to become interested in the question of assimilating the vast numbers of floristic records which were generated years before such standards were erected.

The result of all of this is that, when designing a *Collections Database*, it is necessary to recognize that those who curate the data have little control

over incoming information until it has been received. It is therefore necessary to have a field in the *Collections Database* devoted to the organism
name as it was received [Cco0OrinaA]. That field contains the primary
source of data on the identity of the organism to which the current record
relates, and it should not be subject to editorial control. Both BMS databases have a field for this information, [Name of Fungus] (though at 40
characters it may be a little short), and another field of 30 characters
([Associated Organism] in the Foray Records Database, and [Associated
with] in the BMS/JNCC Database) to store the original name of the associated organism, data which on restructuring would be placed in
[Cco0OrinaA]. In the IMI system, this field is located in the *Collections
Core Subsidiary Table* ([Cco0......]) rather than the *Collections Core Table*
([Ccor......]) because it contains text of variable length.

A text link to the currently accepted organism name, 100 characters, indexed [Cco0AccnaA]

You cannot easily use [Cco0OrinaA] to search for records relating to a particular organism, however, because the information it contains does not
conform to any editorial standards. For that job, a different field is needed,
which stores an editorial opinion on the identity of the current organism.
In its most simple form, this field contains the accepted scientific name of
the organism as a piece of text. If the field is indexed, searching for records
relating to a particular organism becomes a simple matter. Nevertheless,
such a solution is problematic, because of homonyms ('*Hypoderma*' is an
ascomycete and an insect, '*Oenanthe*' is a flower and a bird, '*Hypoderma
brachysporum* Speg.' refers to a different organism from '*Hypoderma
brachsporum* Rostr.'). Names with accented characters ('*Naïs*', '*Elsinoë*',
'*Oïdium*', etc.) present a further problem: a decision has to be made about
whether or not to use these special characters and after that it is very difficult in practice to ensure that keyboarders remember to adhere to the decision.

A more satisfactory solution is to have a field which makes a link
between that *Collections Database* record and the unique and correct
record in the *Nomenclature and Taxonomy Database*. Through such a link
field the user can be provided with access not only to the correct name,
correctly spelt, without confusions over homonyms, but also to a lot of
other information relating to that name and its use, e.g. its authors, the
date and place of publication, its status and its taxonomic position.

There are two options for such a link field. The first is to make the link
using a unique piece of text based on the scientific name of the organism.

Since most scientific names are not homonyms, and do not contain accented characters, almost all of the links will be text comprising the scientific name in an unedited form, and nothing else. In the case of scientific names containing accented characters, the edited form of the name used at IMI to make the link is simply the scientific name with the accent removed from the character (*'Elsinoë'* for example becomes *'Elsinoe'*). For names where the rank is normally indicated, this rank is included in the link data (e.g. *'Entoloma hirtipes* forma *bisporicum'*). In the case of homonyms, the link comprises the scientific name (minus any accents) plus the name(s) of its author(s) in the standard abbreviated form prescribed by Brummitt & Powell (1992) or, for fungi, the subset of those data provided by Kirk & Ansell (1992). Thus *'Hypoderma brachysporum* Speg.' is the link for this fungus, to distinguish it from *Hypoderma brachysporum* Rostr. In the very rare case of homonyms described by the same author(s), the year of publication and, if necessary, a further distinguishing factor are added. In the case of links at ranks higher than species, only the scientific name is used, without the addition of 'sp.', 'gen. indet.' or similar words.

[CcorAccnaA] is the field used at IMI for this link information. At IMI a large number of *Collections Database* records are linked to the *Nomenclature and Taxonomy Database* using this option. The BMS has no Nomenclature and Taxonomy Database to accompany its Foray Records Database (nor should it consider duplicating the efforts of IMI and other bodies to set up such a database), but it would be a comparatively easy task to check and edit [Current Name] (the analogous field, 40 characters long, in the BMS databases) so that it conformed to the link standards used at IMI (terms like 'sp.' would have to be removed). There is no field in the BMS Foray Records Database to store the accepted name of an associated organism, though if the necessity of the major restructuring already pointed out is faced, such a place to store that information would become available. If the BMS databases had data conforming to the link standards in use at IMI, the 2 character long field [Order Code] would become superfluous. In the IMI system, [Cco0AccnaA] is located in the *Collections Core Subsidiary Table* ([Cco0......]) rather than the *Collections Core Table* ([Ccor......]) because it contains text of variable length.

A numeric link to the currently accepted organism name, 8 characters indexed *[CcorAcclkN]*

The second option for a link field is to allocate a unique number to each record in the *Nomenclature and Taxonomy Database*, and to use that number to identify the currently accepted organism name. Provision for this option exists within the IMI database structure, but at present it remains unused. It may be appropriate now to consider the relative advantages and disadvantages of the two options.

The advantage of the numeric option is that the link information is compact, and of a uniform size for all records. Since computers are, ultimately, number-crunching machines, having link information in the form of, say, eight bytes of binary data is simpler for the computer to process, and the presentation of data derived from links is likely to be faster. Since the database software is likely to reserve the same number of bytes for this link in all records, it follows that if the link is changed (perhaps a different name has become the accepted one), the change in the data is unlikely to entail a restructuring of the location of the data on the hard disc. The test option is disadvantageous for precisely the same reason.

The big disadvantage of the numeric option is that it is hard for humans to use. A table composed purely of numbers will never be so easy to use as a table containing data which is meaningful to humans. '*Fagus sylvatica*' has a familiarity to the field biologist which '0009238' will probably never have! Of course it is possible to devise systems for data-entry whereby the user never comes into contact with such numbers, but these systems add a whole tier of complexity to the interactive software providing the user editing and viewing facilities, and in any case, as anyone who has had to deal with corrupted data will confirm, it is far easier to identify that there is something wrong with 'Fagxs syxvatxxa' and to guess what it originally meant than it is to do the same for '0019238'!

With [Cco0OrinaA] and [Cco0AccnaA] or [CcorAcclkN] it is possible to store the primary data, and the present editorial opinion of the identity of the current organism for any *Collections Database* record. The separate storing of these two items has a particular advantage when dealing with organisms with unstable names, in that it is possible to change the current opinion without destroying the original data. It is analogous to drawing a polite pencil line through the original identification on a herbarium packet, and writing your own opinion below. Of course, it may happen that a particular record is controversial, and different scientists may identify the organism differently. Provision for such events is considered later in the *Collections People Cross-reference Table* ([Cpex......]).

A flag to mark fossil records, 1 character [Cco0Foss_A]

Fossil fungi have occupied a dark corner in the past of traditional mycology! Records are infrequent, and compilations have generally treated them separately from those of living fungi. Even the names used for genera have tended to be different, '*Rhytismites*' for example being used for fossils looking like present-day species of *Rhytisma*. This particular habit of traditional mycology does not fit well with other biological disciplines and, since the *Collections Database* deals with records of all organisms, it needs to be scrutinized.

It is well known that taxa originally known as living organisms can also be found as fossils. The examples of the *Ginkgo* and coelocanth, among others, show that the reverse can also be true: taxa originally known as fossils can subsequently be discovered alive. It follows that the practice in mycology of using different scientific names for fossils, just because they are fossils, is suspect. There is no such thing as an intrinsically fossil genus: only individual records in the *Collections Database* can be labelled as fossil or non-fossil.

[Cco0Foss_A] was introduced to deal with this possibility. The default condition of an empty field indicates that the current observation does not relate to a fossil. Each year a few fungal taxa are described with fossils as their types, and this field is used at IMI to distinguish them. Neither BMS database makes provision for such a field.

A text description of the developmental state of the observed organism, 200 characters [Cco0DevtdA]

Organisms are encountered in different developmental states. Fungi are probably no less complex than most in this respect: from symptomless occurrence as, for example, an endophyte, through the symptoms of a plant or animal pathogen, to mycelial features such as stromatic lines, then one or more different anamorphs, and finally a teleomorph, different states can occur, often in more than one combination. This field provides the opportunity for the developmental state of the organism observed to be noted in free text form. It is particularly valuable as a place to store original information about developmental states. Neither BMS database provides this field.

An encoded description of the developmental state of the observed organism, 20 characters [Cco0DevtcA]

This field permits the storing of a standard or editorial opinion of what developmental states were observed. Because the contents of this field are structured, it can be used for mechanical searching and manipulation in a way not possible with [Cco0DevtdA]. There seems to be no universally agreed coding system for recording developmental states in the fungi, and only with the rusts is any system widely used. The BMS Foray Records

Database and the BMS/JNCC Database each have a field 1 character in length ([Morph Code] and [Morph] respectively), which permit the keyboarder to note whether the anamorph ('a'), teleomorph ('t') or holomorph ('h') was present, but that system seems rather simplistic.

At IMI, for rusts, the codes commonly used within that group to note developmental states are used. For ascomycetes, other basidiomycetes and zygomycetes, the following simple annotation is employed: symptomless occurrence ('a'), symptoms ('b'), mycelial features ('c'), anamorph ('d'), and teleomorph ('e'); after 'd' and 'e' three suffixes may be used: not yet sporulating ('1'), sporulating ('2'), and sporulation finished ('3'). Thus 'cd3e2' would indicate that a mycelial feature (for example a stroma), an anamorph, and a teleomorph had all been observed, and that the anamorph appeared to be effete, while the teleomorph was actively sporulating.

It is clear that this code also has deficiencies, particularly for fungi with more than one anamorph. The BMS should recognize that its own field for storing this information is inadequate, and should help to devise either a more effective code which has universal use in the fungi, or a series of codes which have effective application within individual groups of fungi.

A text description of the abundance of the observed organism, 200 characters [Cco0AbundA]

Many observations of fungi are accompanied by a comment by the observer on abundance. These comments are almost invariably subjective ('rare', 'very rare', 'common' etc.), but even in this condition they have a value worth preserving. [Cco0AbundA] provides a location to store these comments in free text form. Like [Cco0DevtdA] this field is particularly valuable as a place to store original information. Neither BMS database provides this field.

An encoded description of the abundance of the observed organism, 2 characters [Cco0AbuncA]

This field permits the storing of a standard or editorial opinion of the abundance of the observed organism. Because the contents of this field are structured, it can be used for mechanical searching and manipulation in a way not possible with [Cco0AbundA]. There seems to be no universally agreed coding system for recording abundance in the fungi, and the subject seems to be rather controversial. An often-quoted criticism of abundance recording in the fungi is that the observation of, say, 50 basidiocarps may genetically represent the observation of only one individual united by mycelium beneath the soil. To get round this problem, the first step is to ensure that anyone scanning the data can see on what basis the assessment of abundance was made.

To do that, the following main factors need to be noted: what was observed (ascocarps, basidiocarps, colonies, conidiophores, pycnidia, stromata, symptoms, zone lines, etc.), how many of each item were observed, and how long it took to observe that number. Thus, the user of the data can distinguish the abundance assessment of a recorder who observed fifty toadstools, from another who observed one colony. Since what was observed is already (at least in potential) the subject of the field for recording developmental stages in code, it remains for the present field to record the number of items observed, and the time taken in a similar codified format. In doing so, the practicalities of gathering data in the field need to be taken into account.

At IMI I use a system experimentally which requires two characters of data. The first is a score from zero to five representing the base 10 logarithm of the approximate number observed ('0', no items observed; '1', 1 to 9; '2', 10 to 99; '3', 100 to 999; '4', 1000 to 9999; '5', more than 10000). The second represents the time taken to make that observation ('1', less than 1 minute; '2', 1 to 5 minutes; '3', 5 to 15 minutes; '4' 15 to 30 minutes; '5', more than 30 minutes). Since chance is a major factor in observing many fungi, particularly those with ephemeral and unpredictable fruiting, a further option is permitted for the second character: 's' denotes a serendipitous observation. The BMS Foray Records Database and the BMS/JNCC Database both have a field 1 character in length [Abundance] which is not yet in use. The idea of using it to record the base 3 logarithm of the approximate number observed has been discussed, but there is no provision to record the time taken to make that observation, and there is no explicit relationship between the information in that field and information about what was observed.

Neither of these two systems is ideal. The system used at IMI represents an adaptation which favours recording microfungi, where far larger numbers of items are likely to be encountered. The system discussed by the BMS favours the recording of larger fungi, with the consequent greater numeric precision at the lower end. In cases where, say, an anamorph and a teleomorph are both recorded, neither system can distinguish the relative amounts of each state. The BMS should recognize that its own field for storing this information is inadequate, and should help to devise either a more effective code which has universal use in the fungi, or a series of codes which have effective application within individual groups of fungi, making sure that such a code distinguishes between deliberate searches and serendipitous discoveries.

Notes relating to the current record, but not intended for publication, 2000 characters [Cco0InnotA]

It is convenient to have space allocated for users to make notes associated with each record. Not all notes, however, are of equal interest to all. In particular, it is often convenient for the user to have a scratchpad in which ephemeral notes (e.g. 'remember to check the spelling of Krets[z?]chmaria') can be placed, without worrying that your orthographically-challenged condition will be exposed to the ridicule of the scientific world! This field serves that purpose.

Notes relating to the current record, and intended for publication, 2000 characters [Cco0ExnotA]

By comparison, there are other notes which quite approximately should be printed out with the record when it is published (e.g. 'apparently the first record from Shropshire'). [Cco0ExnotA] provides that avenue. The field may be entered in free text, but each separate note within the field must be attributed to a source, and where possible a date must be given. Thus, 'apparently the first record from Shropshire (B. Ing, 27/03/1992)' would be an appropriate format. When deriving records from reference collection packets the source can be recorded as '(packet)' without a date. The BMS Foray Records Database and the BMS/JNCC Database each provide one note field [Notes], 250 characters long, without distinguishing between private and publishable information, and no organized attempt is made to record the source of each note.

A flag to mark a foray record, 1 character [Cco0ForayA]

Some Collections records are generated during forays and field meetings of biological societies. [Cco0ForayA] flags such a source. The upper-case letter 'F' is used to indicate records derived from BMS forays. Both BMS databases include a 1 character field [Foray Record] with the same data specifications.

The Collections Locality Cross-reference Table (Clox)

To the average field biologist, only one location is important in recording, and that is the location where the organism was observed. Unfortunately and particularly for fungi other locations may also be important. A fungus may be collected in Borneo, isolated into pure culture in Japan, intercepted by quarantine officials in Australia, the USA and Argentina, and stored in a culture collection in the UK. The fungus has travelled in a living condition to each of these locations, albeit sometimes with the help of that useful symbiont *Homo sapiens* and very often the only locality information available will be that of place of isolation or place of quarantine intercept. It is thus necessary to recognize a many-to-one relationship can exist between relevant locations and an individual *Collections Database* observation. The *Collections Locality Cross-reference Table* permits this many-to-one relationship to be expressed. The BMS databases permit only one location, that of collection, to be recorded for each record, thus enabling them to remain flat-field rather than relational database systems. For their purpose as data-entry or feeder systems, this is adequate, but the need to store resulting information in a proper relational structure should now be recognized by the BMS.

A link to the unique observation identifier, 8 characters indexed [CloxLink_N]

This field stores the numeric link between the *Collections Locality Cross-reference Table* and other tables in the *Collections Database*.

A unique locality identifier, 8 characters indexed [CloxLocnoN]

This field stores a number which is unique for each record in the *Collections Locality Cross-reference Table*. The field is useful for example in cases where more than one quarantine intercept has been made of the current material, since it permits an unambiguous link between the locality of each quarantine intercept and records in the *Collections People Cross-reference Table* with information about who made the interception, and when.

A link to the Geography Database, 8 characters [CloxLoclkN]

This field permits a numeric link for the current locality with the *Geography Database*, to provide access to, for example the boundaries of the area referred to (e.g. 'Rocky Mountains', 'Westerness', etc.), temperal limits to those boundaries (e.g. 'USSR, 1945–1991') and other centralized information on that locality. It is not yet in use.

The category of location, 1 character [CloxCategA]

One character is allocated to this field, to record the category into which the current location falls. At present three categories are in use: 'c', collection;

'i', isolation; 'q', quarantine intercept. Others will be made available as the need for them becomes apparent. Because the BMS databases are flat-field databases, they have no need for this field. All localities in those databases are assumed to belong in the 'c' category. As the BMS keyboards data relating to British fungi from its various publications, it will inevitably encounter records of fungi isolated rather than directly collected, and problems in distinguishing these different categories of records will then arise: you have been warned!

Latitude, 7 characters [CloxLat_A], and longitude, 8 characters [CloxLong_A]

To locate any locality associated with a fungus on a map, it is necessary to have co-ordinates which are as accurate as possible. The cardinal co-ordinates of latitude and longitude, based on the Greenwich meridian, are in almost world-wide use, and information on these co-ordinates should be added to *Collections Database* records wherever possible. [CloxLat_A] is the first of two fields storing this information. It contains information about the latitude of collection in the following format: the first character must be either 'N' or 'S', and identifies the northern or southern hemisphere respectively; characters two and three are reserved for the number of degrees of latitude ('00' to '90'); characters four and five are similarly reserved for the number of minutes ('00' to '59'); characters six and seven are reserved for the number of seconds ('00' to '59').

Information should be entered into this field to the greatest possible accuracy: it should be noted that for example 'N04' implies four degrees north while 'N4' implies somewhere between forty and fifty degrees north. The addition of 'x' to indicate null data (e.g. 'N45xxxx') is also permissible. For making maps, the user will then be able to define the permissible level of accuracy for the current map (e.g. only records accurate to more than ten minutes), and will be able to instruct the computer how to deal with less accurate data (e.g. place dot on the centre of a notional square, or place dot at intersect of lowest common denominator coordinate etc.). [CloxLong_A] is the second of these two fields, and stores the longitude, following similar rules to [CloxLat_A] except that hemispheres are defined as 'E' and 'W', and three characters are reserved for degrees of longitude ('000' to '180'). As the technology for recording exact position by satellite becomes more available, the provision of accuracy only down to the nearest second may become inadequate, and the sizes of these two fields may need to be revised.

The BMS databases have no provision for this information. Most foray records are generated within the British Isles, for which the UK and Irish National Grids are available for mapping. The relationship between these grids and latitude and longitude is mathematical, but rather complex, so for British foray records, the absence of latitude and longitude data may be rectifiable by machine. It seems a shame, nevertheless, that forayers are not being given the opportunity to record latitude and longitude information which is so readily available from Ordnance Survey maps. For foray records held outside Britain, which now number several thousand, the absence of provision to store this information is more serious.

The national grid identifiers, 2 characters, 3 characters and 3 characters, respectively [CloxGridlA] [CloxGrideA] and [CloxGridnA]

Many countries employ a national grid to provide an easy system of identifying locations. The UK and the Irish national grids, and those of several other countries, particularly those with a British tradition of surveying (e.g. Zimbabwe), have similar formats. These three fields enable locality information to be stored in these formats, as accurately as possible, thereby facilitating distribution map production specific for those countries. Two characters are allocated to [CloxGridlA], and they identify the relevant 100 km square of the grid. Three characters are allocated to [CloxGrideA] to enable eastings to be identified to within 100 m. Three characters are allocated to [CloxGridnA] to enable northings to be identified to within 100 m. Consult mapping systems of individual countries for further information. The BMS Foray Records Database and the BMS/JNCC Database provide all three fields to the same specifications ([Grid Ref A], [Grid Ref B], [Grid Ref C] and [Grid reference A], [Grid reference B], [Grid reference C] respectively).

Vice-county (for records from the British Isles only), 3 characters [CloxVc_A]

The British vice-counties are a system devised long ago to divide the British Isles into subnational units for biological recording. Although regarded by many as now archaic, they are still in widespread use among field biologists. This field enables the vice-county of the current record to be noted in the form of the conventional number allocated to that vice-county. The BMS Foray Records Database and the BMS/JNCC Database provide this field to the same specifications ([Vice-County] and [VC], respectively).

A free text description of altitude, 200 characters [CloxAlt_A]

The altitude at which an organism occurred is often of interest from an ecological point of view. Many older collections of fungi contain information about altitude on the packet, and similar data can be encountered in many records in the literature. This older information is often unstructured, or uses measurement systems such as the imperial which are no longer acceptable to science internationally. The present field permits the storage of this original information. The BMS databases provide no equivalent. As records of fungi in literature published by the BMS are computerized, the absence of this field may be an occasional problem.

Structured fields describing altitude spread, 4 characters each [CloxMinalA] and [CloxMaxalA]

These two fields are allocated to store a structured statement or an editorial opinion about the altitude at which the observation was made. In practice, these observations are often made over a range of altitudes, and the provision of two fields enables that range to be reflected. Where the source of information provides only one figure, or where the observation was made at only one altitude, both fields will contain the same information.

Each field contains data representing the number of metres above or below sea-level at which the observation was made. The size of the fields permit observations between –999 and 9999. [CloxMinalA] contains the minimum altitude of the observation: [CloxMaxalA] contains the maximum. The figure in [CloxMinalA] must never be greater than that in [CloxMaxalA]. Where the altitude is not known to the nearest metre, it is possible to enter null data to indicate a level of uncertainty, thus '11xx' would indicate that the collection had been made between 1100 and 1199 metres.

The BMS Foray Records Database and the BMS/JNCC Database each provide one field for altitude ([Altitude] and [Altitude (m)], respectively). For a flat-field database accumulating only new data, and with high quality maps generally available to the field mycologists supplying the data, this is unlikely to be problematic.

Original locality information: [CloxOrmajA], 80 characters; [CloxOrcouA], 80 characters; [CloxOrstaA], 80 characters; [CloxOrplaA], 500 characters

Storing original locality data can be problematic. In practice the information can be divided and allocated to four fields, of which [CloxOrmajA] is the first. This stores any original information identifying the continent or other major area of the current locality. Common major areas encountered are, for example, the Caribbean, Oceania, the Middle-East, Scandinavia, the Andes, the Himalaya. The second field, [CloxOrcouA], identifies the original country. Over the past century, many countries have changed their names and boundaries. Storage of the original name, together with the relevant date (located in the *Collections People Cross-reference Table*), enables the user to have some idea of the boundaries in operation at the time of collection, isolation or quarantine intercept. Much the same comments apply to the third field, [CloxOrstaA], which stores the name of the original state, county, Land, département or other subnational division, and the fourth field, [CloxOrplaA], which stores the original description of the exact area.

The BMS Foray Records Database provides one field 15 characters long for storing all textual locality information [Place Name]. The BMS/JNCC Database provides a field with the same name 30 characters long. Neither has provision to store original information separate from the current editorial opinion. This is an area where the structures used by the BMS need urgent up-grading.

Accepted locality information: [CloxAcmajA], 80 characters; [CloxAccouA], 80 characters, indexed; [CloxAcstaA], 80 characters, indexed; [CloxAcplaA], 500 characters

These next four fields form the editorial counterpart of the previous four fields. They store the current editorial opinion of the continent or major area ([CloxAcmajA]), country ([CloxAccouA]), county etc. ([CloxAcstaA]), and the current description of the exact place ([CloxAcplaA]).

In the fourth of these fields, [CloxAcplaA], where there is a string of localities, wherever possible these localities are stored in decreasing order of size, with each unit delimited by a comma, thus 'Kew, Royal Botanic Gardens, 200 m north of the pagoda' is preferable to '200 m north of the pagoda, Royal Botanic Gardens, Kew'. Abbreviations such as 'Mt' (mountain), 'sw' (south-west), etc. are discouraged as confusing to those whose native language is not English. Imperial and other units used in the original data should be converted to their SI equivalents.

A free text description of the ecosystem, 1000 characters [CloxEcol_A]

A free text account of the prevailing ecosystem is quite frequently encountered when abstracting data from dried reference collection packets, literature and other sources and, if requested, is also quite frequently provided by forayers contributing data. [CloxEcol_A] provides space to store this information. The BMS databases both provide a field of 25 characters [Ecosystem], which it may wish to enlarge.

In older flat-field databases with separate fields for the names of one fungus and one associated organism, this field, or its equivalent, has often been used to store the names of any other associated organisms. For example, in the case of '*Amanita muscaria*, on soil under birch, oak and hazel', the associated organism field might quite arbitrarily contain 'birch', while 'oak and hazel' will have been relegated to this ecology field. Of course this use of the field has been preferable to losing this information altogether, but when such databases are re-structured to a more durable form, editorial time will have to be allocated to deal with this problem.

Encoded description of the ecosystem, 25 characters
[CloxEcocoA]

In the same way that free text and coded fields are provided for developmental state and abundance, this field permits a structured statement of the prevailing ecosystem of the current location. Both BMS databases provide an analogous field 10 characters long [NCC Ecocode]: the difference in length is unlikely to be significant given the codes currently in use.

Both the IMI system and the BMS system use the habitat codes devised by the former British, and now dismembered Nature Conservancy Council (NCC). By and large, for recording in much of north-west Europe, these codes work pretty well, though there are no subterranean or aerial habitats, and the marine habitat is not subdivided. Further afield, the codes tend to fall down: coverage of categories of *Acacia* scrubland, or volcanic slopes, for example tends to be pretty minimal! The BMS should look for far wider habitat codes, but until found, the NCC habitat codes remain quite a good stop-gap.

Notes about localities not intended for publication, 1000 characters
[CloxInnotA]

Space for notes associated with each locality, but not intended for publication, has been provided, as there is frequently a need during editing of problematic locality information to record temporarily possible alternative correct identities for a given place. There is however no provision for notes about localities intended for publication. Neither BMS database has a field for this purpose.

The Collections People Cross-reference Table (Cpex)

Many people can be associated with a *Collections Database* observation: several people may contribute to making the observation or collection, others may make the original identification, yet others may wish to comment on that identification, or redispose the record under a different name, or isolate the fungus into pure culture, or intercept it for quarantine purposes, and so on. The list includes all the possibilities which make the relationship of localities to individual observations many-to-one, and many others besides. To permit the expression of the many-to-one relationship, a *Collections People Cross-reference Table* is necessary. Furthermore, since most events during the existence of an observation only occur because of the action of humans, this table seems as good a place as anywhere to store the many dates associated with the observation.

The BMS databases, having a flat-field design, cannot cope properly with the many-to-one relationship recognized here. Each stores an indication of the identities of some of the people associated with each record, and the date of collection of the current fungus, but no more. Even some of this information is in a form which some observers might find unsatisfactory, and the possibility that more than one person might simultaneously be involved in, for example, the same collection or identification is not recognized.

A link to the unique observation identifier, 8 characters, indexed [CpexLink_N]

This field stores the numeric link between the *Collections People Cross-reference Table* and other tables in the *Collections Database*.

A link to the Collections Locality Cross-reference Table, 8 characters indexed [CpexClox_N]

This field stores the numeric link between the current record and the appropriate locality in the *Collections Locality Cross-reference Table*. For an explanation of its use, see the earlier notes on [CloxLocnoN].

A link to the People Database, 8 characters [CpexPeolkN]

This field permits a numeric link for the current person with the *People Database*, to provide access to shared information about that person, for example, full name, address, interests, status as a referee of identifications etc. It is not yet in use. Before entering use, and before accession of data to the *People Database* the provisions of data protection legislation would need to be taken into account.

The name of the current person as received by the database, 160 characters [CpexPeoplA]

Only one person is permitted for each record in this table, and this field stores the name of that person. It takes the form of original information, so it appears in an unedited form. The editorial opinion of the identity of that person will, when it becomes available, be expressed through the previous field, [CpexPeolkN]. The BMS database fields [Collector], [Determiner] / [Identifier], [Confirmer] and [Current Referee], each 4 characters long, store merely initials. The BMS should urgently consider upgrading its provisions for these fields, so that full names can be stored, as there is a great danger of losing the identities of many of the people behind the less frequent initials.

The category into which the action of the current person falls, 1 character [CpexCategA]

One character is allocated to this field, to record the category into which the action of the current person falls. At present six categories are in use which are of interest in *Collections Database* recording: 'c', collector; 'o', original identifier; 'i', current identifier; 'f' former or other identifier; 's', isolator; 'q', quarantine interceptor. Others exist, but are not of relevance to foray recording, or will be made available as the need for them becomes apparent.

In their current format, the BMS databases have no need for this field. The BMS databases recognize the need to store information identifying the collector, original identifier, a 'confirmer' (this is usually the same person as the original identifier, or is a different identifier with a higher standing as a taxonomist than the original identifier), and a 'referee'. These categories are all catered for in the present structure. When information from the BMS database is restructured to a more durable form, data from these different fields will form separate records in the present table.

Personally-allocated collection and observation numbers, 50 characters [CpexPeonoA]

Many collectors have personal systems for allocating numbers to the collections and other observations they make. These numbers are usually different from the numbers allocated to material when it arrives in, for example, a national herbarium, or from the numbers allocated to observations when they are entered into a computerized database. [CpexPeonaA] provides space for these personal numbers. It is worth noting that, when several different people participate in making one collection, it is quite possible that several different personal collection numbers might get allocated to the same collection. That eventually is no problem for this database structure: each individual collector is allocated a different record in this table in respect of that collection. The BMS databases do not allocate space for such personal numbers.

Date fields: [CpexSday_A], 2 characters; [CpexSmontA], 2 characters; [CpexSyearA], 4 characters; [CpexFday_A], 2 characters; [CpexFmontA], 2 characters; [CpexFyearA], 4 characters

Dates are important things to record. By them we can tell what is the earliest record of an organism, and when that organism was last observed. We can use dates to keep track of changes in distribution or abundance through time, and to relate developmental stages observed to the seasons of the year. There are many things we can do with dates. Not surprisingly, many computerized databases provide specialized date fields to cope with date information. Unfortunately these are generally not suitable for our purposes. Many only provide date services back to the start of the twentieth century, whereas biological recording has a need to store dates going back at least to the middle of the eighteenth century. The other problem with using specialized date fields is that they cannot usually cope with null data, while a very large number of biological records exist in which, for example, the month is known but not the day or year, or the year is known, but not the day or month, etc.

To get round that problem for biological recording, it is necessary to express any date through three fields: one for the day, another for the month, and the third for the year. Even then, the problem of null data has some bearing on the choice of field. There are many records which are known to date from the nineteenth century, but for which the exact year is not known. The fact that the information is over one hundred years old is itself valuable, and needs to be stored. '1800' is unsuitable, since it implies an exact year, as does '18'. What is needed is '18xx'. Similar considerations apply to the month and day fields. As a result, each of these fields needs to be alphanumeric rather than numeric.

The third difficulty is that, for a small (not that small!) but significant proportion of dates associated with records, the original information supplied is a spread of dates: 'May-July 1882' or '23–24 February 1933' etc. To deal with that eventuality, two sets of date fields, a start set

([CpexSday_A], [CpexSmontA], [CpexSyearA]), and a finish set ([CpexFday_A], [CpexFmontA], [CpexFyearA]), are needed. Where (as is usual) the source information comprises only a single date, both sets of fields contain the same data. The BMS databases provide one set of three date fields for the date of collection/observation ([Day], [Month], [Year]). They make no provision for recording any other date.

Text and numeric links for alternative identifications: [CpexTaxnaA], 100 characters; [CpexTaxlkN], 8 characters

Apart from the original name given to the organism when the observation or collection was made, and apart from the currently accepted name provided through the central editor of the database, other people may wish to express an opinion about the identity of the organism, and there may be a need to record the history of earlier currently accepted names which are no longer acceptable as opinion shifts.

[CpexTaxnaA] provides space to make a text link to the *Nomenclature and Taxonomy Database*, thus linking an identification of the organism with the person of the current record. As for [Cco0AccnaA], the contents of this field must conform to the standards for the text link field of the *Nomenclature and Taxonomy Database*. Records with information in this field will all relate to identifications other than the original or current identifications, and so must all contain 'f' in [CpexCategA]. [CpexTaxlkN], which is not yet in use, has the same function as [CpexTaxnaA], but makes the link through a number rather than by text.

The BMS databases have no provision for these fields. Other flat-field collections databases exist which have a field called [Redisposition history] or something similar, in which the histories of different opinions as to the identity of the organisms are stored. Such fields are better than nothing, but represent an inadequate attempt to deal with the informational problem. A history of different opinions will inevitably build up over time: examination of packets in any dried reference collection will show the polite crossings-out and re-identifications which are the evidence that our paper-based databases record such information. For properly structured long-term storage, our computerized databases need the facility to store such histories too.

Text and numeric links to a bibliographic source: [CpexBiblkA], 100 characters; [CpexBiblkN], 8 characters

Revisionary work on the systematics of organisms is being published all the time and, particularly in the case of the fungi, this can result in large-scale name changes which may, or may not be adopted by the editors of the database. Specialists who wish to keep track of the opinions of others about the organisms in their group may wish to record these different opinions, even if they are not accepted. These fields enable a link to be made in such cases between the current record and the record in the *Bibliography Database* representing that revisionary work. They should not be confused with the facility for bibliographic links provided through the *Collections Bibliography Cross-reference Table* which is far more general in its application.

As with links to the *Nomenclature and Taxonomy Database*, there are arguments for and against text versus numeric links. At present, at IMI, links between the *Collections Database* and the *Bibliography Database* are textual rather than numeric though numeric links are a long-term aim. A discussion about the editorial standards used in creating textual links with the *Bibliography Database* can be found later in the discussion on [CbixBiblkA]. The BMS databases have no provision for either of these or the following field.

Identification of the exact page within that bibliographic source, 20 characters [CpexPage_A]

This field is used in the case of a link between the current record and the *Bibliography Database*. It identifies the exact page or page spread on which, within the publication, the identification to which the current record relates was made.

The Collections Substratum Cross-reference Table (Csux)

Dealing with information about the substratum on or in which a fungus is observed is tricky. The material on which the fungus is observed is always its physical location, but it may or may not also be the source of its nutrition. Thus this table has to cope with information about the rocks on which lichens are found, the soil bearing toadstools, the leaves carrying ascomycetes and many other very diverse materials, and it has to cope with them in a way which will allow at least some meaningful information to be extractable mechanically.

In every record in this table, by definition, a substratum must be linked with an observation of an organism: the fungus, insect, or plant etc. which was found on the substratum. An example might be the link: fungus–soil. A further complication, however, is that for many records, and certainly most records involving fungi, the substratum is part of another organism. As a result, a second link to an observation of that organism needs to be made. The result is a double link, e.g. fungus–leaf–plant. These two links, the first mandatory, the second optional, are made with different records in the *Collections Core Table*. One must (arbitrarily) be assigned the 'driver' link, and the second (if it exists) must be the 'slave' link.

It often happens that in a single observation a fungus, for example, is noted on the leaves, twigs and bark of a plant. Under the structure adopted here, that would come out as three records: fungus–leaf–plant, fungus–twig–plant, fungus–bark–plant. Similarly, if a fungus were observed on the thorax of a beetle which was on the leaves of a plant, that would result in two records: fungus–thorax–beetle, beetle–leaf–plant. In that case, the fact that the fungus was also associated with the plant would not be lost. The two observations would share the same unique collection identifier [CcorColnoN], but as the association did not involve a substratum linking the two, there would be no need to create a record for the association in this table. In the third example of a fungus on the thorax of a beetle which was on the leaves of a plant, but where the fungus also spread out over the leaves of the plant, three records would have to be made: fungus–thorax–beetle, beetle–leaf–plant, fungus–leaf–plant.

It may be seen from these examples that a many-to-many relationship exists between substrata and the organisms with which they are linked. The provision of this table and its structure permit very complex ecological inter-relationships to be observed. For long-term structured storage this table is essential. It is impossible to construct a flat-field feeder database which avoids these complications for most records during the data-entry stage, but even then it is likely that there will be a cost in terms of data lost simply because there is no provision for picking it up. The BMS databases each allocate one field ([Medium]), 20 characters long for the whole of this area of data.

'Driver' and 'slave' links to the unique observation identifier: [CsuxDrilkN], 8 characters, indexed; [CsuxSlalkN], 8 characters, indexed

[CsuxDrilkN] stores the numeric link for the 'driver' observation, and [CsuxSlalkN] the numeric link for the 'slave' observation between the *Collections Substratum Cross-reference Table* and other tables in the *Collections Database*.

A combined identifier of 'driver' and 'slave' links, 8 characters, indexed [CsuxComb_N]

This field stores a number calculated using the equation [CsuxComb_N] = ([CsuxDrilkN]*1000000) + [CsuxSlalkN]. This number is unique to the current combination of 'driver' and 'slave' observations, enabling such records to be identified quickly and mechanically through the index. This combination number is needed when outputting data, for example lists of substrata associated with particular species, or indexes of species found on or providing particular substrata.

A free text description of the substratum, 1000 characters [CsuxDesc_A]

In this field, the original text describing the substratum is stored. The best example I have come across to date was 'dead beatle'!

An edited description of the substratum, 160 characters [CsuxAcdesA]

This field, the nearest equivalent to [Medium] in the BMS Foray Records Database, stores an edited opinion of the meaning of the original text in [CsuxDesc_A]. A well-defined structure and a comprehensive thesaurus of accepted terms are necessary. Both the IMI and the BMS database systems attempt to provide these. Both attempts are similar, though not exactly alike, and neither is fully adequate. Both systems recommend the use of a thesaurus term (a noun for the BMS system, a noun or a phrase for the IMI system) first as a descriptor, followed by qualifying adjectives or nouns in apposition (or also phrases in the IMI system). In the BMS system, adjacent elements are separated by a space. In the IMI system they are separated by a comma and a space. Because the IMI system admits as thesaurus items, the comma is a necessary part of the separator: these differences are really quite minor.

The IMI system also contains an added tier of initial descriptors which enable certain otherwise diffuse categories to be brought together for indexing: 'substance' always precedes any naturally occurring non-biological object (e.g. 'substance, soil', 'substance, water'); 'artefact' precedes any object which has been produced through the skill or creativity of a living organism (e.g. 'artefact, concrete', 'artefact, plaster', 'artefact, nest'); 'food' precedes any foodstuff used as a fungal substratum (e.g. 'food, bread', 'food, fruit'). It might be worth pointing out that 'fruit' would imply fruit on, say, a tree, but 'food, fruit' would imply fruit in the greengrocer's shop or something similar. Once at IMI, the term 'artefact, dung' was entered by a keyboarder: most of us felt this was going a little too far, and at IMI we index 'dung' as though it were a part of the organism that produced it!

Without this extra tier, it is impossible to produce a substratum index which gathers together such miscellaneous items into meaningful groups. This extra tier is, furthermore, merely a continuation of a long tradition in paper-based recording systems of such a classification. Efforts should be made to bring together the BMS and IMI systems. These efforts should include a standardizing (and extension) of thesaurus terms, an acceptance that phrases may sometimes be necessary as thesaurus terms, and hence that individual components of a statement should be separated by more than just a space, and the introduction to the BMS system of the extra tier of descriptors.

The position of the 'driver' organism in relation to the 'slave' organism, 8 characters [CsuxPositA]

For some fungi, the position a fruitbody or colony occupies in relation to its substratum is a specific character which remains the same in all collections encountered. For others though, for example *Diatrype stigma*, the position varies from collection to collection: the teleomorph stroma may be on the surface of bare wood here, but under bark there. This field allows the user to specify what position was observed in the case of the current record. Strictly speaking, the field allows for a description of the position of the 'driver' organism in relation to the 'slave' organism: the syntax is 'driver organism' its 'position' the 'slave organism'.

The ecological relationship of the 'driver' organism to the 'slave' organism, 1000 characters [CsuxRelatA]

For records where there is both a 'driver' and a 'slave' organism, there is frequently ecological information about the relationship of those two organisms. The fungus may be parasitic on the plant, or mutualistic with the plant etc. This field is set aside for a free-text description of that relationship. The BMS databases make no formal provision for such recording, nor for the recording of symptoms, and in the IMI system this and the following three fields have been introduced comparatively recently.

An edited description of the relationship between the 'driver' organism and the 'slave' organism, 100 characters [CsuxAcrelA]

This field contains an edited opinion on the meaning of the data in [CsuxRelatA]. Like [CsuxAcdesA] this field is used for mechanical retrieval and processing of information, so the description must be structured. The field therefore stores a thesaurus term describing the ecological relationship of the 'driver' organism with the 'slave' organism. Thus, if the 'driver' organism is *Puccinia oxalidis* and the 'slave' is *Oxalis*, the field might read 'parasitic'. If, however, the 'driver' is *Oxalis* and the 'slave' *Puccinia oxalidis* (there is no logical reason why this should not be the case), the field should read 'parasitized'. There is a need to develop or identify and adopt a thesaurus of ecological relationship terms.

The symptoms caused by the 'driver' organism to the 'slave' organism, 1000 characters [CsuxSymp_A]

For records where there is both a 'driver' and a 'slave' organism and their relationship involves parasitism, there is frequently information about the symptoms caused by the 'driver' to the 'slave'. This field is set aside for a free-text description of those symptoms.

An edited description of the symptoms caused by the 'driver' organism to the 'slave' organism, 100 characters [CsuxAcsymA]

This field contains an edited opinion on the meaning of the data in [CsuxSymp_A]. Like [CsuxAcdesA] and [CsuxAcrelA] this field is used for mechanical retrieval and processing of information, so the description must be structured. The field therefore stores a thesaurus term describing the symptoms caused by the 'driver' organism to the 'slave' organism, following a syntax similar to that of the preceding fields. There is a need to develop or identify and adopt a thesaurus of symptom terms.

The Collections Bibliography Cross-reference Table (Cbix)

In addition to specimens in dried reference collections, isolates in living collections, and field observations not backed by material, the other great sources of data for the *Collections Database* are citations in the literature. This table provides a means of relating *Collections Database* records to the literature. It is quite possible for many different publications all to refer to the same collection or observation: many publications have, for example, referred to Fleming's observation which led to the discovery of penicillin. The relationship between this table and each individual *Collections Database* observation is therefore many-to-one. Being designed originally solely to capture observations generated at forays and other field meetings, the BMS Foray Records Database makes no provision for any bibliographic link. The BMS/JNCC Database has the facility of recording a single bibliographic link for each record. For these two databases to be compatible with each other, re-structuring of the BMS Foray Records Database will be necessary, and that may be an opportunity to upgrade both to the much more flexible structure proposed here and in use at IMI.

A link to the unique observation identifier, 8 characters, indexed [CbixLink_N]

This field stores the numeric link between the *Collections Bibliography Cross-reference Table* and other tables in the *Collections Database.*

Text and numeric links to the Bibliography Database: [CbixBiblkA], 100 characters; [CbixBiblkN], 8 characters

[CbixBiblkA] is the first of two fields, both in current use, permitting a link to be made via this table between the *Collections Database* and the *Bibliography Database.* This link is in text form, and is used to connect *Collections Database* records to the *Bibliography Database.* Although much longer than a numeric link, it nevertheless has many convenient aspects, including a greater chance of recovering a connexion in the event of a corrupt file than with a numeric link.

The link information in [CbixBiblkA] is built up from the surnames of each author of the bibliographic item, the year of publication, and a single character identifier. Adjacent authors are separated by a comma and space, except the last two authors who are separated by a space, ampersand and a second space. The year on the publication and the single character identifier are enclosed within parentheses, and follow a space after the last author. For a given author/year combination, the first identifier is 'a', the second 'b' and so on. Like the text link used for the currently accepted organism name [Cco0AccnaA], information in this field uses no accented characters: thus 'Léveillé' becomes 'Leveille'. Authors with names in non-Latin alphabets are transliterated using the *CABI Database Production Manual* standards. Some examples of typical bibliographic link data are: 'Sutton (1980a)', 'Gayova (1992a)', 'Sutton & Hodges (1978b)', 'Muller & Arx (1950a)', 'DiCosmo, Nag Raj & Kendrick (1979a)'.

Because the link information closely corresponds to the way humans tend to remember bibliographic references, there is no need to remember numbers. If the authors of the paper, and the year of publication are known, the link can be made by constructing the link data and adding, then replacing single character identifiers successively until the correct data are located, or it becomes clear that the desired publication is not represented in the *Bibliography Database*. It is generally accepted, however, that for these bibliographic links, full and alternative availability of a numeric link through [CbixBiblkN] is a long-term goal. That link is already being employed in the BMS/JNCC Database through the field [BSM Link], 8 characters.

Identification of the exact page within the linked publication, 40 characters [CbixPage_A]

This field is used to identify the exact page or page spread with which, within the linked publication, the data of the current *Collections Database* record is connected. This field is twice the length of [CpexPage_A] which performs a closely related function and stores the same category of data: minor inconsistencies of this sort can be very hard to avoid, and have to be rectified as and when noticed! The BMS/JNCC Database, even though it makes a bibliographic link, notes no exact pages.

A place to store unverified bibliographic data, 240 characters [CbixRef_A]

Sometimes a link between the *Collections Database* and the *Bibliography Database* cannot be successfully made. Perhaps the desired data simply do not exist in the *Bibliography Database*, or perhaps the data available to the keyboarder aren't adequate to identify the bibliographic record required: there can be lots of reasons. Under those conditions, it could be undesirable to enter any data, since the link to the bibliographic source cannot be established. To get round that problem, and the keyboarding 'log-jams' that can result, there is a need for a field which can store, in free-text form, the bibliographic data actually available to the keyboarder at the time. This field is thus used only when there is no information in either [CbixBiblkA] or [CbixBiblkN].

The existence, and use of this field implies a further stage of work where a specialist bibliographic editor scans the table for records containing data in [CbixRef_A], and deals with the problems. As might be expected, fields with similar functions have to be located strategically at many points in the different databases used for recording, and the work of reviewing such fields forms a significant part of the routine maintenance of these relational databases.

Replica of typification comment, 1000 characters [CbixTypifA]

Where the current record in the *Collections Database* relates to a specimen in a dried reference collection or more recently a culture in a living reference collection, there is always a possibility that the specimen in question may be a nomenclatural type. Since type specimens are generally designated through publications, a place to store published information about the type status of the current specimen is clearly desirable. This field is devoted to storing, as nearly as possible, a replica of what was actually published, so that the information it contains can function as a primary source of data.

Given the fact that most nomenclaturalists are said to be failed lawyers, it is not uncommon for the type status of a particular specimen to be argued over in many different publications and, while no-one likes to encourage such litigation, that may be another reason why the *Collections Bibliography Cross-reference Table* has a many-to-one relationship with the *Collections Core Table*. Unpublished observations about the type status of a particular specimen should be placed in the general fields [Cco0Exnota] (notes to be published) or [Cco0Innota] (notes not to be published), depending on the commentators assessment of their potential for controversy!

General theme of current bibliographic observation, 500 characters
[CbixThemeA]

This final field in the *Collections Bibliography Cross-reference Table* allows space to store some indication of the general theme of the bibliographic data in relationship to the current Collections observation. For example, the publication may have cited the observation in the context of a study in ecology, biochemistry, genetics, systematics, or any one or several of a wide range of other scientific aspects. Having this field makes it possible, at least in theory, to pull out all the bibliographic records dealing with, say, the ecology of a particular organism. This field is rather new, and rules to give it structure have not yet been devised. It seems likely to represent an area where, in the future, several different fields will be needed.

The Collections Other Collections Cross-reference Table (Coth)

Many of the records in the *Collections Database* refer to specimens residing in reference collections. An important aspect of the structure of this database is therefore to provide an indication of where those specimens may be found. Reference collections do not necessarily last for ever. They may be moved, disbanded, amalgamated or even destroyed. The important private mycological collection of Ted Ellis, for example, was moved to the Royal Botanic Gardens, Kew after his death. Mycologists, furthermore, have a habit of dividing a single specimen and sending the parts to different reference collections. One specimen may therefore be at different times in different collections.

The *Collections Other Collections Cross-reference Table* provides a place to store reference collection information which recognizes this need for a many-to-one relationship. Only some of the available fields are considered here. The BMS databases each provide one field, 3 characters long ([Location of Specimen] and [Loc of specimen]), to identify one collection associated with the current record, the BMS Foray Records Database also provides another field, 7 characters long [Recond Number], with data which may belong here, identifying the current item in the BMS Recorder's database.

A link to the unique observation, 8 characters, indexed [CothLink_N]

This field stores the numeric link between the *Collections Other Collections Cross-reference Table* and other tables in the *Collections Database*.

Free-text and structured information identifying the current reference collection: [CothColl_A], 200 characters; [CothColcoA], 20 characters

The publication *Index Herbariorum* (Holmgren, Holmgren & Barnett, 1990) lists internationally recognized dried reference collections. Each collection is allocated a unique code which identifies it. Where such structured information is available for the current specimen, it is included in [CothColcoA]. Many collections, however, are personal or otherwise not internationally recognized, and no code is available. Information identifying such collections is placed in the free-text field [CothColl_A]. The BMS database fields [Location of Specimen] and [Loc of specimen] contain data belonging here.

Accession number in current reference collection, 50 characters [CothColnoA]

Many reference collections issue accession numbers to items received. The number issued by a particular collection to the current specimen is stored in this field.

Dates of entry to and exit from the collection: [CothSday_A], 2 *characters;* [CothSmontA], 2 *characters;* [CothSyearA], 4 *characters;* [CothFday_A], 2 *characters;* [CothFmontA], 2 *characters;* [CothFyearA], 4 *characters*

These fields store the date of arrival of material in a collection and, if necessary, the subsequent date of departure, in a format similar to the date fields in the *Collections People Cross-reference Table.*

The Collections Administration Table (Cadm)

Lastly, and regrettably necessary, is the *Collections Administration Table.* This separates all the administrative aspects associated with the record from the scientific aspects, but has a one-to-one relationship with the *Collections Core Table* and the *Collections Core Supplementary Table.* Apart from the fields relevant to biological recording, this table also contains fields providing the following (and other information): an assessment of any hazards presented by the current item; a note on ownership of the rights to the current item; a note on the confidentiality of current data. In this table, information relating to the administration of records as they progress through an identification service may also, for example, be kept.

Another important, but large and difficult topic, which cannot be discussed here, is the problem of keeping track of exchanged records. If the BMS were to donate a copy of 20000 records to a sister society, how would the BMS be able to tell they were not getting their own records back as duplicates if, at some later point, the sister society made a return donation? The worrying prospect raises itself of duplicated mycological data mushrooming as records circle between different databases on an electronic merry-go-round! It is certainly already happening doubled in trumps over the internet.

A link to the unique observation identifier, 8 characters, indexed [CadmLink_N]

This field stores the numeric link between the *Collections Herbarium Table* and other tables in the *Collections Database.*

Keyboarding dates: [CadmAcdatD], 8 *characters;* [CadmUpdatD], 8 *characters*

It helps to know when a record arrived in the system, and it is also useful to know when it was last altered: without the first date it may be impossible to know at what rate records are being accessed to the database, the second date may be needed when, for example, a recurrent error has been attributed to a keyboarder who started work on a particular day. These two fields store that information. Both are 'date' fields, thus relying on the database software's services for handling date information. [CadmAcdatA] stores the date of accession of the record. [CadmUpdatD] stores the date of most recent update. Neither BMS database provides space to record this information.

Proof-reading flags, 300 characters [CadmRecstA]

The final example is a field category which has to be located at strategic points throughout databases. All databases need to be proof-read. A problem is that, for many records, it is easy to validate the contents of some fields, while those of other fields may present problems which require further investigation. [CadmRecstA] and similar fields note the data-quality of each field in the current record. One character is allocated to each field and what fills that character depends on where that field is in the process of proof-reading. The BMS databases have no fields to record proof-reading status.

The system in use at IMI (which has not yet been introduced for the *Collections Database*, but is in regular use for the *Bibliography Database*) is to recognize three levels for each field, marked by '0', '1' and '2' respectively. The first indicates that the field has not passed the first stage of proof-reading. The second, that the field has passed the first stage, but not the second. The third, that both stages have been passed. A surprisingly large amount of data quality can be checked mechanically, and at IMI the first stage is a mechanical scan of each record.

Only fields for which the proof-reading flag is set to '0' are examined. If the information in the field passes the mechanical test, the flag is automatically reset to '1'. If the information fails the test, the proof-reading software assesses the reason for the failure. The software is designed to correct some problems automatically, and then reset the flag to '1'. When this happens, a warning that mechanical correction of data has occurred is printed in the report at the end of the proof-reading operation. For problems which the software cannot correct, the flag remains set at '0', and attention is drawn to the problem often with an indication of a proposed correction. The second stage is human proof-reading, after which the flag for each field reaches an acceptable standard is reset to '2'.

Some problems with these data structures

With a need to store complex data, it is more or less impossible to devise faultless data-structures. There are many problems inherent in the fields just discussed. What has been presented is not an ideal final format, but is offered in the hope that it represents a significant step in the right direction. These fields enable a lot more information to be stored in a way which reflects its true nature than can be done with a simple flat-field database. It is easy to set up a simple database which will deal with 80% of all records adequately, particularly if the information is used only by browsing humans. It is much more difficult to set up a system which will be correct more than 99.9% of the time, and adequate 99.99% of the time. For most aspects, the structures suggested here work to such tolerances. To eliminate the remaining problems would require a disproportionately large effort. A selection of some of the more obvious problems which remain will now be briefly discussed.

Other alphabets

A problem scarcely catered for by the structures just described is the storage of original data in scripts other than the Latin alphabet. More precisely, the problem exists of original data in scripts other than those provided for in the standard extended ASCII character set. Even more precisely, you need to know which character set you are using within your extended ASCII code. And there are lots of them. At the fourth International Mycological Congress in Regensburg, I proposed that mycologists should at least agree what character set they would use as a default, but the need for even this base line was not recognized at the time. As a result, world wide, mycological databases are using many different character sets, and the users are often not even aware that there are alternatives.

In a joint project between IMI and the N.G. Kholodny Institute of Botany, Kiev, funded by the UK Darwin Initiative, the mycological records of the Kholodny dried reference collection are being computerized. When this project began, the problem of different alphabets and character sets suddenly become acute. On encountering thousands of records with information like, '*Erysiphe cruciferarum* on *Sinapis arvensis*, Україна, Одеська обдасть Овідіопольський р-н, село Барабой, 1/07/1978, leg. В.II. Гелюта' it is necessary to consider how they should be stored. And it is not just the Cyrillic alphabets (there are several), or the

Greek. It is also the specialized letters in the Latin alphabet family: the 'g<breve>' (ğ), the 'd<hacek>'(d'), the 'c<acute>' (ć), the 'u<circle over>' (ů), the 'e<dot over>' (ė), the 'a<hook>' (ą), the 'i<dotless>' (ı), the 'o<double acute>' (ő), and many many more.

To handle the keyboarding of such data in an MS DOX operating environment, it is very convenient to have a small TSR ('terminate and stay resident') utility file which sets up an alternative character set for display on screen, and re-maps the scan codes delivered by the keyboard so that those characters can be easily typed. DRUG409.COM produced by Yuri Shkolnikov (formerly at the N.G. Kholodny Institute of Botany, Kiev) provides this service superbly for the Russian and Ukrainian alphabets, and is in widespread use. In WINDOWS, 'PARAWIN' is widely available throughout the former USSR for similar purposes. While it is possible to purchase cyrillic keyboards for PCs, they tend to be in a layout rather unfamiliar to the western European user. In addition, therefore, it can help to paint in the appropriate letters on your keyboard (it is better to paint them on the vertical front of each key, rather than the part touched by the fingers, so that they don't become eroded so quickly).

As long as the byte (8-bits) forms the basis of computer processing, the 256 character extended ASCII sets will form the basis of data storage. The 16-bit character set (which will apparently include all the letters of all alphabets and many pictograms known to man, and more besides) is, at the time of writing, said to be on the way, but it may be some years before it becomes widely available, while the much advertised 'Windows 95' handling of different alphabets is merely another 'kludge'. In the meantime, the problem of storing information in different character sets is being tackled at IMI in three ways. The first is to add fields each one character in length, at appropriate locations within the tables to note the character set in use for the current record. The second is to add fields, storing parallel information in the original language where that differs from English. The third is to mark the beginning and ending of a character set different from the default within a text field (e.g. '<cyrillic>é.Å. âÑ½εΓá<latin>' which would be output as 'В.II. Гелюта'). These extra facilities, while solving this problem, also, regrettably, add to the complexity of the tables.

Notes

The provision of space for notes, and the rules governing their entry, are inadequate in the system outlined in this chapter. In particular, the field length provides unnecessarily small physical limits, and there is no ability to record notes in different languages, even though that could be easily done in the paper-based systems we are emulating: scientists simply wrote on the collection packet in their own language. The best that can be said for the present notes fields ([Cco0ExnotA] and [Cco0InnotA]) is that they provide a convenient place to store notes made up to the present.

An upgrade under consideration at the moment is to provide two new fields within the *Collections People Cross-reference Table* ([CpexExnotA] and [CpexInnotA]). Each time a person wished to make a note about an observation in the *Collections Database*, a new record would be created in the *Collections People Cross-reference Table*, storing the date of the note, the name of the person making the note, and the note itself. An extra character 'n' (for 'notes') would be added to those available for the field noting the category of the current record ([CpexCategA]).

Doubtful records

The one field in the BMS databases not catered for in the structures reviewed earlier, is the 1 character flag to mark a doubtful record. Initially, the absence of this field reflected a feeling that all records are intrinsically doubtful (it is merely a question of extent), but the need to mark those which are held to be especially doubtful by some experienced authority may be a reasonable requirement. Even so, it seems undesirable to allow a record to be questioned without noting who questioned it, when and why.

The provision of notes fields in the *Collections People Cross-reference Table* would enable all of that information to be stored properly. An extra character 'd' (for 'doubting') would be added to those available for the field noting the category of the current record ([CpexCategA]). The name of the person casting doubt on the record, and the date, together with reasons, could then be stored.

UTM map references

In many countries, neither latitude and longitude, nor a system similar to the UK national grid is used to provide co-ordinates identifying a collection location. The most common other system is Universal Transverse Mercator (UTM). While this is unfamiliar to collectors in the British Isles (except the Channel Islands, where it is used for Ordnance Survey maps), it is widely used in other parts of Europe and North America. There may, therefore, be grounds for including fields to record the UTM reference to localities in the *Collections Locality Cross-reference Table*.

Complexity

In his introduction to the BMS *Guide to Recording Fungi*, Marriott (1994) observes that the 'sheer size (28 fields) of the BMS database has a daunting effect on many people'. In the preceding analysis, over 80 fields were identified as needed to store *Collections Database* information about fungi and the organisms with which they are associated. Even those 80 represent a selection of those thought to be most relevant to the recording needs of the BMS: the full set of fields currently recognized in IMI as being needed in an ideal *Collections Database* is several times that number. And the *Collections Database* is only one of seven databases identified as being needed for field mycology. Since each of those databases is roughly as complex as the one just described, it means that somewhere over seven hundred different fields, perhaps even more than one thousand, are potentially needed for fungal recording. This is indeed daunting.

It is instructive to examine what structures are being considered by other groups looking at computerized systems for biological recording. One relevant venture, directed mainly to higher plants, is the 'Common Datastructure for European Collections Databases' (CDEFD) project. The sort of data structures being considered by that project relate to collections, and so can be compared quite easily with those identified in this chapter. Even a short perusal reveals that certainly no fewer, and quite possibly many more fields are under consideration in their model than are suggested here. Many of their fields cover similar areas to those of this chapter, others mycologists may well need to adopt, but equally there are some fields and ideas here which, being suited for recording fungi, appear not to have been considered closely by that group.

Another is the 'Global Plant Checklist Project' of the International Organization for Plant Information (IOPI), which is aimed at producing a world-wide nomenclature and taxonomy database for vascular plants. The fields under consideration by that group can easily be compared with the fields of the *Nomenclature and Taxonomy Database* referred to, but not considered in detail in this chapter. Again, there is broad similarity over wide areas, though specialist fields for fungal nomenclature are lacking, and the greater experience of vascular plant systematics shows in much better provision for the nomenclature and taxonomy of higher plants. The general impression from comparison with both of these, and other groups, therefore, is that the complexity of fields recognized at IMI for recording fungi is probably still on the modest side of what would be ideal: there is little danger of over-shooting the mark.

If databases for biological recording are moving towards complex structures, their maintenance will become increasingly a specialized and highly skilled job; or rather, a series of specialized and highly skilled jobs. Control of the hardware and the underlying commercial software is one skilled area. Database design and field structuring is another which is likely to remain important for the foreseeable future: the perfect database structure for biological recording has yet to be developed, and until then, there will be constant adjustments as needs for new fields and new tables are identified. The provision of specialist software to service these structures will also be important for some time to come: having decided on particular structures, software is needed to enable the data they contain to be presented in a format easily assimilated by the human user.

Then there is the need for specialist editors, and groups to establish and maintain data standards. In the past, for production of commodities like distribution maps, generalist editors were required: people with a spread of knowledge about fungal names, host names, taxonomy, literature and geography. The job of these editors was to ensure output was of a good quality. But with structured information, output can be purely mechanical. What is needed is to ensure the quality of the input. This requires specialist editors. Associated with their work is the need for simple 'feeder' databases which exist solely to gather data (they will be considered next), and for conversion programs which manipulate the information in these feeder databases, and read it into the appropriate fields of the main storage databases. It is clear that if, as seems likely, the databases of the future have such complex structures, they will require greater resources to maintain them.

Data-entry (feeder) systems

Data-entry systems for gathering observations from participants of forays and other field meetings, and from donors of independently generated records must be simple and friendly to use: people who want to contribute shouldn't need a degree in computing first. The data-entry structure must accordingly be non-relational. It could use database software and be a simple flat-field database, or could take the form of text entered using a word-processor, with the start of each new record marked by a reserved character, and each field by a carriage return. The number of fields in use should be kept small, and should not be subject to frequent changes. This simplicity is important to ensure that data can be received from many different sources, using contributors, machines and software of a wide-ranging calibre.

Most important, instructions on how to set up this simple data-entry system, together with rules on what fields it should contain, and advice about how to export data to the central database should all be documented in an easily-accessible and helpful guide, while the rules for entering data in those fields should be as simple and forgiving as possible. This is one thing the BMS does really well: its guide (Marriott, 1994) is an excellent example of how this can and should be done.

In practice, for a big database, records will come from a variety of sources: from professional and amateur mycologists carrying out field work, from plant pathologists, from the public, from people or organizations requesting assistance with identifying fungi, from quarantine services, and by keyboarders working on backlog data from existing literature and reference collections. Each of these (and many other sources) may wish to contribute slightly different types of data. For example, the locality of interest to the forager is the place of collection, to the quarantine officer it is the place of interception, while the experimental mycologist may wish to store the locality where the fungus was isolated into pure culture. It is therefore quite normal and necessary to have to set up different data-entry systems, each customized to the needs of a particular group.

Another important point to bear in mind is that those generating records will often have little or no concept of what sort of things can go wrong, what sorts of misunderstandings can occur, and what complexities exist in a fully structured database system. They may be unaware that the main storage database has to cope with information from sources quite unlike theirs, and they may as a result sometimes have difficulty in appreciating the need for 'all those extra fields'. It is thus important that they should not be presented with more fields than they need, and that they should perceive that the data being entered is of direct use to them.

Although much of what is keyboarded at data-entry stage cannot easily be checked, some parts of the information can be validated at that stage. For example, it should be possible to check keyboarded dates and point out any which are invalid. It is surprising how many fungi have been observed on 31st February and similar dates, and it is surprising how many were observed, if the data were to be believed, before the Norman conquest or in a time scale more appropriate to science fiction than science! Similarly, if many records are being keyboarded from a single country, the names of the different administrative regions (counties, states, oblasts, départements, Lands, etc.) could be checked at this point. It is often useful to write and distribute simple data-checking software in such cases, or to embed such software in the data-entry system used.

Mapping

Although just one of many forms in which information can be output from these databases, the computerized production of fungal distribution maps does present interesting problems. Undoubtedly the future possibilities for computerized mapping are enormous. At the time of writing, some highly sophisticated pieces of mapping software are coming onto the market. At the affordable end, however, most of this software is designed to present data internal to the software itself (and of often suspect quality), in an 'educational' format (e.g. a map of the British Isles with the caption 'Country: Great Britain. Capital: London'), or to be purchased by the gullible in the unthinking belief that it will provide them with an alternative road route between their home and their in-laws'. The ability to read a data-file of the user's choice, in a format specified by the user, and display its information on the maps is not provided by this software. To get such features, it is necessary to go to the top end of the market, and there the prices are beyond what most scientific societies are prepared to pay, let alone individuals.

The current realities for mapping fungi by computer are therefore more mundane: at the end of its first century, the BMS does have the capability to produce simple computerized distribution maps, but that capability is restricted to a very small number of individuals. Rather than running before we can walk, the purpose of this second part of the chapter is to try to spread that rather primitive capability a little further, by looking in lay terms at some of the basic building blocks which can be used to produce simple computerized maps, and for this, the best place to start is by reviewing the sort of maps which have been produced, mainly manually, over the last few years.

Existing published maps

Distribution maps of fungi can be found in a number of works published by or in association with the BMS. The maps of British species of *Hypoxylon* produced manually by Watling & Whalley (1977), and of British species of *Gasteromycetes* produced by computer by Cooper, in Pegler, Læssøe & Spooner (1995) are two excellent examples (Fig. 12.1). In each map in both of these sets, information is presented on the distribution of a single fungal species in the form of black circles set inside notional 10 km squares based on the UK and Irish Ordnance Survey grids, with the coastline of the British Isles as a background. In presenting their

Handkea utriformis

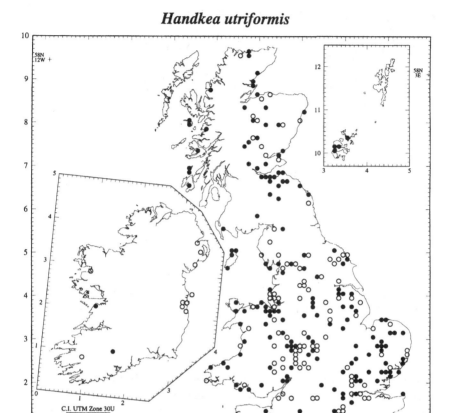

Fig. 12.1. Distribution map based on 10 km squares of the UK and Irish National Grids: open and closed circles indicate different categories of records (J. Cooper).

information in this manner, they were adopting the most widely used format for presenting distribution maps of lichens, higher plants and many other organisms in Britain. Similar maps based on 10 km square or larger or smaller grids have been published for fungi in a number of other countries (examples from the Netherlands may be found in various back issues of *Coolia*). The grid system is, however, not the only way to present distributional information, nor is it necessarily the best. The manually produced but valued and long-lived series of maps produced at IMI depicts distribution by drawing lines around the areas where the organism is

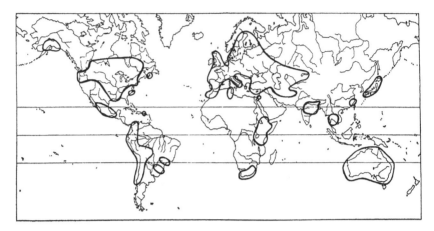

Fig. 12.2. Map indicating distribution by circling appropriate areas (IMI).

known to occur (Fig. 12.2). A similar alternative which is occasionally used is to shade in such areas. All of these maps are monochrome.

Assessment of styles

A grid system is of enormous value when setting out to survey an area for organisms: it means that adequate checks can be made to ensure that every part of the region gets surveyed and, with the selection of an appropriate-sized grid, that the number of localities to survey can be matched to the resources of the surveying organization. These two needs are probably the main reasons why most biological societies in the British Isles have adopted the 10 km square grid. Before computerization, when production of each map was manual, adding a black circle by hand for every record of occurrence was very labour intensive: many older field mycologists will have memories of long hours spent with dry-transfer lettering sheets. Presentation of distributional data in the form of a grid seemed a natural solution: it limited the maximum number of circles to the number of grid squares, visually emphasized the manner in which the survey was carried out, protected the rare species from undue attention by adding an element of deliberate inaccuracy to the presentation, and looked 'scientific' with its nice straight rows of orderly data!

Some of these reasons for presenting data in the form of a grid can look a little old-fashioned in the computer age. Any half-decent mapping system should permit users the choice between producing one black circle

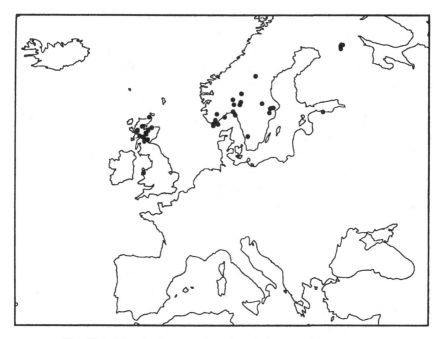

Fig. 12.3. Distribution map based on latitude and longitude coordinates, with precise collection locations indicated (IMI).

per grid square, and placing a black circle exactly over every place of collection (Fig. 12.3): the difference in time and effort between the two options is negligible for a computer. And given that choice, the second option is surely almost always preferable. It reveals whether the grid-based surveyors did their job properly, or merely visited one site at the point where four squares of the grid meet. It reveals whether the survey was carried out from the convenience of a roadside. It reveals the true scientific results of the work.

Using symbols (black circles, squares, triangles, etc.) to mark individual locations is particularly advantageous where the distributional information is precise compared with the scale currently in use to present the coastal or political outline data. Such symbols can however look ridiculous on a map of, for example, North America if a single dot in the middle of the USA indicates that the organism has been recorded from somewhere within that huge country. Even more misleading examples, such as records of quarantine intercepts located all exactly on capital cities, are not known.

In such cases, so long as the convention is properly understood, shading the whole region from which the imprecise record originated can be a more satisfactory way of presenting the information. A better solution still is to combine the two methods of presentation. For a monochrome map produced by printing hard-copy on paper, this could take the form of marking regions represented by very imprecise records with a pale grey shading, marking those more precisely represented with a darker grey, and placing exact records on top using a small black symbol. On screen, the greys and black could easily be replaced by colours.

This format enables us to use records with information in the latitude and longitude or grid square fields for the precise records, and records lacking that information, but having progressively less precise data in the state and country fields. In a nutshell, it means our data storage and output systems are dovetailed.

Specifications for a computerized mapping system

The specifications for a simple computerized system producing distribution maps should therefore include the ability to create and store simple maps at a variety of scales permitting a resolution of down to, say 20 km, for a world map, and 1 km for a national map. The system should be able to present the maps using a variety of different projections, showing the basic boundaries (coastal and/or political) of the area being mapped. It should provide the option to superimpose lines of latitude and longitude, or of national grids. It should be able to find and read distributional data direct from the *Collections Database* and other data (such as nomenclatural data) direct from other databases, and to use that information to shade in appropriate areas at different levels of intensity and to superimpose small symbols to indicate the position of precise records. It would need to be able to display each map on screen, and to output it as hard copy printed on paper, or in some standard graphics file format for incorporation in a word-processed document. For screen displays, another very desirable feature would be the ability to display full information about any collection selected using a mouse pointer. The ability to edit such data and return it to the database would also be useful.

Hardware considerations

In setting up such specifications, with the view of extending map-making capabilities more widely among mycologists, it would be important to take into account the requirements of the hardware and operating systems likely to be available: such considerations are more ephemeral, but at the time of writing, a system designed for MS DOS using a VGA 640 × 350 pixel screen would probably be pretty universally applicable and, if rather conservative, would at least permit the use of assembly language code to handle some graphics display to screen, at the point where speed is essential. Although, at the time of writing, the signs are that MS DOS will eventually be replaced as a disk operating system, the current alternatives of WINDOWS and WINDOWS95 are great if you don't object to dying of old age before they load, but they are very slow in handling graphics, and both rather difficult programming environments unless you are a specialist. For monochrome hard-copy output, a printer with a resolution of 300 dpi (dots per inch) capable of receiving instructions using the POST-SCRIPT language would be ideal, but a driver to talk to a bottom-of-the-range dot-matrix-printer would also be necessary.

Boundary data (coastlines and political boundaries)

Latitude and longitude co-ordinates of coastlines and political boundaries are available as shareware in text file format. At least some take the form of a sequence of many thousands of paired coordinates representing individual points along the boundary, each pair being separated by a carriage return and line feed. For data on a world scale, sequences accurate to about ten minutes (15 km in temperature regions) are available. As text files, they are unsuitable for use in creating maps, and need to be converted to numeric format. In making that conversion, it is necessary to decide whether to express minutes unchanged as sixtieths of a degree, or to convert them to a decimal figure.

Since latitudes extend from 90° north to 90° south, any latitude can be expressed, accurate to one minute, using the numbers −9000 to +9000. Similarly, since longitudes extend from 180° east to 180° west, any longitude can be expressed, accurate to one minute, using the numbers −18000 to +18000. In computing terms, therefore, any latitude and longitude can be expressed accurate to one minute, in four bytes using two signed integers: a processed copy of shareware data of the whole world's coastline at ten minute's accuracy can be stored in this way in just under 350K of disc space. To gain greater accuracy, it would probably be necessary to progress to using single-precision numbers which would double the storage requirements: for mapping on a global scale or at national level for most countries this precision is probably not worth the bother.

UK and Irish Ordnance Survey grid data can be handled similarly (because they are based on different projections, amalgamating data from the two sets of grids provides an additional and challenging difficulty), except that the integers are treated as unsigned. Using unsigned integers, one pair of eastings and northings accurate down to 100 m, if desired, can be stored in four bytes. The whole coast of Great Britain and associated islands, but excluding Ireland can, for example, be stored at a resolution of 500 m in less than 160K of disc space. This means that it is possible to pack the basic data necessary for producing maps of the world as a whole, and the country of your choice in particular, on to a single floppy disc, with plenty of space left for software.

For large-scale maps of small areas using the UK Ordnance Survey grid, or some similar system, an interesting alternative which might be worth exploring could be to use a scanner to produce a graphics image of the map, to the corners of which coordinates could be assigned. This might just work if the scanning were carefully done, though it is hard to see how such a short cut could be reliably applied to the more complex projections used with latitude and longitude. For smaller local projects, in the absence of any other source of data, it is always possible to draw in grid lines or lines of latitude and longitude on a map, and keyboard the coordinates oneself.

Placing a simple map on screen

This is not a particularly difficult stage: in one absorbing weekend, the mathematics of a bright sixteen-year-old, with some cookbook formulae and some facility with a programming language in the BASIC family are quite enough to put together code for displaying maps in a variety of simple projections on screen. Maps can be displayed by reading each pair of coordinates from the map data file, converting them to screen coordinates, and drawing a line from one pair to the next using the language's screen-drawing commands. The conversion formula should take into account the fact that computers address the screen from top to bottom, whereas, for example the UK national grid increases from south to north. A reversing factor may have to be put in to ensure the map is the right way up! Furthermore, because a VGA screen is 640 pixels wide, but only 350 pixels high, a width/height conversion factor has to be built in so that the map does not look distorted. For latitude and longitude maps, each pair of coordinates has to be put through the mathematical formula necessary to obtain the desired projection – one useful source is Steers (1956); it is also important to take into account whether the fine detail of your coordinates is expressed as sixtieths or hundredths of a degree. Finally, and for all projections, factors have to be applied to ensure the map is placed at the correct magnification in the right place on screen.

The main difficulty with such a program is likely to be the length of time it takes to draw the map. Presentation of a map of the whole world on screen using a Mercator-style projection, for example, might take the computer 5–20 minutes, depending among other things on processor speed. At least the user has the mild entertainment of watching the map evolve on screen. Using the same data set, however, to produce a map of, say, Africa is visually more tedious: without some form of indexing of the raw mapping data – a whole order of programming complexity greater – every pair of coordinates for the whole world has to be crunched, most resulting in numbers beyond the bounds of the VGA screen coordinate limits. One is led to the conclusion that the creation of maps *de novo* for screen display is a time-consuming job.

Saving a map to a library

One solution to this is to enable the user to save maps produced at such a time cost in a map library. In that way, the time overhead of preparing the map and saving it to disc is incurred only once. It is not enough just to save the screen image. If it is to be used to display distributional information at a later date, additional information has to be saved. At a minimum this would have to include the scale of the map, the type of boundary data (e.g. latitude and longitude, or grid), the type of projection, and reference coordinates to either the central point of the map, or to its sides. This information could be saved either as a supplementary file, or as a header to the screen information itself.

Building a zoom facility

Experience rapidly shows that, given the facility of a map library, users then express the wish to 'zoom in' on a particular area of the map. If the area desired is wholly a subset of an existing map, this can be done quite easily. One option would be simply to magnify the screen data for the area selected by an appropriate factor. This would be quick, but has the serious disadvantage that the resulting magnification would have no enhanced resolution. An alternative would be to redraw the selected area of the map by going back to the original data set. This has the disadvantage of being time consuming. Effectively you are creating a brand new map for your library: there is no zooming.

One way round this is to generate additional information about each new map at the time of creating it. Consider, for example, a map of Africa based on the data-set for the whole world. Only those latitude and longitude coordinates which result in meaningful screen coordinates will be displayed. If, during creation of this map, the computer notes in an index every time the transition is made from meaningful to meaningless screen coordinates, and every time the reverse transition is made, then in any future reference to the data-set it need only access those coordinates which had proved relevant to the present screen display. As with the other information which has to accompany the saved map, this index can either be an ancillary file or a header within the saved map file.

Restoring a map to screen

This is the procedure likely to be most frequently used, and for which speed is therefore essential. Although it is easy to restore a screen using graphics commands of a programming language such as a member of the BASIC family, such commands are very slow, even for compiled languages, and especially so for interpreted ones, because they work by calling the BIOS screen services which are notoriously slow. Restoring a map to screen in this way could take up to 45 seconds. Unfortunately, the way information is transmitted to a VGA screen can seem as though it was designed to discourage direct access through the processor. Even so, it is not too difficult to devise an assembly language program which will reduce the time of restoring a map to about one third of a second. Useful suggestions as to how this can be achieved may be found in Stevens (1988).

Placing distributional data on screen

An easy way to achieve this would be, using the database software, to prepare in advance a file composed of 4 byte pairs of coordinates, each pair representing one distributional record. Placing such data on screen is then merely a matter of reading in each pair of coordinates, crunching them to screen coordinate form, and putting a coloured circle centred on those coordinates up on screen. Probably many computerized maps prepared for the BMS so far use this or a similar method. What this gains in ease of handling, has the balancing disadvantage that the source data is once removed from the mapping software, making it impossible to access and display full information about a selected record.

A more sophisticated solution would be to derive the desired coordinates directly from the database itself. At present, using an MS DOS-based system, this is quite difficult. Although WINDOWS- and WINDOWS95-based systems permit such access through object linking, their current slowness with graphics remains a deterrent. To access the database itself in an MS DOS-based system, a detailed knowledge is required of how the database software organizes information for each record. An understanding of how it runs its physical indexes and key fields is also particularly important. For much commercial database software, this sort of information is not easily available. The BMS has used the SMARTWARE database system for the last ten years and, fortunately, it is easy to hack into its database and index files. Even the key fields files, which are rather more tricky, can with patience be deciphered.

Given that detailed knowledge, the mapping software should locate each desired record in the data file, note its unique observation identifier [CcorLink_N], then use that to find and read the fields containing the necessary coordinates (stored as text information) in the appropriate record in the *Collections Locality Cross-reference Table*. It should then convert those coordinates to numeric format, and display them on screen. At the same time, it should construct within itself a 'ghost' database with three fixed length fields, each record storing the unique observation identifier and the two screen coordinates. This 'ghost' database (which is erased on quitting the program, or on restoring a different map from the library, etc.) provides the information needed to find additional data about a selected record. If the distribution of more than one organism is to be displayed in different colours on the same map, each will require its own 'ghost' database, and the software will need to be able to distinguish between them.

Displaying imprecise records by shading areas is a more complex operation, necessitating quick access to boundary data which is often political or a mixture of political and coastline. On locating an imprecise record in the main database, the mapping software would have to find the appropriate boundary data, then draw and fill that area on the map, in the appropriate colour. The most imprecise records would have to be drawn before the more precise, and both would have to be done before displaying records for which coordinates were available. A 'ghost' database could be made for each level of precision of data. Individual records within each database would have to be identified by some means of storing the collective coordinates of the area on the displayed map shaded because of that record.

Selecting records for display of additional data

For precise records, the mouse can be used to point to individual circles displayed on the current map. By reading the screen coordinates of the mouse pointer at the time of clicking, allowing for a certain level of inaccuracy of pointing, and comparing them with those in the 'ghost' database, it is possible to identify the appropriate record or records for which further information is desired. The main database can then again be accessed from the outside to supply full information about each desired record. For less precise records, access to the correct 'ghost' database could be made on the basis of screen colour at the point of clicking the mouse. The screen coordinates where the mouse was clicked would then have to be compared with the collective coordinates for each record in the 'ghost' database.

Gathering additional data about the selected record

To fetch the full information, something like the following operation would be necessary. Using the unique observation identifier stored in the current 'ghost' database record, the correct record in the *Collections Core Table* would have to be found, to determine whether the current record related to a successful search, and to pick up the unique collection identifier [CcorColnoN]. Then the correct record in the *Collections Core Supplementary Table* would have to be found, from which would be gained information about the developmental state and abundance of the current fungus, and confirmation that it was not a fossil. While reading that record, the information in [Cco0AccnaA] would be picked up and used to find the name of the organism, its authors and taxonomy from the *Nomenclature and Taxonomy Database*, and information about where it was published from the *Bibliography Database*.

An index would then have to be made of all the records in the *Collections Locality Cross-reference Table* for which [CloxLink_N] matched the unique observation identifier stored in the 'ghost' database. Each of the records in the index would have to be examined to find which one contained the information on the locality of collection, and that information would have to be read, together with information on the prevailing ecosystems. A similar index would then have to be made for the *Collections People Cross-reference Table* to pick up the names of all collectors, together with any numbers they may have allocated the current observation, the date of collection, and the names of the original and current identifiers.

Next, an index would have to be made of all the records in the *Collections Core Table* sharing the same unique collection identifier [CcorColnoN]. In that way all associated organisms would be selected. It would then be necessary to find the name, authors and taxonomy of each associated organism, using data in [Cco0AccnaA] in the appropriate record for each. Next, the *Collections Substratum Cross-reference Table* would have to be examined for the current fungus in combination with every different associated organism, to find what substrata were being colonized, together with any ecological relationships or symptoms. For British fungi, where it is normal to note only one associated organism, this is not too bad, but it can be slow for Brazilian rain forest fungi which may have over 25 different associated organisms!

If desired, information about reference collections where the current item is held and numbers allocated to that item by the collection may also be gathered for display by making an index of the *Collections Other Collections Cross-reference Table*. Similarly bibliographic sources for the current item can be gathered from the *Collections Bibliography Cross-reference Table*. Although it might seem complicated, this sequence is, of course, no more than a human operator would do if all the facts were to be checked meticulously. In fact this method of gathering data about a single observation from a suite of different tables is actually pretty routine and forms the bulk of work done in many of the common output programs one is likely to need. The only difference is that, for mapping, it has to be done from outside of the database software.

Displaying additional data about a selected record

Once gathered into memory, all of the information needs to be processed into a format suitable for presentation to the user. Those with a taste for graphics programming might amuse themselves by presenting the fungus name, host name and place of publication on screen in italics. Those with humbler aims might be satisfied with merely outputting them in a different colour. One system at IMI uses pale green to indicate italics, and pale cyan to indicate roman type.

To display that information on screen, it may be necessary to overwrite part of the map, especially if the map takes up a large area of screen. The sort of screen saving and restoring facilities already used can be adapted for this purpose too. It may be worth noting that, if the same area of

screen is always used for displaying full information, at times it may overwrite the locality on the map of the currently desired record. It may be
worth programming the computer to select one of two areas for over-
writing to ensure this does not happen. For those who are confident they
understand the mechanics of the database software, an editing interface
can then be constructed.

Exporting a map as a graphics file

Merely exporting screen data tends to be rather unsatisfactory. A map can
look great on screen, can be exported to a word-processor file, and then
can look disappointingly small, distorted, and with a discernible 'jerky'
outline when printed. The slower but more satisfactory alternative would
be to use the header information saved with that map to access the raw
source coordinates, and to build up a graphics image extending beyond the
limits imposed by VGA resolution. To provide an export service, knowledge of one or more of the common graphics file formats ('.PCX' or
'.BMP') would be necessary.

Printing a map on paper

Although quick, merely transferring the screen data to the printer should
be avoided for the same reasons as make exporting the map as a graphics
file disappointing. The slower, but more satisfactory alternative would
again be to use the header information saved with that map to access the
raw source coordinates, and to build up a graphics image appropriate to
the resolution of the printer in use. The POSTSCRIPT language is probably the easiest format in which to output maps to printers though, being an
interpreted language, it can be very slow. Some printers have their own
built-in printer language which it is possible to use, but there are often
memory and other problems when larger and more complex maps are
being produced. For those with a dot-matrix-printer it is worth remember-
ing the cheap and dirty trick that a map printed at 60 dpi at five times the
scale of a laser-printed map can then be reduced by photocopying until
the resolution is more or less the same at that of a laser printer.

Conclusions

The biggest problems in recording and mapping fungi still remain in gathering data. For most species, even for a comparatively small area like the

British Isles, not enough is known to make very meaningful comments about distribution. The absence of an agreed data structure in which to store mycological data is encouraging a proliferation of many, often inadequate, mycological databases all differing slightly from each other, while discouraging the development of software to access mycological information. Because the tasks which it is mechanizing are complex, such software has to be pretty sophisticated. When a machine is doing the job, the instructions usually have to be more precise and more detailed than for humans: it is not possible to wave a magic wand and the machine get things correct.

Mycologists appear to have no formal representation in the various groups trying to set up internationally agreed data structures for biological recording, yet to be recorded adequately, fungi by their nature require specialist data structures not needed by other groups. The BMS should consider ways of involving itself and other mycological societies in such developments. The BMS computerized databases have provided a useful service, but they are now showing their age, and urgently require considerable upgrading if they are to store information which can be manipulated mechanically for a range of purposes rather than merely be browsed by humans or output in just a limited number of forms. Computerized mapping has not progressed very rapidly within the BMS, and may need more encouragement in the next few years.

Acknowledgements

Various people have provided helpful advice during the preparation of this manuscript. I am particularly grateful to Dr Jerry Cooper (IMI), who has done so much to promote mapping in the BMS, for supplying the formulae for some mapping projections used during my studies. Dr Paul White (Rolls-Royce Aero-engines) provided valuable advice on other formulae. My son, James Minter made sure I remained reasonably up to date with the worlds of computer hardware and software. Dr Yuri Shkolnikov (formerly of N.G. Kholodny Institute of Botany, Kiev) collaborated in deconstructing the SMARTWAREII database system, particularly in working out how its key field system operates.

References

Brummitt, R.K. & Powell, C.E. (1992). *Authors of Plant Names.* 732 pp., Kew: Royal Botanic Gardens.

Da Silva, M. & Minter, D.W. (1995). Fungi from Brazil recorded by Batista and co-workers. *Mycological Papers*, **169**, 1–585.

Holmgren, P.K., Holmgren, N.H. & Barnett, L.C. (eds) (1990). *Index Herbariorum, Part 1: The Herbaria of the World*. Eighth edition. 693 pp., USA, New York, Bronx: New York Botanic Garden.

Kirk, P.M. & Ansell, A.E. (1992). Authors of fungal names. *Index of Fungi*, supplement. 95pp., Kew: International Mycological Institute.

Marriott, J.V.R. (1994). Guide to recording fungi. *Guides for the Amateur Mycologist*, **3**, i–ii, 1–32. British Mycological Society.

Pegler, D.N., Læssøe, T. & Spooner, B.M. (1995). *British Puffballs, Earthstars and Stinkhorns: an Account of the British Gasteroid Fungi*. 255 pp., Kew: Royal Botanic Gardens.

Sankaran, K.V., Sutton, B.C. & Minter, DW. (1995). A checklist of fungi recorded on *Eucalyptus*. *Mycological Papers*, **170**, 1–376.

Steers, J.A. (1956). *An Introduction to the Study of Map Projections*. 324 pp., London, University of London Press Ltd.

Stevens, R.T. (1988). *Graphics Programming in C*. 639 pp., California, Redwood City: M&T Publishing, Inc.

Watling, R. & Whalley, A.J.S. (1977). *Provisional Maps of Xylariaceous Fungi*. 24 pp., Huntingdonshire, Monks Wood: Biological Records Centre & British Mycological Society.

Index

Note: page numbers in *italics* refer to figures and tables